Impact of Global Warming and Climate Change on Human and Plant Health

The Editors

Prof. (Dr.) Arun Arya is currently working as Senior Professor, Department of Botany and Head of Environmental Studies Department in The Maharaja Sayajirao University of Baroda. He is a Botanist, Phytopathologists, Philatelist and popular Science Writer. He served Head, Department of Botany during 2008-2013, after obtaining his Doctorate Degree from the Universtiy of Allahabad in 1983, he joined as a Lecturer in The Maharaja Sayajirao University of Baroda in 1985. Dr. Arya is a recipient of Young Scientist award of DST, New Delhi in 1986. He has guided numerous research projects funded by DST, CSIR, DBT, ISRO, MoEF and UGC. He has supervised 8 Ph.D. programmes published more than 130 research papers and 10 books. He presented his research findings in International Conferences at Japan, Canada, Italy and Morocco. He had a honor to co-chair the session on eco regions in 12th World Forestry Congress at Quebec city, Canada in 2003.

Prof. Vinod S. Patel, Consulting Engineer and Environmental expert. He has been working in this field for last 35 years. He has worked on various engineering projects in USA, East Africa and India in different capacities. He was the Fellow of Indian Association for Air Pollution Control (IAAPC) and guided various activities of IAAPC as its President. He is an expert in Radial well and groundwater supply management. He served as Visiting Professor in the Faculty of Technology and Engineering and regularly delivers lectures to the masters students in the Science Faculty, The Maharaja Sayajirao University of Baroda. He has published more than 40 technical papers and three books on various aspects of Pollution and Environment.

Impact of Global Warming and Climate Change on Human and Plant Health

Editors

Dr. Arun Arya

Prof. V.S. Patel

2015

Daya Publishing House®

A Division of

Astral International Pvt. Ltd.

New Delhi – 110 002

Cataloging in Publication Data--DK

Courtesy: D.K. Agencies (P) Ltd. <docinfo@dkagencies.com>

National Seminar on Air Pollution Management in Changing Global Environment Scenario (2010 : Maharaja Sayajirao University of Baroda)

Impact of global warming and climate change on human and plant health / editors, Dr. Arun Arya, Prof. V.S. Patel.

pages   cm

Includes bibliographical references and index.
ISBN 9789351306856 (International Edition)

1. Global warming--Health aspects--Congresses. 2. Climatic changes--Health aspects--Congresses. 3. Air--Pollution--Control--Congresses. 4. Crops and climate--Congresses. 5. Public health--Congresses. 6. Plant health--Congresses.   I. Arya, Arun, editor. II. Patel, V. S. (Vinod S.), editor. III. Maharaja Sayajirao University of Baroda. Department of Botany.organizer. IV. Maharaja Sayajirao University of Baroda. Environment Science Programme.organizer. V. Title.

DDC 613.11   23

Published by   : **Daya Publishing House®**
                            A Division of
                            Astral International Pvt. Ltd.
                            – ISO 9001:2008 Certified Company –
                            4760-61/23, Ansari Road, Darya Ganj
                            New Delhi-110 002
                            Ph. 011-43549197, 23278134
                            E-mail: info@astralint.com
                            Website: www.astralint.com

Laser Typesetting   : **Classic Computer Services**, Delhi - 110 035
Printed at   : **Thomson Press India Limited**

PRINTED IN INDIA

सत्यं शिवं सुन्दरम्

The Maharaja Sayajirao University of Baroda
Office : University Road, Fatehgunj, Vadodara 390002 Gujarat, India

Prof. Yogesh Singh
PhJ. (Computer Engineering)
Vice-Chancellor

Foreword

I feel pleasure in writing the Foreword for the book entitled "**Impact of Global Warming and Climate Change on Human and Plant Health** ," edited by Prof. Arun Arya and Prof. V.S. Patel.

Nature is self-sustaining. Beauty of nature is flooded with a number of biological and mineral wealth. We are supposed to conserve these with better technologies and proper use. Increase in CO_2 level will result in change of behavior of a number of birds, insect and microbial pests. Honey bees are needed for pollination of flowers. Von Frisoln on the basis of his experiments done in nineties' proved that these are not totally color blind and have a very high level of intelligence. Direct sunlight and gravity helps in finding their path to return to beehive.

Only one per cent of the earth's water is drinking water. Unlike Western world we should follow the practice of reuse, reduce, and recycle in our daily life. Botanists estimate that approximately 17 trees are needed to produce one tone of paper. Every tone of recycled paper will save 1728 litres of oil. Computers and use of solar power is now in use extensively throughout the world. Silicon produced from one tone of sand when used in photovoltaic cells could produce electricity equivalent to burning of 500,000 tons of coal. Efforts should be made to increase the yield of the crops, and find out alternative sources of energy.

The present book encompasses 36 chapters written by experts in the field. I hope that the knowledge seeking students, and experts in the field of pollution and climate change will be benefitted by the volume. Recycling and clean development mechanism will reduce the global warming.

I congratulate both the editors for taking up the task and enhancing our knowledge in the field of climate change and its effect on plants and human beings.

(Yogesh Singh)

Preface

Give love; and love to your life will flow.
A strength in your utmost need.
Have faith, and a score of hearts will show
their faith in our work and deed.

We are facing a jigsaw puzzle that is called 'global change'. The greenhouse effect means that the average temperature on Earth is 33 degrees Celsius higher that it would have been in the absence of what is a natural phenomenon. Average earth temperature is 18°C and without greeenhouse effect it will be minus 15 degree Celsius. According to IPCC the temperature on Earth will rise by an average of 1.5 to 4.5 degrees over the next century.

United Nations Conference on Human Environment was held in Stockholm. UNEP and WHO are dealing with the environmental problems on a global scale. Recent event all across the globe-melting glaciers, monstrous storms and unstoppable floods, continental scale forest fires, scorching heat and droughts are pointing to an urgent need for scientific communities to pay attention to issues arising out of global climate change

The science of climate change is rapidly evolving as new models; new predictive capabilities, new satellite observations and new analytical tools are brought to bear on the core questions of climate change: How is the climate of our planet changing? What changes are attributable to natural variability? What is the role of human activities in climate change? How can developing economies mitigate the deleterious impacts of climate change? What are the options for the developed world to reduce the climate change related impact on global ecology? Research on the implications of Climate change on India's vast ecological regions is of paramount importance to gain insights into these complex processes. Interdisciplinary dialogue and discussion are vital as climate change science covers many domains and disciplines.

To make the artificial leaf, the MIT team spread its catalysts on opposite sides of a silicon wafer. The silicon absorbs sunlight and passes energetic, negatively charged electrons and positively charged electron vacancies (holes) to the catalysts on the opposite sides that use them to make H_2 and O_2. When the device is placed in a clear jar and exposed to sunlight, the setup converts 5.5 per cent of the energy in sunlight into hydrogen fuel, which can be stored and used in fuel cell to generate electricity. The new artificial leaf uses nickel and cobalt as catalysts, which are relatively cheap, and has so far operated continuously for at least 45 hours. The research was reported by Daniel Nocera, a Chemistry Professor at MIT, who did the work, at the biannual meeting of the American Chemical Society at Anaheim, California, USA on 28th March 2011.

The book contains research papers presented in National Seminar on **Air Pollution Management in Changing Global Environment Scenario**, which was organized on 7th February, 2010 jointly by Department of Botany and Environment Science Programme of the M. S. University of Baroda. The seminar was supported by Indian Association for Air Pollution Control, Indian Science Congress Association – Baroda Chapter, Indian Environmental Association. **Shri Bharatsinh Solanki,** Union Minister for Power, Govt. of India, was chief guest and was attended by dignitaries like **Dr. (Mrs.) Mrunalini Devi Puar**, Chancellor of MSU, **Dr. S.K. Nanda**, IAS, Principal Secretary, Forests and Environment, Govt. of Gujarat. **Shri Solanki** stated that with the increasing demands of people more and more industrialization is required at the same time necessary measures should be taken to minimize the air pollution. Leading a healthy long life without any disease is still a dream.

"Satam Jivam Shardah Satam"

We should live for more than 100 years without suffering from diseases.

In cities like Delhi and Mumbai health costs due to pollution are as much as Rs.120 crores per year. By reducing the pollution from industries and vehicles a drastic cut can be made in health budget of citizens of National capital or in Mumbai. NOx and SO_2 when breathed in can lodge in our lungs and can cause lung damage and respiratory problems. Pollutants can damage not only human life but also the life of animals and plants as well by preventing the use of diesel by buses and using CNG in roads of Delhi the change in pollution levels can be seen. A number of record show that our common house Sparrow (Chakli) has gone from near by trees and we wake up not with the sweet melodious voice of the bird but with the sound of call bell or our mobile phones. We have forgotten our old life style and children see the scene of rising sun in the televisions or in internet.

Uplift of society was conceived by the **Great Visionary Shrimant Maharaja Sayajirao Gaekwad III** and we should try to infuse in young minds full determination, human values, Indian culture and a Scientific temper, as has been stated by **Dr. A.P.J. Abdul Kalam**, our ex President.

Beautiful hands are those that do work
that is earnest and brave and true
Moment by moment
the long day through.

We thank all the authors for their valuable contributions and IAAPC for bringing this volume and hope that the documented knowledge will definitely be a source of novel ideas to be implemented to combat the pollution and reduce the global warming.

Dr. Arun Arya
Prof. V.S. Patel

Contents

Part I
Global Warming: Health Effects

अनमोल जीवन

हे मानव! तू संभल जा
बहुत हो चुका अत्याचार
अब तो संभल जा।
न मिलेगा यह जीवन दोबारा,
न होगी यह धरती कल्याणकारी
न मिलेगा सुन्दर बसेरा।
तुम्हारा अस्तित्व शायद न रह पाये
बिना स्वच्छ जल के।
है हरी भरी धरती और
सुन्दर मखमली घास,
वृक्ष जहां फैलाते हैं हरीतिमा,
और भरते हैं माँग
सिन्दूरी पलाश के मनमोहक फूल।
बुद्धि और विवेक का रखों ध्यान विकास में
धरा की लाज बचाने को
न आयेगा अब कोई कृष्ण
हे मानव अब तू सँभल जा।।

2015, Impact of Global Warming and Climate Change on
        Human and Plant Health

Pages 3–11

Editors: Dr. Arun Arya and Prof. V.S. Patel
Published by: DAYA PUBLISHING HOUSE, NEW DELHI

Chapter 1

Global Warming: An Overview from Human and Animal Health Perspectives

A. Basu[1], S.K. Basu[2]* M. Asif[3] and X. Chen[4]

[1]*Calcutta Medical College, 88 College Street,*
Kolkata – 700 073, West Bengal, India
[2]*University of Lethbridge, Lethbridge, Alberta, Canada T1K 3M4*
[3]*University of Alberta, Edmonton, Alberta, Canada T6G2P5*
[4]*Yunan Academy of Agricultural Sciences, Kunming, Yunan, China 650205*

ABSTRACT

If we go through the pages of history, we find that there had always been a tug-of-war between man and nature. Primitive men, though depended a lot on nature for procurement of food and shelter, would often find it difficult to tame the wildness of Her. Be in the time of floods or massive volcanic eruptions, man would have to rely on the mercy of Mother Nature for sparing their weak helpless lives. But with the gradual evolutionary changes among human beings, the strength for dominating over the invincible natural powers grew steadily among men and this led to several challenging discoveries *viz.* fire, metal tools and weapons and other alike and practices of deforestation, farming and domestication of wild animals. With the advent of science and technology, these challenges became almost an obsession among men and they planted their flags of victory, time and again, into the heart of Mother Nature. Although such victories helped in making human beings the most supreme species among all members of the animal kingdom, the condition of the natural environment started being on the downhill. An economic boom in the different nations followed the historic

* *Corresponding author.* E-mail: saikat.basu@uleth.ca

Industrial Revolution of England in the 1800s. But this paved the way for the forthcoming deterioration of Earth's atmosphere caused by the gaseous and other chemical by-products liberated off the factories and industries. Widespread use of air-conditioners and refrigerators (containing chlorofluorocarbons) to serve our modern day house-holds and working places seemed to be an excellent opportunity for the till filtered UV rays to enter the Earth's open field through the disrupted Ozone layer and lead to an increased incidence of skin and ocular cancers. All these and several other similar so-called fantastic scientific developments have together resulted in severe pollution of the Earth's environment and, as a consequence, rise in its temperature, a condition termed as 'Global Warming' by the environmentalists. This has brought about serious impacts on the health of human beings and also on his natural fellows *i.e.* animals and plants. The following piece of text will deal with the adverse effects of global warming on man and natural world and a few possible means to curb this particular situation.

Keywords: *Biome, Environment, Greenhouse effect, Global warming, Food security, Ozone.*

Abbreviations

UV: ultraviolet; **IPCC:** Intergovernmental Panel on Climate Change; **RPM:** Respiratory Particulate Matter; **VOC:** Volatile Organic Compounds; **UN:** United Nations; **ppm:** parts per million

1.1 Introduction

So, what is Global Warming? It simply means the rise in atmospheric temperature of the planet Earth (Naik, 2011). However, this rise in temperature does not take place on its own. There are several factors that contribute to this phenomenon (Naik, 2011). And this rise is significant enough to bring about detrimental effects to the Earth's environment; both physical and biological (Bounds, 2010). The major cause behind Global Warming is 'Greenhouse Effect', a term that has been a hot topic of discussion for several years (Mitra *et al.,* 2007). Now to elaborate, Greenhouse effect is not always harmful. It is the man-made component of it that makes it hazardous for the planet (Ansari, 2010). The natural component of Greenhouse effect is essential for maintaining an optimum temperature of our planet and making it a suitable place to live in (Mitra *et al.,* 2007).

A greenhouse is a place where plants growing in temperate climate are kept inside a glass house, the temperature inside of which is kept high (Mitra *et al.,* 2007). The glass of a greenhouse allows the short wave radiations *i.e.* visible light and UV rays coming from the sun to enter inside but traps the long wave radiations *i.e.* infrared radiations that leads to heating up of the greenhouse (Chugh, 2006). Now there are certain gases like carbon di-oxide and methane that are produced in significant amount in the Earth as a byproduct of several chemical reactions, both in nature and by human causes (Ansari, 2010). These gases act like the glass of a greenhouse (Chugh, 2006). They filter out the short wave radiations from the longer ones and allow the former to escape from the Earth's surface, thus trapping the longer wave radiations to warm up the atmosphere (Chugh, 2006). Hence, these gases are termed the Greenhouse gases (Bounds 2010). Without them, the global temperature would have been 60 per cent colder (Ansari 2010). But when such gases are produced in excess, they trap greater amount of the long wave radiations. As a result, the temperature of the Earth goes up several times higher than normal (Ansari, 2010). It has been recorded by the IPCC that the global temperature has increased by one degree Fahrenheit in the last century (Naik, 2011).

Greenhouse gases are produced both in nature and by human activities. Natural sources of these gases are forest fires, volcanic eruptions and from the marshes (Naik, 2011). The anthropogenic sources of Greenhouse gases are the excessive burning of fossil fuels, automobile pollution, mining, massive deforestation and several others (Naik, 2011). The total amount of carbon di-oxide emitted from energy systems can be calculated by the Population Multiplier formula of Ehrlich and Holdren (Schneider, 1989):

$$CO_2 \text{ emitted} = CO_2 \text{ emission} \times \frac{\text{Technology}}{\text{Technology}} \times \frac{\text{Total Population Size}}{\text{Capita}}$$

1.2 Effects on Human Health

Global warming is likely to bring about drastic changes in the human health, both from within and outside. The probable health hazards are respiratory ailments, heat stroke and water borne and water bred diseases (Roy Britt, 2005). The main agents behind causing respiratory tract illness are the ground level ozone and RPM (Roy Britt 2005). The ground level ozone occurs in the lower atmospheric strata compared to the mush familiar ozone that occurs in the upper atmospheric strata *i.e.* Ozonosphere (Rastogi and Kishore, 2006). The stratospheric ozone layer is responsible for preventing the harmful UV rays from entering the Earth's atmosphere (Rastogi and Kishore, 2006). But the ground level ozone is formed as a result of photochemical reaction between nitrogen oxides and VOC (Kumar *et al.,* 2010). These reacting chemicals are released from industries and automobiles. This ground level ozone is responsible for significant decrease in the lung function and thus, the exercise tolerance. It also increases the airways resistance and causes lung inflammation (Kumar *et al.,* 2010). The population that is more susceptible to the adverse effects of ground level ozone are the children, outdoor workers and people with impaired airway function (*e.g.* asthmatics or emphysematous patients) (Kumar *et al.,* 2010).

RPM in the range of 0.5-3.0μ in diameter (*i.e.* less than 10μ) are the most harmful agents causing respiratory ailments (Park 2011A). These particles can easily reach the lungs upon inhalation and lead to an inflammatory reaction (Kumar *et al.,* 2010). This inflammation ultimately leads to fibrosis of the lungs and reduction in the lung function (Park, 2011C). It has also been documented the increased exposure to RPM may lead to thickening of blood (Roy Britt, 2005). Heat stroke occurs when people are exposed to greatly elevated temperature during outdoor working hours, especially in the tropics. Due to the intense heat, the individual experiences headache, giddiness and excessive thirst. The characteristic feature in heat stroke is the absence of perspiration for which the skin appears hot and dry. There is associated hyperpyrexia with the patient losing his/her consciousness. The pulse becomes feeble, breathing is labored, blood pressure falls and cyanosis develops. There is scanty urination leading to disturbed fluid-electrolyte balance in the body. This results in cardiac arrhythmias, convulsions and death if medical intervention is not done timely. Even if the person survives, in case of severe heat stroke, there may be life-long neurological disorders like loss of memory, epileptic episodes or insanity (Karmakar, 2007).

The global climate change is likely to bring about an increase in the incidence of water-borne and water-bred diseases. Increased frequency of storms and rainfall are likely to result in severe flooding of the plain areas and this will eventually be followed by outbreaks of water-borne diseases (*e.g.* diarrhea, dysentery, cholera, hepatitis, etc.) (Roy Britt, 2005). These infectious diseases will lead to deterioration of man's health, leading to malnutrition (Park, 2011B). In turn, malnutrition itself will predispose the individual to further infections. Thus a vicious cycle is set up that will lead to severe consequences in the human health (Park, 2011B). At the same time, flooding will create water stagnation and this will provide a welcome site for infectious disease vectors to breed and multiply, thereby creating a potential source for spread of vector borne diseases.

Among the notable diseases are malaria, dengue and other mosquito and fly borne diseases (Roy Britt, 2005). There are different mosquito vectors for malaria, each with a different characteristic breeding habit. For example, *Anopheles* sp. of mosquitoes prefer clean collected water for breeding purpose, while *Culex* sp. prefer dirty stagnant water. Hence, admixture of collected water with human or animal filth and wastes will allow the latter variety of mosquitoes to breed unabated (Park, 2011D). And not only flooding and rainfall, but a rise in the temperature itself will allow a change in the breeding habit of the mosquitoes, thereby bringing about different varieties of these insects to reproduce under favorable circumstances and cause outbreak of diseases. Flies act mostly as mechanical vectors, unlike mosquitoes that are biological vectors. So their body parts easily pick up the infectious agents from the contaminated water sources and transmit them to food and drinks, leading to severe epidemics (Park, 2011D).

1.3 Effects on Flora and Fauna

Plants and animals form an indispensable part of nature and hence, our life. Therefore, degradation of nature resulting from this global climate change is likely to affect them as well. Both men, animals and plants are interconnected to each other in a web where even if a single member is affected, the entire balance will be lost with a severe consequence to follow eventually (Rajeev, 2012).

A place where extremely high concentrations of flora and fauna are seen, including the endemic species, is tagged as a biodiversity 'hot spot'. These hot spots have attracted global attention in recent years as the foci where effects global climate change seems to be more pronounced. It has been reported that tropical hot spots are least susceptible to global climate change, while those in the Mediterranean and the savanna regions are much more vulnerable. The Cape Floral Kingdom and its neighboring Karoo of South Africa are the most prominent examples. It has been estimated that an average annual rise in temperature by 3-4°C is sufficient to wipe out the entire floral biodiversity of Karoo (Smith *et al.,* 2001).

Apart from the hot spots, importance is also laid down upon the ecotones, coral reefs and mangrove vegetations as potentially vulnerable sites to climate change (Smith *et al.,* 2001). Ecotones are the transition zones connecting one or more biomes, thereby becoming a rich species diversity and genetic diversity zone (Smith *et al.,* 2001). Biomes, in turn, are ecologically diverse groups of animals and plants spread over a significantly measurable area, with each group comprising its characteristic flora and fauna in par with the suitable environment *viz.* rainfall, temperature, altitude (Rastogi and Kishore, 2005). Ecotones being a transition zone, the effect of climate change will be noticed there much earlier than changes affecting the larger biomes become visible (Smith *et al.,* 2001). This is because gene flows occur through these ecotones from one biome to the other and hence a significant gene pool exists in these ecotones (Smith *et al.,* 2001). For example, a semiarid dry ecotone exists between a dry biome and an arid biome. Rising temperature will lead to increased desertification in the ecotone with loss of valuable biodiversity both in the ecotone and in the adjacent biomes. Another example could be coral reefs, though occupy only 1 per cent of the oceans of this world, are the homes of several thousand varieties of fishes and crustacean and other forms of marine life (Smith *et al.,* 2001).

Increased temperatures have been found to cause bleaching of the corals that are responsible for the building of these reefs. Raised CO_2 concentration also significantly delays the calcification of the corals, thereby leading to degradation of these reefs (Smith *et al.,* 2001). In certain instances, near total elimination of a particular coral species from the reef has been seen owing to these climate changes. In other conditions of shorter duration thermal changes, massive coral species like the *Porites* sp. have survived while the weaker acroporids and pocilloporids have been eliminated (Walther *et al.,* 2002). Mangrove vegetations are extremely susceptible to the rise in sea levels following global climate change, the significant being in the Sunder bans (Smith *et al.,* 2001). These vegetations protect the coastline from the effect of floods, cyclones and tsunamis (Mitra *et al.,* 2007). Their destruction therefore will bring about inundation of the coastlines (Smith *et al.,* 2001).

Now let us come to certain specific animals whose existence has become critically endangered due to the global warming. Recent researches indicate that amphibians and reptiles are steadily declining in their global population due to these temperature changes (Lind, 2011). However, reptiles being more mobile than amphibians will be able to withstand the climate changes to some greater degree. These two classes of animals are poikilothermic and hence highly sensitive to changes in the temperature of air and water, precipitations (*i.e.* rain and snow) and the hydro-period. These changes become more important for the amphibians laying eggs in water, their larval stage of development including metamorphosis and post-metamorphosis stages (Lind 2011). Moreover, certain reptilian species exhibit temperature dependant determination of sex at the time of incubation of eggs. So it is quite evident that climate changes are likely to bring about a drastic reduction in the growth and survival of these animals (Lind, 2011).

An interesting issue that lizards may be starved to death by the global warming has been brought to light by a team of researchers headed by Barry Sinervo of the University of California (Than, 2010). Dummy lizards created using temperature gauges and painted pipes were placed at four different zones in the Mexican Yucatán Peninsula, two at sites where the blue spiny lizards (*Sceloporus serrifer cyanogenys*) are still present and the remaining two in the other two zones where the same lizards have been locally wiped out (Than, 2010). It was recorded that the former two zones experienced intense heat only for a period of 4 hours during which the lizards would remain inactive and then move out to gather food. But in the latter two zones, the temperature was very high for almost entire day. This meant that these high temperatures prevented the lizards from moving out to hunt and this scarcity of food led them to die (Than, 2010).

It is important to note that not only reptiles and amphibians, even the pollinating insects are experiencing the same fate. Extinction of the pollinating insects would lead to serious shortage in the human food source in the long term and subsequently a mass extinction of life (Than, 2010).

Now moving to the Poles, it has been documented that the Arctic has seen a tremendous rise in temperature in the past few years. Melting of ice and rise in sea levels have led the polar bears (*Ursus maritimus*) to seek shelter in the interior lands. During summer, forest fires have burnt down their dens while during the spring time, heavy rainfall has collapsed them. Moreover, as the bears have moved out of the sea ice, their hunting has suffered a lot. There has also been a severe reduction in the bear cub survival rate along with certain anatomical changes in the bears and thus seem to be one of the worst hit animals by the global warming.

In the Antarctic, penguins form a prominent group of animals that is suffering tremendously from the global climate change. In the Antarctic Peninsula, four species of penguins co-inhabit and of them the Emperor penguins and Adélie penguins suffer the most. The Emperor penguins lay their eggs on packed ice. Due to rising temperature, the packed ice melts and the eggs or the chicks are carried off to the ocean before they hatch or mature, thus facing the inevitable death. On the other hand, the Adélie penguins lay eggs in nests made of pebbles in the snow free areas of Antarctic coastline. In the western half of the peninsula, the temperature is comparatively higher. This allows the air to expand and hold greater amount of moisture, leading to excess precipitation *i.e.* snowing that wipes out the nesting grounds (Wolf, 2009).

1.4 Effects on the Environment

Global warming has brought about significant changes in the Earth's climate and much more remains to be changed if the condition is allowed to persist. The rising temperature is resulting in melting of the glaciers and elevation of the sea levels. This melting of glaciers is more significant in the tropics than elsewhere because of their high sensitivity to temperature changes (Smith *et al.,* 2001). Elevation of sea levels will result in flooding of the coastlines and the island nations. According to Smith *et al.* (2001) the major island nations under threat are Marshall Islands, Kiribati, Tuvalu, Tonga, the Federated States of Micronesia, and the Cook Islands of the Pacific, Antigua and Nevis of the Caribbean Sea and the island of Maldives in the Indian Ocean. Melting of glaciers will also affect the water supply to the people in certain parts of Asia and South America who depend largely on the mountain water for their daily need (NOAA Satellite and Information Services, 2011).

Rising global temperatures have increased the occurrence of natural disasters like floods, cyclones, hurricanes and volcanoes. Tsunami has been recently added to the list. All these natural hazards result in severe loss of life and property of millions every year (Kaur, 2012). An important issue is the impairment of food security due to the global climate change. Global warming creates water scarcity and this affects the irrigation in the fields. It also creates altered environmental conditions like temperature, soil changes, etc. that become difficult for adaptation by the agricultural plants. However, this problem of adaptation is far more pronounced in the developing nations than the industrialized countries. Such problems faced in raising crops leads to sky-rocketing of food prices that makes both selling the foods in the local market and importation difficult. Since the people of developing nations cannot afford such costly crops, most of them remain under-fed or consume the cheaper crops with poor nutritive values. As a result, they suffer from malnutrition and infections. This adds to the global burden of nutrition related disorders (Quaye *et al.,* 2012).

A conveyor belt like system of heat transport exists in the oceans of this planet that is responsible for the global heat transport and thus influences the different climatic patterns in the different regions of the planet (Smith *et al.,* 2001). This conveyor belt system carries warmer surface waters to the Poles while bringing in cooler waters to the Equator (Roach, 2005). An important component of this heat distribution system exists in the Atlantic Ocean (Smith *et al.,* 2001). Recent researches have shown that melting of glaciers and collapse of packed ice sheets has increased the amount of freshwater runoffs into the North Atlantic waters (Roach, 2005). If this continues unabated, then the Atlantic conveyor belt system of heat transport is likely to be shut down completely within one or two centuries (Roach, 2005). This will severely affect the marine life and also cause a significant alteration in the global heat budget (Smith *et al.,* 2001).

1.5 Future Strategies to Control Global Warming

From the discussion so far, it is quite evident that we have reached a point from where turning back seems to be a pretty difficult task. But it's not at all impossible. If we take proper measures to curb these ongoing insults upon Nature from this very moment, then we can hope of saving our planet. To maintain a stable environmental temperature, the global rise in temperature must not be more than 2°C annually. But the severe increase in the global CO_2 concentration in the atmosphere poses a serious threat to the achievement of this goal. So the UN leaders have decided to limit the atmospheric concentration of greenhouse gases at the level of about 450 ppm of CO_2 equivalent (IEA, 2011).

Combustion of waste materials produces heavy amount of CO_2 that escapes into the atmosphere and adds to the load of greenhouse gases. So adoption of more eco-friendly methods of waste disposal (like controlled tipping or composting) can curb down this environmental hazard. The principle of 3 R's: Recycle-Reduce-Reuse has become the heart of environmental protection measures in the recent years. Similarly, practicing reforestation and cutting down automobile emissions seem to be other effective means of reducing the burden of greenhouse gases (Malicdem, 2010). Though adoption of these afore-mentioned measures is easy said, yet their implementation is all the more difficult. But we must not forget that dominance of a mere single form of life can never ensure the viability of this planet. It is the harmonious existence among all the living forms that will make this planet a safe place to live in. So it is high time that we start implementing the proverb in our lives: *Live and let live*.

References

Ansari, A.H. 2010 Povert Link to the Environment: International and National Perspectives. *Indian Journal of International Law* 50: 13.

Chugh, K.L., 2006. *Environmental chemistry*. Kalyani Senior Chemistry. Kalyani Publishers, Ludhiana. pp. (I): 10/1-10/31.

Karmakar, R.N., 2007. *Injury and its medicolegal aspects.* In: Karmakar (ed.) J. B. Mukherjee's *Forensic medicine and toxicology*. Academic Publishers. pp. 319-570.

Kaur, U. 2012. Environmental hazards and global warming. *Assoc Geog Std*. pp. 1-48 Available online at: http://ags.geography.du.ac.in/Study per cent 20Materials_files/Ushvinder per cent 20Kaur_SSN.pdf (accessed August 30, 2012)

Kumar, V., Abbas, A., Fausto, N. and Aster, J. 2010. *Environmental and nutritional diseases*. In: Kumar *et al.* (eds.) Robbins and Cotran's pathological basis of disease. 8th ed. Elsevier. pp. 399-445.

Lind, A. 2011. *Amphibians and reptiles and climate change.* Available online at:

http://www.fs.fed.us/ccrc/topics/amphibians-reptiles.shtml interaction/climate-change

Malicdem, L. 2010. *5 Environmental hazards and their simple eco-friendly solutions for Environmental protection*. Available online at: http://seniordebutante.blogspot.in/2010/01/5-environment-hazards-and-their simple.html

Mitra, A., Guha, S. and Chanda, S. 2007. *Environmental biology*. In: Dr. Chattopadhyay (ed.) *A text book of biology*. Book Syndicate Pvt. Ltd. (I): 498-565.

Naik, A. 2011. *Global warming*. Available online at: http://www.buzzle.com/articles/causes- of-global-warming.html

NOAA Satellite and Information Services: National Environmental Satellite, Data, and Information Service, 2011. *Global climate change indicators*. Available online at: http://www.ncdc.noaa.gov/indicators/

Park, K. 2011A. *Environment and health. Park's Textbook of preventive and social medicine.* Banarsidas Bhanot Publishers. pp. 652-729.

Park, K. 2011B. *Nutrition and health. Park's Textbook of preventive and social medicine.* Banarsidas Bhanot Publishers. pp. 561-617.

Park, K. 2011C. *Occupational health. Park's Textbook of preventive and social medicine.* Banarsidas Bhanot Publishers. pp. 744-759. Bounds, B. 2010. Facts about the greenhouse effect. Available online at: http://www.brighthub.com/

environment/science-environmental/articles/93925.aspx International Energy Agency, 2011. Prospect of limiting the global increase in temperature to 2°C is getting bleaker. Available online at: http://www.iea.org/index_info.asp?id=1959

Park, K. 2011D. *Principles of epidemiology and epidemiologic methods. Park's Textbook of preventive and social medicine.* Banarsidas Bhanot Publishers. pp. 49-123.

Polar Bears International. *Climate change.* Available online at: http://www.polarbearsinternational.org/polar-bears/bear-essentials-polar-style/human-interaction/climate-change

Quaye, W., Yawson, R.M., Ayeh, E.S. and Yawson, I. 2012. Climate change and food security: The role of biotechnology. *AJFAND.* 12(5): 6354-6364.

Rajeev, L. 2012. *Global warming effects on animals.* Available online at: http://www.buzzle.com/articles/global-warming-effects-on-animals.html

Rastogi, V.B. and Kishore, B. 2005A. *Ecosystem. A complete course in ISC biology.* Pitambar Publishing Company Pvt. Ltd. (I): 59-95 (Unit-III).

Rastogi, V.B. and Kishore, B. 2005B. Environmental pollution. *A complete course in ISC* biology. Pitambar Publishing Company Pvt. Ltd. (I): 159-193 (Unit-III).

Roach, J. 2005. *Global warming may alter Atlantic currents, study says.* Available at:http://news.nationalgeographic.com/news/2005/06/0627_050627_oceancurrent_2.html

Roy Britt, R. 2005. *Caution: Global warming may be hazardous to your health.* Available online at: http://www.livescience.com/168-caution-global-warming-hazardous-health.html

Schneider, S. 1989. *The greenhouse effect: science and policy. Science* 243:771-81. Available online at: http://www.ciesin.org/docs/003-074/003-074.html

Smith, J., Schellnhuber, H. and Mirza, M. M. 2001. *Vulnerability to climate change and reasons for concern: A synthesis.* In: McCarthy *et al.* (eds.) Climate change 2001: Impacts: Adaptation and vulnerability. Cambridge University Press. pp. 913-970.

Than, K. 2010. *Mass lizard extinctions looming; global warming blamed.* Available online at: http://news.nationalgeographic.com/news/2010/05/100513-science-environment-lizards- global-warming-extinctions/

The consequences of global warming on glaciers and sea levels. Available online at: http://www.nrdc.org/globalwarming/fcons/fcons4.asp

Walther, G., Post, E., Convey, P., Menzel, A., Parmesank, C., Beebee, T., Fromentin, J., Guldberg, O. and Bairlein, F. 2002. Ecological responses to recent climate change. *Nature.* (416): 389-395.

Wolf, S. 2009. *Climate change threatens penguins.* Available online at: http://www.actionbioscience.org/environment/wolf.html

2015, Impact of Global Warming and Climate Change on
    Human and Plant Health
Editors: Dr. Arun Arya and Prof. V.S. Patel
Published by: DAYA PUBLISHING HOUSE, NEW DELHI

Pages 12–21

Chapter 2

Air Pollution and their Effects on Human Health

Hitesh A. Solanki* and Brijesha A. Shah**

*Department of Botany, University School of Sciences,
Gujarat University, Ahmedabad – 380009, Gujarat, India*

ABSTRACT

Earth is surrounded by air cover up to the hight of about one mile from the surface of earth. The World Health Organization (WHO) defines air pollution as -the presence of materials in the air in such consentration which are harmful to man and his environment. Air pollution also can be defined as the presence of undesirable substances in the air in such quantities and of such durations that tends to be injurious to human health or welfare, animal or plant life, or property and would unreasonably interfere with the employment of life or property. The air may become polluted by a number of other sources such as power and heat generating systems, industrial and transportation process, use of insecticides and fungicides.Nuclear energy programmes may be sources of air pollution. Air pollution causes lot of adverse effects on human health, animals and plants and also can affect weather and bring climate change.

Keywords: Air pollution, Effects, Human health, Asthma, Cancer.

2.1 Introduction

Air pollution is the introduction of chemicals, particulate matter, or biological materials that cause harm or discomfort to humans or other living organisms, or

---

E-mail: *husolanki@yahoo.com; **sweetyshah2121@gmail.com

damages the natural environment, into the atmosphere. The atmosphere is a complex, dynamic natural gaseous system that is essential to support life on planet earth. Stratospheric ozone depletion due to air pollution has long been recognized as a threat to human health as well as to the Earth's ecosystems.

An air pollutant is known as a substance in the air that can cause harm to humans and the environment. Pollutants can be in the form of solid particles, liquid droplets, or gases. In addition, they may be natural or man-made. ("Reports". WorstPolluted.org. Retrieved 2010-08-29) Pollutants can be classified as either primary or secondary. Usually, primary pollutants are substances directly emitted from a process, such as ash from a volcanic eruption, the carbon monoxide gas from a motor vehicle exhaust or sulfur dioxide released from factories.

Secondary pollutants are not emitted directly. Rather, they form in the air when primary pollutants react or interact. An important example of a secondary pollutant is ground level ozone – one of the many secondary pollutants that make up photochemical smog.

Smog hanging over cities is the most familiar and obvious form of air pollution. But there are different kinds of pollution some visible, some invisible that contribute to global warming. Generally any substance that people introduce into the atmosphere that has damaging effects on living things and the environment is considered air pollution. Carbon dioxide, a greenhouse gas, is the main pollutant that is warming Earth. Though living things emit carbon dioxide when they breathe, carbon dioxide is widely considered to be a pollutant when associated with cars, planes, power plants, and other human activities that involve the burning of fossil fuels such as gasoline and natural gas. In the past 150 years, such activities have pumped enough carbon dioxide into the atmosphere to raise its levels higher than they have been for hundreds of thousands of years.

Other greenhouse gases include methane which comes from such sources as swamps and gas emitted by livestock and chlorofluorocarbons (CFCs), which were used in refrigerants and aerosol propellants until they were banned because of their deteriorating effect on Earth's ozone layer. Another pollutant associated with climate change is sulfur dioxide, a component of smog. Sulfur dioxide and closely related chemicals are known primarily as a cause of acid rain. But they also reflect light when released in the atmosphere, which keeps sunlight out and causes Earth to cool. Volcanic eruptions can spew massive amounts of sulfur dioxide into the atmosphere, sometimes causing cooling that lasts for years. In fact, volcanoes used to be the main source of atmospheric sulfur dioxide;

Industrialized countries have worked to reduce levels of sulfur dioxide, smog, and smoke in order to improve people's health. But a result, not predicted until recently, is that the lower sulfur dioxide levels may actually make global warming worse. Just as sulfur dioxide from volcanoes can cool the planet by blocking sunlight, cutting the amount of the compound in the atmosphere lets more sunlight through, warming the Earth. This effect is exaggerated when elevated levels of other greenhouse gases in the atmosphere trap the additional heat. Most people agree that to curb global warming, a variety of measures need to be taken. On a personal level, driving

	MAJOR SOURCES	HEALTH EFFECTS	ENVIRONMENTAL EFFECTS
SO_2	Industry	Respiratory and cardiovascular illness	Precursor to acid rain, which damages lakes, rivers, and trees; damage to cultural relics
NO_x	Vehicles; industry	Respiratory and cardiovascular illness	Nitrogen deposition leading to over-fertilization and eutrophication
PM	Vehicles; industry	Particles penetrate deep into lungs and can enter bloodstream	Visibility
CO	Vehicles	Headaches and fatigue, especially in people with weak cardiovascular health	
Lead	Vehicles (burning leaded gasoline)	Accumulates in bloodstream over time; damages nervous system	Fish/animal kills
Ozone	Formed from reaction of NO_x and VOCs	Respiratory illness	Reduced crop production and forest growth; smog precursor
VOCs	Vehicles; industrial processes	Eye and skin irritation; nausea; headaches; carcinogenic	Smog precursor

Figure 2.1
Source: **Matt Kallman, World Resources Institute, August 20, 2008.**

and flying less, recycling, and conservation reduces a person's "carbon footprint"–
the amount of carbon dioxide a person is responsible for putting into the atmosphere.

On a larger scale, governments are taking measures to limit emissions of carbon
dioxide and other greenhouse gases.

2.2 Sources of Pollution

1. Anthropogenic Sources

☆ Related to burning different types of fuels.

e.g. Stationary sources, mobile sources

2. Natural Sources

☆ Dust

☆ Methane

☆ Radon

☆ Smoke

☆ Carbon Monoxides

☆ Volcanic Activity

There are two types of Air Pollutions,

1. Indoor Air Pollution
2. Outdoor Air Pollution

1. Indoor Air Pollution

Biological sources of air pollution are also found indoors, as gases and airborne particulates. Pets produce dander, people produce dust from minute skin flakes and decomposed hair, dust mites in bedding, carpeting and furniture produce enzymes and micrometre-sized fecal droppings, inhabitants emit methane, mold forms in walls and generates mycotoxins and spores, air conditioning systems can incubate Legionnaires' disease and mold, and houseplants, soil and surrounding gardens can produce pollen.

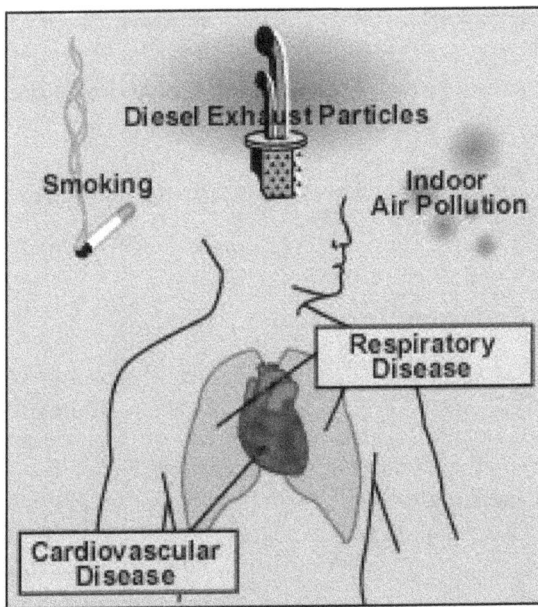

Figure 2.2: Showing Cardiovascular Diseases due to Smoking.
Source: http://www.home-air-purifier-expert.com/lungs.html.

2. Outdoor Air Pollution

Effects of Air Pollution

☆ Global Warming

☆ Acid Rains

☆ Ozone Depletion

☆ Climate Change

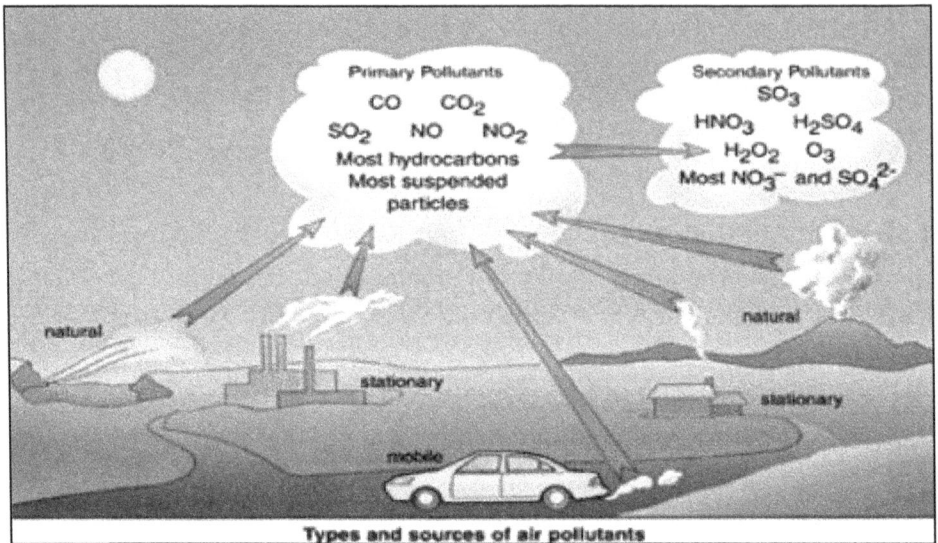

Figure 2.3: : Showing Sources of Air Pollution.
Source: **http://ecourses.vtu.ac.in**

2.3 Effects of Air Pollution on Human Health

Mostly it affects:

☆ Human respiratory system

☆ Human cardiovascular system

Air pollution is a significant risk factor for multiple health conditions including respiratory infections, heart disease, and lung cancer, according to the WHO. The health effects caused by air pollution may include difficulty in breathing, wheezing, coughing and aggravation of existing respiratory and cardiac conditions. These effects can result in increased medication use, increased doctor or emergency room visits, more hospital admissions and premature death. The human health effects of poor air quality are far reaching, but principally affect the body's respiratory system and the cardiovascular system. Individual reactions to air pollutants depend on the type of pollutant a person is exposed to, the degree of exposure, the individual's health status and genetics.

The most common sources of air pollution include particulate matter, ozone, nitrogen dioxide, and sulfur dioxide. Both indoor and outdoor air pollution have caused approximately 3.3 million deaths worldwide. Children aged less than five years that live in developing countries are the most vulnerable population in terms of total deaths attributable to indoor and outdoor air pollution ("Air quality and health". www.who.int. Retrieved 2011-11-26).

The World Health Organization states that 2.4 million people die each year from causes directly attributable to air pollution, with 1.5 million of these deaths attributable to indoor air pollution.("Estimated deaths and DALYs attributable to selected environmental risk factors, by WHO Member State, 2002". Retrieved 2010-08 29.) "Epidemiological studies suggest that more than 500,000 Americans die each year from cardiopulmonary disease linked to breathing fine particle air pollution.("Newly detected air pollutant mimics damaging effects of cigarette smoke". www.eurekalert.org. Retrieved 2008-08-17) Worldwide more deaths per year are linked to air pollution than to automobile accidents. Published in 2005 suggests that 310,000 Europeans die from air pollution annually. Causes of deaths include aggravated asthma, emphysema, lung and heart diseases, and respiratory allergies. The US EPA estimates that a proposed set of changes in diesel engine technology (Tier 2) could result in 12,000 fewer premature mortalities, 15,000 fewer heart attacks, 6,000 fewer emergency room visits by children with asthma, and 8,900 fewer respiratory-related hospital admissions each year in the United States.

2.3.1 Effects on COPD and Asthma

Chronic obstructive pulmonary disease (COPD) includes diseases such as chronic bronchitis and emphysema. (Zoidis, 1999). Researchers have demonstrated increased risk of developing asthma and COPD from increased exposure to traffic-related air pollution. Additionally, air pollution has been associated with increased hospitalizations and mortality from asthma and COPD (Andersen *et al.,* 2011)

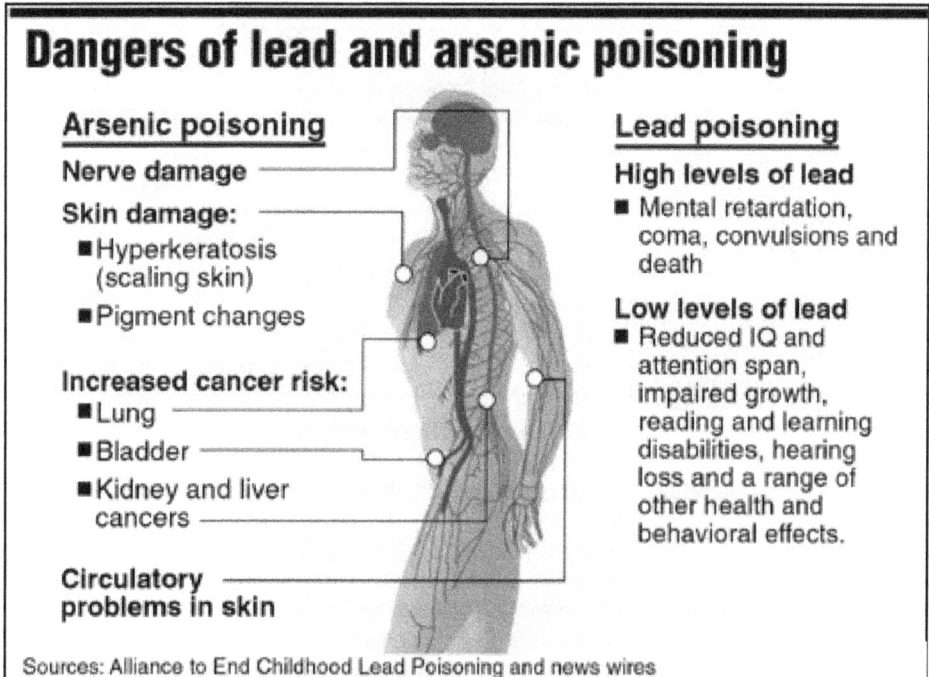

Dangers of lead and arsenic poisoning

Arsenic poisoning

Nerve damage

Skin damage:
- Hyperkeratosis (scaling skin)
- Pigment changes

Increased cancer risk:
- Lung
- Bladder
- Kidney and liver cancers

Circulatory problems in skin

Lead poisoning

High levels of lead
- Mental retardation, coma, convulsions and death

Low levels of lead
- Reduced IQ and attention span, impaired growth, reading and learning disabilities, hearing loss and a range of other health and behavioral effects.

Sources: Alliance to End Childhood Lead Poisoning and news wires

Figure 2.4: Damages of Pb and As Poisoning.
Source: http://colonelspeaks.wordpress.com/tag/who/

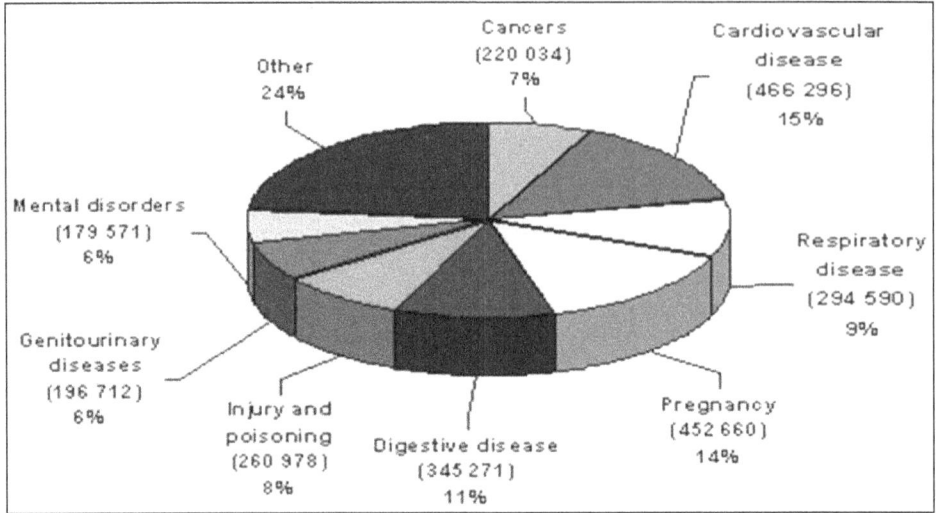

Figure 2.5: Leading Causes of Hospitalizaton.

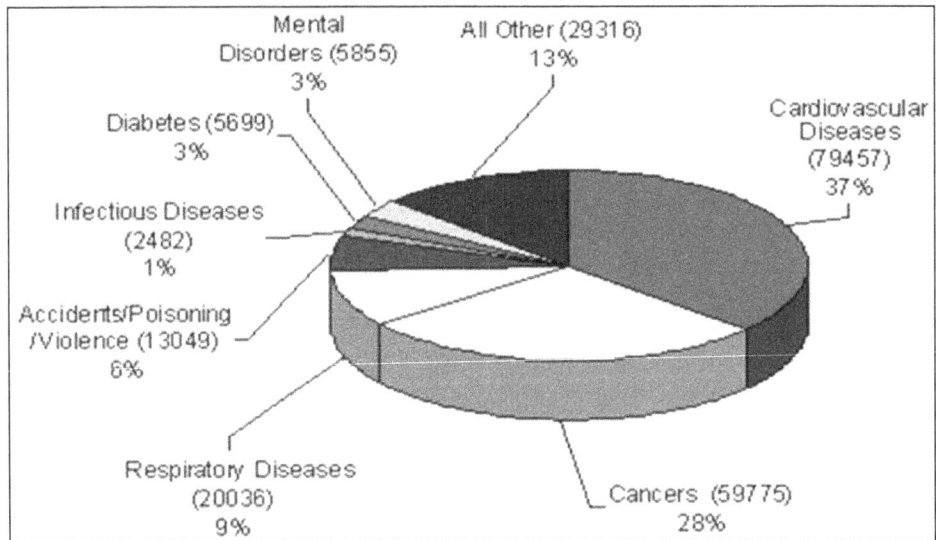

Figure 2.6: Leading Causes of Death.

Source: http://sivanandammsw.blogspot.in/2010/10/health-effects-of-air-pollution.html

We can control air pollution by decreasing carbon dioxide, increasing forest area, reduction of poisonous gases. Our good and healthy future is in our hands, we all have to take care of our environment by reducing the air pollution.

(1) Greenhouse effect, (2) Particulate contamination, (3) Increased UV radiation, (4) acid rain, (5) Increased ground level O_3 concentration, (6) Increased levels of NO_x.

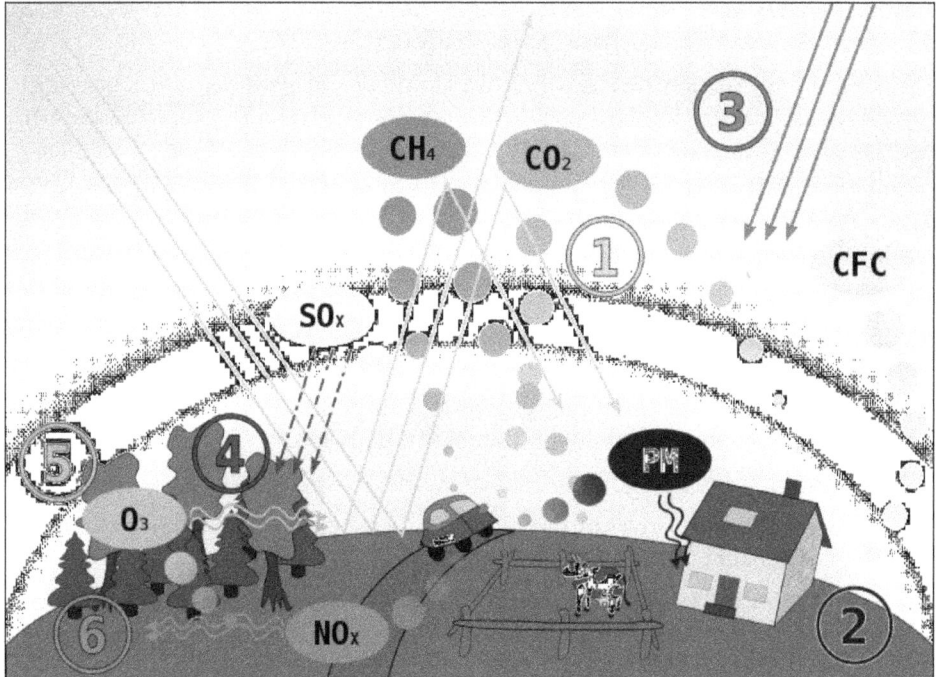

Figure 2.7: Effect of Greenouse on Plants and Animals.
Source: http://en.wikipedia.org/wiki/File:Air_Pollution-Causes per cent 26Effects.svg

2.3.2 Possible Links to Cancer

A large Danish epidemiological study found an increased risk of lung cancer for patients who lived in areas with high nitrogen oxide concentrations. In this study, the association was higher for non-smokers than smokers. There are also possible associations between air pollution and other forms of cancer, including cervical cancer and brain cancer (Raaschou-Nielsen *et al.,* 2010).

References

Air quality and health". www.WHO.Int. Retrieved 2011-11-26.

Andersen, Z. J., Kristiansen, L. C., Andersen, K. K., Olsen, T. S., Hvidberg, M., Jensen, S. S.,. Raaschou-Nielsen, O. (2011). Stroke and Long-Term Exposure to Outdoor Air Pollution From Nitrogen Dioxide: A Cohort Study. Stroke; a *Journal of Cerebral Circulation*. doi: 10.1161/STROKEAHA.111.629246 AP 42, Volume I". Epa.gov. Retrieved 2010-08-29.

Christopher H. Goss, Stacey A. Newsom, Jonathan S. Schildcrout, Lianne Sheppard and Joel D. Kaufman (2004). "Effect of Ambient Air Pollution on Pulmonary Exacerbations and Lung Function in Cystic Fibrosis". *American Journal of Respiratory and Critical Care Medicine* 169 (7): 816–821.DOI:10.1164/rccm.200306-779OC. PMID 14718248.

David P. and Smith K. "Biomass Pollution Basics". WHO.

Davis, D. (2002). *When Smoke Ran Like Water: Tales of Environmental Deception and the Battle Against Pollution*. Basic Books. ISBN 0-465-01521-2.

Diesel exhaust inhalation increases thrombus formation in man† Andrew J. Lucking1*, Magnus Lundback2, Nicholas L. Mills1, Dana Faratian1, Stefan L. Barath2, Jamshid Pourazar2, Flemming R. Cassee3, Kenneth Donaldson1, Nicholas A. Boon1, Juan J. Badimon4, Thomas Sandstrom2, Anders Blomberg2, and David E. Newby1

Duflo, E., Greenstone, M., and Hanna, R. 2008 "Indoor air pollution, health and economic well-being". ''S.A.P.I.EN.S." '''1''' (1)''. Sapiens.revues.org. Retrieved 2010-08-29.

EMEP/EEA air pollutant emission inventory guidebook – 2009

Environmental Pollution Evidence growing of air pollution's link to heart disease, death//American Heart Association. May 10, 2010

EPA: Air Pollutants". Epa.gov. 2006-06-28. Retrieved 2010-08-29.

Estimated deaths and DALYs attributable to selected environmental risk factors, by WHO Member State, 2002''. Retrieved 2010-08-29.

Farrah J. Mateen and Robert D. Brook 2011 "Air pollution as an emerging global risk factor for stroke" JAMA 2011 23;305(12):1240-1.http://www.ncbi.nlm.nih.gov/pubmed/21427378

Gehring, U., Wijga, A. H., Brauer, M., Fischer, P., de Jongste, J. C., Kerkhof, M.,.Brunekreef, B. 2010. Traffic-related air pollution and the development of asthma and allergies.

Goldstein, A. H., Charles D. Koven, Colette L. Heald, Inez Y. Fung 2009. "Biogenic carbon and anthropogenic pollutants combine to form a cooling haze over the southeastern United States". *Proceedings of the National Academy of Sciences*. Retrieved 2010-12-05.

http://www.sfgate.com/cgi-bin/article.cgi?f=/c/a/2008/11/13/MNQP143CPV.DTL

Indoor air pollution and household energy". WHO and UNEP. 2011.

Infant Inhalation Of Ultrafine Air Pollution Linked To Adult Lung Disease". Sciencedaily.com. 2009-07-23. Retrieved 2010-08-29.

Kymisis, M. Hadjistavrou K. 2008. "Short-Term effects of Air Pollution Levels on Pulmonary Function Of Young Adults". *The Internet Journal of Pulmonary Medicine* 9(2).

Miller, K. A., Siscovick, D. S., Sheppard, L., Shepherd, K., Sullivan, J. H., Anderson, G. L., and Kaufman, J. D. (2007). Long-term exposure to air pollution and incidence of cardiovascular events in women. [Research Support, N.I.H., ExtramuralResearch Support, U.S. Gov't, Non-P.H.S.]. *The New England Journal of Medicine*, 356(5), 447-458. doi: 10.1056/NEJMoa054409.

Newly detected air pollutant mimics damaging effects of cigarette smoke" (PDF). Retrieved 2010-08-29.

Newly detected air pollutant mimics damaging effects of cigarette smoke". www.eurekalert.org. Retrieved 2008-08-17.

Persistent Endothelial Dysfunction in Humans after Diesel Exhaust Inhalation Ha°kan To¨rnqvist1*, Nicholas L. Mills2*, Manuel Gonzalez3, Mark R. Miller2, Simon D. Robinson2, Ian L. Megson4, William MacNee5, Ken Donaldson5, Stefan So¨derberg3, David E. Newby2, Thomas Sandstro¨m1, and Anders Blomberg1

Reports". Worst Polluted.org. Retrieved 2010-08-29. Revised 1996 IPCC Guidelines for National Greenhouse Gas Inventories (reference manual)". Ipcc-nggip.iges.or.jp. Retrieved 2010-08-29.

Chakrabarti. S."20th anniversary of world's worst industrial disaster". Australian Broadcasting Corporation.

Study links traffic pollution to thousands of deaths". *The Guardian* (London, UK: Guardian Media Group). 2008-04-15. Retrieved 2008-04-15.

United Kingdom's emission factor database". Naei.org.uk. Retrieved 2010-08-29.

Zoidis, J. D. 1999. "The Impact of Air Pollution on COPD". *RT: for Decision Makers in Respiratory Care*.

2015, Impact of Global Warming and Climate Change on
Human and Plant Health
Editors: **Dr. Arun Arya and Prof. V.S. Patel**
Published by: **DAYA PUBLISHING HOUSE, NEW DELHI**

Pages 22–29

Chapter 3

A Step Towards a Greener Tomorrow

M. Dutta[1], X. Chen[2], M. Asif[3] and S.K. Basu[4*]

*[1]Department of Microbiology, Gurudas College,
University of Calcutta, Kolkata – 700 054, W.B., India
[2] College of Life Sciences, Northeast Agricultural University,
Harbin 150300, Heilongjiang, P. R. China
[3]Department of Agriculture, Food and Nutritional Sciences,
University of Alberta, Edmonton, AB, Canada T6G 2P5
[4]Department of Biological Sciences, University of Lethbridge,
Lethbridge, AB, Canada T1K 3M4*

ABSTRACT

Due to the increase in vehicular fuel price, high environmental pollution and non- renewable nature of fossil fuels, an alternative resource is being searched currently and biofuel appears to be the only viable and sustainable answer. Biofuel is a cost efficient, environment-friendly and renewable energy source generated from easily available agricultural materials. Bioethanol produced in bio-refineries by fermentation of sugar, starch or cellulose containing biomass can serve as a viable substitute or can be blended with petrol. Biodiesel extracted from oil crops can perform as an alternative for diesel or can be mixed with diesel. Biogas obtained by anaerobic digestion or gasification of agricultural or animal wastes can be used to generate electricity and also used for cooking purposes. Researches are being conducted all over the globe to find out effectively and efficiently

* *Corresponding author.* E-mail: saikat.basu@uleth.ca

produce biofuels at an industrial scale. Governments across the globe are taking initiatives to encourage and promote the production of energy rich crops that can be used as raw material sources for biofuel generation in a sustainable manner.

Keywords: *Biomass, Biodiesel, Biofuel, Bioethanol, Biogas, Fossil fuel, Feedstock, Bio-refinery.*

3.1 Introduction

Since known times, man has been dependent on petroleum as a source of energy. It is essential for transportation, agriculture and also for aircrafts (Basu *et al.,* 2011). Fossil fuels as coal, petroleum, natural gas have been the primary source of energy up to 80 per cent in the world, whereas, merely, 20 per cent of the electricity comes from other resources like hydraulic, nuclear, wind, solar, geothermal and biogas (Erdem *et al.,* 2009). The world's source of fossil fuel is limited and is depleting on daily basis (Bala, 2005; Soetaert and Vandamme, 2009, Basu *et al.,* 2010). With the advancement of technology, industrial and green revolution the demand for petroleum and its consumption has kept on increasing. Their combustion also aids in serious environmental pollutions and global warming (Soetaert and Vandamme, 2006; Basu and Datta Banik, 2009). Global warming is growing at an alarming rate in countries that are facing a sustained socioeconomic development and are among the front series user of these fossil fuels. Furthermore, these fossil fuel resources are distributed mainly in politically unstable regions (Soetaert and Vandamme, 2006).

Though, some new oil reserves have been found but these are located in poorly accessible areas where extraction cost is extremely high. The cost of extraction and demand is rising and so is the price of crude oil. The supply of fossil fuel is limited and it has been predicted that our global supply of petroleum will come to an end in next 50 years, natural gas around 65 years and coal in about 200 years if current consumption rate continues (Soetaert and Vandamme, 2009). Today USA is the world's largest consumer of petroleum with China and India ranked second and third respectively. With the ever increasing need of petroleum in the economically advancing nations the price of petroleum is also soaring gradually. The question that is currently arising worldwide is what will happen when our supply of fossil fuel runs out in due course of time? Researchers all over the planet are seeking answer through the use of renewable energy resources such geothermal, solar, wind, tidal, hydro energies and biomass based resources (Basu *et al.,* 2011). With the improvements in genetic engineering techniques and development of sophisticated equipment's, the synthesis of bio-based energy systems from agronomic and domestic wastes is rapidly increasing and turning popular and is regarded as one of the convenient and cost-efficient method of obtaining biofuel (Basu *et al.,* 2010).

Crops such as corn, sugarcane and sugar beets can be efficiently processed in bio-refineries into carbohydrate feedstock and can be easily used as starting material for fermentation process to generate bioethanol (Puppan, 2002). Biodiesel can be produced form oilseed crops like rapeseed or canola, soybeans and palm seeds and also from waste vegetable oils and animal fats. Biogas can be made by processing agricultural and domestic wastes (Soetaert and Vandamme, 2006). The major advantages of biofuels over other competing conventional fuel sources are that biofuels are easily biodegradable, carbon dioxide neutral fuel resource, renewable, environment-friendly and can be easily produced from easily available agricultural raw materials (Puppan, 2002). The cheap cost, easy availability and continuous supply of agricultural materials (Puppan, 2002) and new emerging techniques of biotechnology and genetic engineering (Bhowmik *et al.,* 2009; Basu *et al.,* 2010) thus provide the possible answer to the question rose earlier. Biofuel can serve as a promising substitute for fossil fuel and the world population has to rely on these fuels once the conventional non-renewable energy sources runs out (Bhowmik *et al.,* 2009; Basu *et al.,* 2010, 2011).

3.2 Biofuel Generating Crops

Biomass is produced by burning biological materials to generate heat. Biofuels in turn are mainly produced by processing domestic biomass feedstock, or through regular processing of agricultural and/or agricultural food products, or through reprocessing of used cooking and vegetable oil sources (Erickson and Carr, 2009). Biogas is produced using anaerobic digestion to generate methane from biodegradable materials and wastes. Biofuel and bio gas serves as renewable sources of energy. The cultivation, processing, and use of these liquid biofuels cause less pollution than the fossil fuels. Biofuels are also biodegradable in nature. Biodiesel and bioethanol are environment-friendly fuel sources. As a result of incomplete burning carbon monoxide is generated from the petroleum fuels that lacks in molecular oxygen. But since oxygen is present in ethanol, the process of chemical combustion is more complete for synthesizing carbon dioxide in the automobile engines that result in substantial reduction of carbon monoxide emissions (Erickson and Carr, 2009). Ethanol can be blended with gasoline to save the use of fossil fuels that cause serious environmental pollution. In close comparison to petroleum only a minor concentration of sulphur is present in ethanol (Bala, 2005). Blending ethanol with petrol thereby reduces the sulphur content of the mixed blend. This helps greatly in lowering the toxic sulphur oxide emissions to the atmosphere. Biodiesel in comparison with conventional diesel has lower sulfur content, higher biodegradability and is also better with respect to fuel flash point and aromatic (Bala, 2005).

Bioethanol and biodiesel are the two most important biofuel sources for the transport sector (Basu and Datta Banik, 2009). Bioethanol can be processed from crops such as sugarcane, corn, wheat and sugar beet (Pandey, 2009). It can perform as an additive or substitute for petrol (Basu and Datta Banik, 2009). Biodiesel, on the other hand is the methyl or ethyl ester of the fatty acid generated directly from vegetable oils, (both edible and non-edible), recycled waste vegetable oils and animal fat (Mittelbach, 2009). Important sources of biodiesel production are non-edible oil generated from species of plants like *Jatropha curcas* L., *Pongamia pinnata* L., *Calophyllum inophyllum* L. and *Hevea brasiliensis* Müll. Arg. (Pandey, 2009). Biodiesel can be mixed with diesel to generate a specific biodiesel blend or can also be directly used in its pure and unmixed form. The first biofuel produced from plant was bioethanol. It was produced from sugarcane in Brazil. Later USA produced bioethanol using corn as the raw material. It is important to that fuel ethanol production in US has been almost exclusively from corn source (Mousdale, 2008). In case of neighboring Canada the total 2004 demand for fuel ethanol was met from 2,025 million liters of wheat, barley, corn and potatoes (EFSPP, 2004). A country wise break up listing different crops and agricultural raw materials used in biofuel production is presented in Table 3.1.

Table 3.1: Country-wise Raw Materials Used for Biofuel Production

Countries	Raw Materials
Bioethanol	
USA	Corn (sorghum barley, wheat, cheese whey and brewery wastes are also used)
Brazil	Sugarcane and cassava
Canada	Wheat, barley, corn and potatoes
Texas	Wheat, rice, Sorghum, barley and oats
California	Wheat, rice, Sorghum, barley and oats
Africa	Sugarcane molasses
India	Sugarcane molasses and sweet sorghum
Biodiesel	
USA	Soybean
Brazil	Soybean
Malaysia	Oil palm
Indonesia	Oil palm

3.3 Production of Biofuels

3.3.1 Bioethanol

Bioethanol can be produced from carbohydrate containing crops such as sugarcane, beet and sweet sorghum by fermentation and distillation. However, with starch containing crops like corn, wheat, rice, barley, potato and cellulosic materials such grasses scarification precedes fermentation and distillation. In case of sugar containing biomasses like sugarcane, sugar beet and sweet sorghum, commercial yeast (*Saccharomyces cerevisiae*) ferments the sucrose. The sucrose is first broken down into its monosaccharide units which then undergo fermentation. Enzymes like invertase hydrolyse sucrose and next zymase ferments glucose and the fructose into ethanol (Gnansounou *et al.*, 2005). With starchy biomass like corn, wheat, rice, barley and potato, enzymes like glucoamylase hydrolyses the polysaccharide starch into its monomeric units of d-glucose (Liaqat, 2010). Cellulosic materials like grasses can also be converted to bioethanol. The lignocellulose found in these kind of materials first has to be delignified then a number of treatment follows namely steam explosion, dilute acid pre-hydrolysis, hydrolysis using an enzyme and finally fermentation into bioethanol (Erickson and Carr, 2009). Hydrolysis of the hydrogen bonds in the fractions containing hemicellulosic and cellulosic contents directly produces pentoses and hexoses. Bioethanol is then produced through the process of fermentation of the pentoses and hexoses generated as described above (Liaqat, 2010). Ligno-cellulosic materials are hydrolyzed using concentrated sulfuric or hydrochloric acids. The economic utility of this process depends on important factors like optimizing sugar recovery and recovery of the acid for recycling. The acid and sugar are separated via ion exchange and then acid is re-concentrated through multiple effect evaporators (Demirbas, 2008).

3.3.2 Biodiesel

Biodiesel chemically consists of long chain fatty acid alkyl esters synthesized from vegetable oil and cooking oil or from animal fats. It is also synthesized from oilseed crops such as sunflower, safflower, soybean, cottonseed, rapeseed and peanut oils (Demirbas, 2009). Biodiesel closely resembles diesel fuels. Biodiesel however has very high fuel viscosity. This difficulty is overcome by diluting, micro-emulsifying, pyrolysis and transesterification. The transesterification reduces viscosity of the oil. While conducting transesterification ethanol preferred to methanol as it can be obtained easily from agricultural raw materials and is renewable and biologically less harmful (Demirbas, 2003).

3.3.3 Biogas

Biogas can be synthesized through the anaerobic digestion of food or animal waste by microbes (such as anaerobic bacteria). The biogas produced contains a mixture of methane and carbon dioxide (Erickson and Carr, 2009). Biomass can be put into several uses such as heating, cooking, lighting or for electricity production. Wood pellets and briquettes which are compacted forms of biomass can also be used for chemical combustion. The biomass can also be burnt with coal and the methane gas generated can be prevented from escaping to avoid the greenhouse gas impacts and can be utilized. Heat generated during the bio-digestion process kills and eradicates possible harmful pathogens present in the manure and the left over products can be used as valuable fertilizer. There exists another process known as gasification (Erickson and Carr, 2009). Here biogas can be produced from solid biomass. The operating conditions in biomass gassifiers include low oxygen concentration and high temperature. Undesirable chemicals can be removed from biogas through filtration. The biogas formed in gasification process can be used in highly efficient combined cycle power-generation systems that integrate a combination of steam and gas turbines to generate electricity (Ramage and Scurlock, 1996; Demirbas, 2008). Different production techniques and corresponding biomasses used for generating biofuels have been presented in Table 3.2.

3.4 Future Perspective

Opportunities

Government all over the globe especially in developing countries are chalking plans and making policies to promote and encourage bio-fuel. Steps are being taken to communicate poor and marginal farmers in arid lands in different parts of the world, especially the new world countries, with the global biofuels revolution without food security at a stake (Basu and Datta Banik, 2009). The objective is to benefit these farmers. The rising price of the fuel could provide opportunity to serve the purpose. Innovative research on bioethanol from sweet sorghum and biodiesel from *Jatropha* and *Pongamia* ensures energy, livelihood and food security to farmers in arid regions, as well as reduces the use of fossil fuel, lowering environmental pollution (Basu and Datta Banik, 2009). The requisites for cultivating these energy crops are quite simple. These crops do not have much requirement for water, are stress resistant, and cheap, and are thus suitable crops to be cultivated by poor farmers in dry lands. However there is still a necessity to develop partnerships between the public and private sectors in such initiatives so that they can be economically beneficial to poor farmers (Basu *et al.*, 2011).

Table 3.2: Processes Involved in Production of Biofuels

Biofuel	Biomass Used	Production Techniques
Bioethanol	Sugar containing biomass: sugarcane, sugar beet, sweet sorghum	Fermentation and distillation
Bioethanol	Starchy biomass: corn, wheat, rice, barley, potato	Enzymatic hydrolysis followed by fermentation, distillation and dehydration
Bioethanol	Cellulosic biomass: grasses	Delignification, steam explosion, dilute acid pre-hydrolysis, hydrolysis: (1) chemical hydrolysis (dilute and concentrated acid hydrolysis) and (2) enzymatic hydrolysis fermentation
Biodiesel	Rapeseed, canola, oil palm, sunflower, peanut oil	Extraction followed by esterification
Biogas	Agricultural, food or animal waste	Anaerobic digestion, gasification

Using biofuel as a substitute to fossil fuel does not deplete future resources. As these biofuels are produced from energy rich crops and are renewable in nature unlike fossil fuels. According to economic predictions (Ramage and Scurlock, 1996) by the year 2050, per cent of the global populations will be residing in developing and under developed countries of the world. The highly industrialized and developed nations are expected to move towards direct burning of agricultural and domestic wastes for generating electricity, use liquid fuels (bioethanol and biodiesel) from energy rich crops (Demirbas, 2008). Integrated biomass gasification/gas turbine technology has much potential in the future of electric current production due to their high energy conversion efficiencies. Biomass will compete favorably with conventional fuels in the international oil market as a renewable, environment-friendly and adaptable resource (Demirbas, 2008). More and more energy rich crops can be grown to meet the growing demands of changing consumer needs. In future, biomass can supply a cost-effective and continuous, renewable source of energy, and combustion of bio fuel will also assist in reducing the pollution world wide. India is in rich biomass resource. This biomass can yield renewable energy (Demirbas, 2003, 2008). The soaring price of petroleum, lack of efficient distribution system, unstable political scenario across the planet, threats of global warming and ever increasing environmental pollution caused by combustion of fossil fuels are a few factors which provide a promising market for biofuels in the coming days. It is predicted that the demand for bioethanol and biodiesel in the automotive fuel market will grow rapidly in the years ahead (Demirbas, 2009).

3.5 Challenges

In spite of the fact, biofuel holds tremendous opportunities in the not so distant future as a green and sustainable alternative to the global environmental pollution; however there are also some pitfalls to it as well. The issue is critical and quite debatable; while some researchers have supported the promotion of biofuel whereas others have opposed it vehemently. The main reason behind this is the worry that global food production may be disrupted severely due to over emphasis on biofuel crop production (Epp, 2012). Several producers in the developed countries are shifting their production from food crops towards biofuel crops due to higher profit margins. As a result of this practice several economists and sociologists are worried that this may in fact raise the cost of food crops; and as consequence may lead to social instability, political unrest, global food crisis, hunger and poverty in many parts of the world (Hein, 2012). On the other hand, supporters of biofuel production have suggested that the impact of biofuel crop production has been quite marginal and has never impacted the price of food crops in a significant manner. They have also criticized that the overwhelming fear, anxiety and concern regarding the biofuel crop production on the food markets have been over estimated and wrongly projected.

The current recession has hit the global food market and the shrinking strength of the US dollar has also been an important factor that contributed towards food prices not just biofuel production (Epp, 2012). With the current downward trend in global market economy, food shortages, challenges of efficient distribution of food commodities equally in different parts of the world, crop failures due to fluctuating weather patterns, the increasing divide between the global rich and poor and rapid expansion of human population both socio-economic and environmental challenges reasonably anticipated in the next 10-15 years are huge. However, we are optimistic that biofuel production will stood the test of time to emerge as a sustainable green technology for the betterment of our shared future from a holistic and global perspective.

References

Bala, B. K. 2005. Studies on biodiesels from transformation of vegetable oils for diesel engines. *Energ. Edu. Sci. Technol.* 5: 1-45.

Basu, S. K. and Datta Banik, S. 2009. Environment education-Global issues and policies. Volumes I and II. APH Pub Corp, New Delhi, India.

Basu, S. K., Datta Banik, S., Roy Choudhury, S. and Chen, X. 2011. Green technology and environment: Efforts for building a sustainable planet. The Science Association of Bengal, Kolkata, India.

Basu, S. K., Dutta, M., Goyal, A., Bhowmik, P. K., Kumar, J., Nandy, S., Scagliusi, S. M. and Prasad, R. 2010. Is genetically modified crop the answer for the next green revolution? GM Crop. 1(2): 1-12.

Bhowmik, P. K., Basu, S. K. and Goyal, A. 2009. Advances in Biotechnology: E- Book. Bentham Science Publishers, Sharjah, UAE and Oak Park, Illinois, USA.

Demirbas A. 2008. The importance of bioethanol and biodiesel from biomass. *Energy Sources* Part B. 3:177–85.

Demirbas A. 2009. Biorenewable liquid fuels. In: Demirbas, A. (ed.) Biofuels: securing the planet's future energy needs. Springer-Verlag, Berlin, Heidelberg, New York.

Demirbas, A. 2003. Biodiesel fuels from vegetable oils via catalytic and non-catalytic supercritical alcohol trans-esterifications and other methods: a survey. *Eneg Convers. Mgmt.* 44: 2093-2109.

EFSAPP/Economic, Financial, Social Analysis and Public Policies for Fuel Ethanol Phase 1, Natural Resources Canada, Ottawa, November 2004.

Epp, M. 2012. Food versus fuel: The debate is over. Top Crop Manager (East). April, 2012 issue. pp. 20-21.

Erdem H.H., Akkaya A.V., Cetin B., Dagdas A., Sevilgen S.H., Sahin B., Teke I., Gungor C., and Atas S. 2009. Comparative energetic and exergetic performance analyses for coal-fired thermal power plants in Turkey. *Int. J Therm. Sci.* 48: 2179-2186.

Erickson, B. and Carr, M. T. 2009. Bioethanol development in the USA. In: Soetaert, W. and E. J. Vandamme (eds.) Biofuels. John Wiley and Sons, Hoboken, NJ, USA.

Gnansounou, E., Dauriat, A. and Wyman, C. E. 2005. Refining sweet sorghum to ethanol and sugar: economic trade-offs in the context of North China. *Biores. Technol.* 96: 985-1002.

Hein, T. 2012. Are high commodity prices are here to stay? Top Crop Manager (East). April, 2012 issue. pp. 25-26.

Liaquat,A. M., Kalam, M. A., Masjuki, H. H., and Jayed, M. H. 2010. Potential emissions reduction in Road transport sector using biofuel in developing countries. *Atmospheric Environment.* doi:10.1016/j.atmosenv.2010.07.03.

Mittelbach, M. 2009. Process technologies for biodiesel production. In: Soetaert, W. and E. J. Vandamme (eds.) Biofuels. John Wiley and Sons, Hoboken, NJ, USA.

Mousdale, D. M. 2008. Biofuels: Biotechnology, Chemistry, and Sustainable Development. CRC Press, Boca Raton, FL, USA.

Pandey, A. (ed.) 2009. Handbook of plant-based biofuels. CRC Press, Boca Raton, FL, USA.

Puppan, D. 2002. Environmental evaluation of biofuels. Periodica Polytechnica Ser. *Soc Mgmt Sci.* 10: 95-116.

Ramage, J. and Scurlock, J. 1996. Biomass. In: Boyle, G. (ed.) Renewable energy: Power for a sustainable future. Oxford University Press, Oxford, UK. pp. 137-182.

Soetaert, W., and Vandamme, E.J. 2006. The impact of industrial biotechnology. *Biotechnology* J. 1(7– 8): 756–69.

Soetaert, W. and Vandamme, E. J. (eds.) 2009. Biofuels. John Wiley and Sons Ltd, Hoboken, NJ, USA.

2015, Impact of Global Warming and Climate Change on
    Human and Plant Health

Pages 30–44

Editors: **Dr. Arun Arya and Prof. V.S. Patel**
Published by: **DAYA PUBLISHING HOUSE, NEW DELHI**

Chapter 4

Air Pollution and Health: A Case Study of Lucknow City of U.P. (India)

Alok Chantia[1] and Preeti Misra[2]

[1]*Sri Jai Narain Post Graduate (KKC) College, Lucknow, U.P., India*
[2]*Babasaheb Bhimrao Ambedkar University, Lucknow, U.P., India*

ABSTRACT

Man has traveled a long path of civilization where he made numerous efforts for cozy living and happy life. He made big and small dams, technological developments including factories, vehicles etc which caused air pollution and ultimately led to many health problems. Different individuals are affected in different ways by air pollution. The sensitive ones suffer the most as do the elderly or the very young ones. People already afflicted with asthma or those with heart problems suffer more.

Air pollution affects our health in many ways with both *short-term* and *long-term* effects. Some individuals are much more sensitive to pollutants than others. Young children and elderly people often suffer more from the effects of air pollution. The extent to which an individual is harmed by air pollution usually depends on the **total exposure** to the damaging chemicals, *i.e.,* the *duration of exposure* and the *concentration of the chemicals* must be taken into account.

Present study has been done in Lucknow, the state capital of Uttar Pradesh, where fast growing urbanization is adversely affecting the health of its people. In this paper an attempt has been made to examine the correlation between air pollution and health among Lucknowites. It is an exploratory work where 300 people of both sex above 18 years of age have been taken on the basis of random sampling. Present work was done during 15 May 2009 to 14 June 2009. This Study shows that air pollution is causing both short and long term effects to the people of Lucknow. In **short-term effect it is causing** irritation to the eyes, nose and throat, and upper respiratory infections such as bronchitis and pneumonia. Other symptoms include headaches, nausea, and allergic reactions. Short-term air pollution is also aggravating the medical conditions of individuals with asthma and emphysema. In **long-term** it is causing chronic respiratory disease, lung cancer, heart disease, and even damage to the brain, liver, or kidneys. Present study shows that continual exposure to air pollution affects the lungs of growing children and aggravates or complicates medical conditions in the elderly citizens.

Keywords: Air pollution, Human health, Lucknow, Vehicular pollution.

4.1 Introduction

Deteriorating air quality is a major environmental problem in many large urban areas in both developed and developing countries. The environmental impact of emissions from motor vehicles include global climate change from greeenhouse gases, acidification of soil and surface water, adverse effect on plant and animal species and damage to building structures. In India both rural and urban population is affected due to polluting gases. Automobile exhaust is a significant source of air pollution in the urban context. Air pollution is aggravated because of increasing traffic, growing cities, rapid economic development. Increasing vehicles and industries are mainly responsible for the deterioration of air quality as both create noise and emit air pollutants. Technological up gradation and the development of scientific knowledge has resulted in reduced overall pollution levels in Urban areas especially the concentration of gaseous pollutants and mass concentration of particulate matter. However more efforts are needed to achieve the satisfaction level.

Motor vehicles emissions result from fuel combustion or evaporation. The most common type of transport fuels are petrol for light duty vehicles and diesel for heavy duty vehicles. Emission from motor vehicles with spark ignition engines for *e.g.* gasoline- fuelled vehicles are from the exhaust, engine crank-case and fuel system (carburetor, fuel tank etc.). The major pollutants emitted from fuelled vehicles are PM, CO, HC, NOx, etc.

The air quality in urban area depends on the number of vehicles plying on the road, types of fuel (diesel or gasoline), vehicle speed, the meteorological condition (wind speed, wind direction, temperature, relative humidity, etc.). The reason for double-digit growth of automobile-population especially the motorcycle and passenger car segments are that people are gradually becoming attached towards personal means of conveyance and the cities are becoming automobile dependent. This automobile dependence creates several environmental, problems *e.g.*

☆　Oil vulnerability

☆　Photochemical smog

☆　High greeenhouse gas contribution

☆　Increase in the fine and ultra fine particles in the ambient air

☆　Urban sprawl

4.2 Health Hazards of Air Pollutants

Particulate Matter

Particles known as PM10 have a diameter less than 10 µm and when inhaled would penetrate beyond the larynx. Particulate air pollution is associated with a range of effects on health including effects on the respiratory and cardiovascular systems, asthma and mortality. In addition, constituents of particulate matter, such as acid sulphates, may irritate the upper airway and deep lung, reduce bronchial clearance, and modify the lung's resistance to infection.

Effects

☆ Small particles can penetrate deeply into the lung and result in bronco-constriction and an alteration in respiratory mechanisms.

☆ Ultra fine particles ranging from 0.001 to 0.1 micron in diameter are able to penetrate deep down into the lung and to the alveolar sacs where gaseous exchange occurs.

☆ Small particles penetrate deeply into sensitive parts of the lungs and can cause or worsen respiratory disease such as emphysema and bronchitis, and aggravate existing heart disease.

☆ They work by increasing both the rates of blood flow and vascular permeability to white blood cells, elevating clotting activity, constriction of the airways and fever induction.

Sulfur Dioxide (SO_2)

SO_2 is a colorless water-soluble gas. It smells like burnt matches sticks. It can be oxidized to sulphur trioxide, which in the presence of water vapour is readily transformed to sulphuric acid mist. Sulphur dioxide is detectable to the human nose at concentrations of around 0.5–0.8 ppm (1400–2240 µgm^{-3}).

Effects

☆ Exposure to concentrations of 10 to 50 parts per million for 5 to 15 minutes causes irritation of the eyes, nose and throat, choking and coughing.

☆ This causes a reflex cough, irritation, and a feeling of chest tightness, which may lead to narrowing of the airways, particularly likely to occur in people suffering from asthma and chronic lung disease, whose airways are often inflamed and easily irritated.

☆ For nasal breathing with low to moderate volumes the penetration into the lungs is negligible.

☆ For oral inhalation and larger volumes, doses may reach the segmental bronchi.

☆ Exposure of the eyes to liquid sulfur dioxide, (from, for example an industrial accident) can cause severe burns, resulting in the loss of vision.

☆ Repeated or prolonged exposure to moderate concentrations may cause inflammation of the respiratory tract, wheezing and lung damage.

☆ Other health effects include headache, general discomfort and anxiety.

Oxides of Nitrogen (NO_x)

NO_x causes a wide variety of health and environmental impacts because of various compounds and derivatives in the family of nitrogen oxides, including nitrogen dioxide, nitric acid, nitrous oxide, nitrates, and nitric oxide.

NO_2 is a reddish-brown gas with a pungent and irritating odour. It transforms in the air to form gaseous nitric acid and toxic organic nitrates. Nitrogen dioxide (NO_2) can have both acute (short term) and chronic (long- term) effects on health, particularly in people with asthma. Its toxicity relates to its ability to form nitric acid with water in the eye, lung, mucus membrane and skin.

Effects

☆ Eye, nose, and throat irritation.

☆ NO_2 causes inflammation of the airways.

☆ Long term exposure to NO_2 may affect lung function.

☆ May increase the level of respiratory infections in children.

☆ Enhance the response to allergens in sensitized individuals.

☆ Lowering the resistance to diseases such as pneumonia and influenza.

☆ Extremely high-dose exposure (as in a building fire) to NO_2 may result in pulmonary edema and diffuse lung injury.

☆ Continued exposure to high NO_2 levels can contribute to the development of acute or chronic bronchitis.

☆ It can cause collapse, rapid burning and swelling of tissues in the throat and upper respiratory tract, difficult breathing, throat spasms, and fluid build-up in the lungs.

☆ It can interfere with the blood's ability to carry oxygen through the body, causing headache, fatigue, dizziness, and a blue color to the skin and lips.

☆ Industrial exposure to nitrogen dioxide may cause genetic mutations, damage a developing fetus, and decrease fertility in women.

☆ Industrial exposure to nitric oxide can cause unconsciousness, vomiting, mental confusion, and damage to the teeth.

☆ Exposure to low levels of nitrogen oxides in smog can irritate the eyes, nose, throat, and lungs and can cause coughing, shortness of breath, fatigue, and nausea.

Noise

Any unwanted sound can be called as noise. Noise level is more in industries or jet engines. The sound can be measured in Decibel meter.

Effects

☆ Adverse effects varying from hearing loss to annoyance.

☆ Noise produces both temporary and permanent hearing loss. Noise can range from the bursting of the eardrum to permanent hearing loss.

☆ Cardiac and cardiovascular changes, stress, fatigue, dizziness, lack of concentration.

☆ Cause of accident, irritation, inefficiency, deterioration in motor and psychomotor functions, nausea, interference with work tasks and speech communication, headaches, insomnia and loss of appetite and many others.

☆ Continuous noise causes an increase in cholesterol level resulting in constriction of blood vessel making prone to heart attack and stress.

Air pollution is a mixture of solid particles and gases in the air that causes the quality of the air to reduce drastically and become lethal to organisms breathing in it. These gases are nitrogen, oxygen, carbon dioxide, neon, argon, methane, helium, krypton, nitrous oxide, hydrogen, xenon and organic vapours. Particulate matter includes a wide range of pollutants like diesel soot, fly ash, wood smoke, nitrates in fertilizers, sulfate aerosols, lead, arsenic, metals, dust and allergens. Air pollution can have large negative effects on human health.

4.3 Major Components of Air Pollution

Ozone at Ground Level

Ground-level ozone - not be confused with the "good" ozone layer in the upper atmosphere - is one of the major constituents of air pollution in most cities. It's made by a chemical reaction between pollutants such as nitrogen oxides and volatile organic compounds (VOCs) in the presence of heat and sunlight. Exhaust from vehicles, industrial emissions, gasoline vapors, and chemical solvents are major sources of nitrogen oxides and VOCs. The highest levels of ozone pollution occur during summer time. Excessive ozone in the air, especially during summertime, can be dangerous for people with respiratory illnesses. It can cause breathing problems, trigger asthma, reduce lung function and cause lung diseases.

Particulates

Particulates, also referred to as particulate matter (PM) or fine particles, include dust, soot, dirt, smoke and liquid droplets suspended in the air. Some particulates occur naturally, originating from volcanoes, dust storms, forest and grassland fires, others are man made. Chronic exposure to particles contributes to the risk of developing cardiovascular and respiratory diseases, as well as of lung cancer. They can also cause decreased lung function, chronic bronchitis, difficulty in breathing, and irritation in eye, nose and throat.

Carbon Monoxide

Carbon monoxide is a colourless, odourless, non-irritating gas which is comes from vehicle exhaust, wood burning, forest fires, manufacturing processes and cigarettes. Carbon monoxide is a very poisonous gas which can reduce the body's ability to deliver oxygen to tissues and organs, such as the heart and brain. For people exposed to extremely high levels of this gas, it can be fatal. Carbon monoxide poisoning can lead to headache, loss of consciousness, difficulty performing complex tasks and even aggravation of heart problems like angina and coronary artery disease.

Nitrogen Oxides

Nitrogen oxides are a group of highly reactive gases. They, especially nitrogen dioxide, are emitted from burning of fuels by vehicles and industrial plants. They can often be seen as reddish brown layer over many urban areas. Long term exposure to nitrogen dioxide has been shown to increase symptoms of bronchitis in asthmatic children. Nitrogen oxides and sulfur dioxide react with other substances in the air to form acids which then fall to earth as rain or fog.

Sulfur Dioxide

Sulfur dioxide is a colourless gas with a sharp odour. The main source of sulfur dioxide is the burning of sulfur-containing fossil fuels (petroleum and coal) for domestic heating, power generation and motor vehicles. Other sources include process of extracting gasoline from oil, or when metals are extracted from ore. Sulfur dioxide can affect the respiratory system and the functions of the lungs, and causes irritation of the eyes. Inflammation of the respiratory tract causes coughing, mucus secretion, aggravation of asthma and chronic bronchitis. Further oxidation of sulfur dioxide in the presence of a catalyst such as nitrogen dioxide forms sulfuric acid. This is the main component of acid rain which is a cause of deforestation.

Lead

The main sources of lead emissions are metals-processing facilities. Earlier leaded petrol used to be the main source of lead in the air, but it has been phased out. Lead can cause damage to liver, kidneys, brain and other organs. Excessive exposure can cause behavioral disorders, memory problems and seizures.

4.4 How Air Pollution is Damaging to our Health?

Air pollution is generally considered to be potentially deadly as it can cause following complication in our body:

Problems

In a study of the effects of air pollution on children, researchers reported that children and teenagers in Southern California communities with higher levels of air pollution were more likely to have diminished lung function.

Increase the Risk of Diabetes

Researchers at Ohio State University Medical Center have found a strong connection between diabetes, diet, and air pollution. As part of the research, the research team exposed mice to the same dirty air that many of us breathe every day. The tests showed that air pollution increased body fat, caused inflammation and interferes with how the mice processed insulin, a hallmark of diabetes. The effects were strongest when combined with poor diet. Another study reported that the risk of death from diabetes can be more than two times higher in areas of high air pollution.

Increase Incidence of Bronchitis

Increased particulate exposure enhances the incidence of bronchitis in exposed population.

Increase Risk of Lung Cancer

According to a huge scientific study done on around half million people, long-term exposure to the air pollution can significantly increase the risk of lung cancer. The increase in risk is from combustion-related fine particulate matter - soot emitted by cars and trucks, coal-fired power plants and factories.

Increase the Risk of Heart Disease

American Heart Association (AHA) recently suggested that short- and long-term exposure to air pollution directly increases the likelihood of heart attack, stroke, or other cardiovascular problems. Fine particulate matter is again behind this increased risk factor.

Respiratory Diseases

Respiratory problems are a very natural and scientifically established result of air pollution. Problems like asthma, emphysema, pneumoconiosis are quite common in heavily polluted cities. A new study also links air pollution with the increased risk for breathing problems during sleep. More specifically, air pollution increased the risk of sleep-disordered breathing in which breathing stops briefly during sleep.

Cause other Problems

Higher levels of air pollution result in dizziness, mental confusion, severe headaches, nausea, and fainting on mild exertion. Increasing levels of air pollution in the state is also proving to be a major contributor to the formation of dense fog and low day temperatures in winters.

Aerosols or Suspended Particle Matter (SPM) emitted by vehicles and industries in the air combine with the moisture content in the atmosphere to increase the density of the fog, which lasts for a longer period. The fog thus formed bars penetration of sun rays leading to fall in temperatures during the daytime.

4.5 Effect of Air Pollution on Child Birth

According to new research examined by the World Health Organization (WHO), air pollution in cities around the world causes roughly 2 million premature deaths each year (6 October, 2006). A study at Queensland University in Brisbane, Australia warns that women exposed to air pollution during pregnancy have smaller fetuses than those living in areas with cleaner air.

Women should seriously consider reducing their exposure to this form of air pollution during pregnancy. Those who live near major roads are most at risk. A study compared fetus sizes of 15,000+ ultrasound scans to levels of air pollution in different areas within a 14 km radius of downtown Brisbane. The fetuses were between the 13th and 26th weeks of development. The study revealed that "if the pollution levels were high, the size of the fetus decreased significantly." Quantitative measurements of head and abdominal circumference and femur length were smaller in areas of higher pollution (http://www.ehponline.org/docs/2007). These results are important because fetus size is a positive determinant of birth weight and bigger babies have been shown to be healthier and have higher IQs.

Nitrogen dioxide has been linked to Sudden Infant Death Syndrome (SIDS) by University of California, San Diego researchers in a 2005 study. SIDS is characterized by the sudden and unexplainable death of a seemingly healthy infant aged one month to one year (http://adc.bmj.com/cgi/content/abst).

Furthermore, a 2004 study from Sao Paulo University, Brazil found that fewer boys were born in polluted areas. Female fetuses tend to be more apt at surviving harsh conditions in the womb and during birth.

The researchers in Brazil did another study with male rats in filtered and unfiltered ambient air. After four months of these conditions, the male rates were mated with female rats not exposed to pollution. Males from the filtered air produced young of a 1.34 male/female ratio. Males from unfiltered air produced young of a 0.86 male/female ratio. Some hypotheses that can be drawn from these results are that pollution increases the ratio of female/male sperm produced, pollution damages male sperm more than female egg, or that both are damaged the same amount but males are less able to withstand the damage. The exact mechanism is unknown (http://en.wikipedia.org/wiki/Sex_ratio).

4.6 Air Pollution in Lucknow and its Effect

The cities mainly responsible for high vehicular and industrial pollution in Uttar Pradesh are Lucknow, Kanpur, Agra, Varanasi, Jhansi, Allahabad, Meerut, Ghaziabad and Noida. According to the UP Pollution Control Board, the SPM level in major cities of the state is three to five times than the standard limit of 60 μg per cubic metre. The number of vehicles in UP have increased five times in last two decades; from 21 lakh in 1990 to 1.02 crore in 2010. The number of vehicles is said to be increasing at the rate of 5-10 per cent per year.

According to state meteorological director "definitely, air pollution is fast becoming a major reason for a dense fog. The smoke and fog combine to become smog, which can be seen even when temperatures are moderate." Aerosols absorb moisture and blend with the fog formed due to condensation of moisture due to cold air. This combination is more thick and dense in comparison to the normal fog, hence it lasts for a longer period and gets cleared only when sun rays are assisted by the moderate wind.

The change in the weather conditions has been more severe in last few years, particularly in January after heavy snowfall in the hill areas. The SPM are present in the atmosphere in summers also, but do not pose much problem because moisture content is the atmosphere is very low. In monsoon season, the pollutants are cleared by rainfall. However, in winters, as temperatures come down due to icy winds coming to plains from the snow clad hills, the moisture content is condensed into fog. The presence of SPM expedites the process of fog formation, leading to drop in day temperatures and chilly weather (TOI, 2011).

In view of above facts, it is need of the hour to have a look at the Lucknow city having a population of 22,45,509 (Municipal corporation + Cantonment) as per 2001 census and an area of 310 sq. km. In Lucknow alone, over one lakh vehicles are registered with the regional transport office. Total vehicle registered with RTO, Lucknow during 2006-2007 were 9,04,831 as against 8,24,003 during 2005-2006. The overall growth registered is 9.81 per cent during 2006-2007. Number of registered vehicles with RTO Lucknow in different categories during last two years is given in Table 4.5 and details of vehicles plying as public transport (non government) on different routes in Lucknow is shown in Table 4.6. Table 4.7 details concentration of SPM, RSPM, SO2, and NOx in different localities of Lucknow, whereas, Table 8 shows average concentration of SPM, RSPM, SO2 and NOx in different areas of Lucknow.

300 respondents of both sex were selected on the basis of random sampling to see the effect of pollution on the residents of Lucknow. The results are shown in following Tables.

Table 4.1: Result of Questions Asked
What is your opinion about pollution in Lucknow?

Sex	Nos.	Good		Bad		Worst	
		No.	Per cent	No.	Per cent	No	Per cent
Male	143	8	5.59	62	43.36	73	51.05
Female	157	5	3.49	70	44.59	82	52.22
Total	**300**	**13**	**4.33**	**132**	**44.0**	**155**	**51.67**

Table 4.1 shows that majority of respondents felt that environment was not good in Lucknow. Only 8(5.59 per cent) males and 5(3.49 per cent) females thought that environment without pollution was good but it is remarkable that these respondents were living in the out skirts of the city. Majority of respondents showed a resonance between bad and worst environment. The respondents prefer to say worst instead of bad. But actually it may be in category of bad. Table 1 revealed that 43.36 per cent males said bad where as 51.05 per cent said worst. The 44.59 per cent females say bad where as 52.22 per cent said worst. It was found that women were more susceptible to pollution or bad environment.

Table 4.2: How the Pollution Affects Environment ?

Sex	Nos.	It Affects Health		It Affects Air		It Affects Water	
		No.	Per cent	No.	Per cent	No	Per cent
Male	135	97	71.85	27	20.00	11	8.15
Female	152	112	73.68	29	19.08	11	7.24
Total	**287**	**209**	**72.82**	**56**	**19.51**	**22**	**7.67**

Table 4.2 showed that majority of respondents 72.82 per cent were of the opinion that polluted environment affects health. Whereas 19.51 per cent respondents said that it affects air. 22 (7.67 per cent) said that bad environment affects water. So it is crystal clear that almost all respondents correlated bad environment with polluted air and water which ultimately affects their health.

Table 4.3: What Problem you Face in Polluted Environment ?

Sex	Nos.	Breathing Problem		Skin Burn/Irritation		Eyes Irritation		Blood Pressure	
		No.	Per cent	No.	Per cent	No	Per cent	No	Per cent
Male	143	63	44.06	47	32.87	21	14.69	12	8.39
Female	157	72	45.86	59	37.58	15	9.55	11	7.01
Total	**300**	**135**	**45.00**	**106**	**35.33**	**36**	**12.0**	**23**	**7.67**

Consistency in health problem may give an idea to an individual that it might be the cause of pollution or bad environment and Table 4.3 explores this. 135(45 per cent) respondents say that they feel breathing problem. Here percentage of women for complaint are little bit higher (45.86 per cent). 106(35.33 per cent) respondents report about skin burn or irritation, whereas 36(12 per cent) say that they suffer from irritation in eyes. Only 23(7.67 per cent) say that when they remain out from their homes/office for a long time, they feel high blood pressure. Exposure to urban pollution may be a cause of high blood pressure, according to a study conducted by researchers from the University of Dusiburg-Essen in Germany and presented at a meeting of the American Thoracic Society. "Our results also show that living in areas with higher levels of particle air pollution is associated with higher blood pressure." This finding points out that air pollution does not only trigger life-threatening events like heart attacks and strokes, but that it may also influence the underlying processes, which lead to chronic cardiovascular diseases.

Prior studies have shown that short-term spikes in the levels of air pollutants lead to equivalent increases in blood pressure, but little research had been done on the effects of longer-term exposure. In the current study, the researchers recorded air pollution levels over the course of a year, and compared this with blood-pressure readings among 5,000 residents of those areas over the course of four years. The researchers found that after adjusting for potential confounding factors such as age, gender, smoking and weight, increased exposure to fine particulate matter was significantly correlated with higher blood pressure. The effect was stronger in women than in men." There is extensive ongoing research into the link between air pollution and heart disease," said Judy O'Sullivan of the British Heart Foundation. "This will help us understand what needs to be done to minimize harm to heart health and protect people most at risk from pollution."(http://news.bbc.co.uk/2/hi/health)

Air is an ultimate factor for our smooth survival but due to population growth and unavailability of facilities in small districts, Lucknow city is expanding rapidly which has caused an adverse effect on air density and health of an individual too which is shown in Table 4.4. 147(49 per cent) respondents think that air pollution has increased due to increasing number of vehicles in the city. (24.33 per cent) respondents show the tendency of no use of public transport But 38(12.67 per cent) respondents think that number of vehicles have increased because of expansion of boundary of the city. Apart from these responses, the most important point is mismanagement of work of an individual. 42(14 per cent) respondents say that an individual does not make ay schedule when he/she leaves home/office. So he/she uses vehicle arbitrarily to and fro to complete a routine work which ultimately causes pollution.

Table 4.4: What do you think is the Cause of Polluted Environment?

Sex	Nos.	Increased No. of Vehicles		Extension of City Boundary		Minimum Use of Public Transport		Mismanagement of Work	
		No.	Per cent	No.	Per cent	No	Per cent	No	Per cent
Male	143	68	47.55	21	14.69	40	27.97	14	9.79
Female	157	79	50.32	17	10.83	33	21.02	28	17.83
Total	**300**	**147**	**49.0**	**38**	**12.67**	**73**	**24.33**	**42**	**14.0**

Table 4.5: Registered Vehicles with R.T.O. Lucknow During 2005-06 and 2006-07

Sl.No.	Type of Vehicle	Number of Registered Vehicles		Per cent Rise
		On 31st March, 2006	On 31stMarch, 2007	
1.	Multi Axial	917	1365	48.85
2.	Medium and Heavy weight vehicles	8014	8232	2.72
3.	Light Commercial Vehicles (Three Wheelers)	2930	3362	14.74
4.	Light commercial Vehicles (Four wheeler)	5365	6217	15.88
5.	Buses	3978	4198	5.53
6.	Taxi	5979	8012	34.00
7.	Three Wheelers and Auto Rickshaw	12502	15154	21.21
8.	Two wheelers	660093	720378	9.13
9.	Car	94222	105674	12.15
10.	Jeep	12428	13000	4.60
11.	Tractor	13385	13923	4.02
12.	Trailers	991	1062	7.16
13.	Others	3199	4254	32.97
	Total	**8,24,003**	**9,04,831**	**9.81**

Source: RTO, Lucknow, 2007.

Table 4.6: Present Status of Public Transport Available on Different Routes as on 31-03-07

Sl.No.	Type of Vehicles		Number
1	Nagar Bus		110
		CNG Bus	43
2	Tempo Taxi		1776
		Diesel	578
		CNG	1153
		Battery	45
3	Auto Rickshaw		2129
		Diesel	26
		CNG	2103

Source: RTO, Lucknow, 2007.

Table 4.7: Concentration (μg/m³) of SPM, RSPM, SO$_2$, and NOx.

Location	Days	SPM	RSPM	SO$_2$ A	B	C	Mean	NOx A	B	C	Mean
Aliganj	I	305.3	148.8	15.1	18.5	13.2	15.9	18.4	33.8	24.1	25.5
	II	372.1	196.7	20.9	27.5	15.41	21.3	28.7	30.2	28.2	29.0
	Avg	**338.7**	**172.7**	**18.0**	**23.0**	**14.3**	**18.4**	**23.6**	**32.0**	**26.2**	**27.2**
Vikas Nagar	I	336.0	195.4	18.8	25.1	15.3	21.2	23.5	30.1	23.0	25.7
	II	288.0	160.7	20.7	30.3	21.1	24.0	20.7	25.3	15.8	20.7
	Avg	**312.0**	**178.0**	**19.7**	**27.7**	**18.2**	**22.6**	**22.1**	**27.7**	**19.4**	**23.1**
Indira Nagar	I	355.9	181.7	14.2	30.2	20.1	21.5	23.0	36.9	30.6	30.2
	II	361.4	188.3	26.4	29.3	19.1	24.9	28.2	32.2	29.2	29.8
	Avg	**348.4**	**185.0**	**20.3**	**29.7**	**19.6**	**23.2**	**25.6**	**34.5**	**29.9**	**30.0**
GomtiNagar	I	355.3	174.4	21.9	22.7	19.3	21.3	26.7	27.7	23.5	25.9
	II	317.6	155.3	20.1	24.5	15.2	19.9	24.5	26.2	18.5	23.1
	Avg	**336.4**	**164.9**	**21.0**	**25.6**	**17.3**	**20.6**	**25.6**	**26.9**	**21.0**	**23.5**
Hussainganj	I	405.3	202.7	31.8	44.2	25.8	33.9	27.7	49.5	31.3	36.2
	II	434.1	238.8	27.4	33.3	28.1	29.6	36.8	52.2	28.5	39.2
	Avg	**419.7**	**220.7**	**29.6**	**38.8**	**26.9**	**31.7**	**32.3**	**50.9**	**29.9**	**37.7**
Charbagh	I	358.9	181.4	25.2	37.7	31.3	31.4	38.4	50.5	33.4	40.7
	II	421.9	211.2	28.3	31.8	24.3	28.1	40.3	48.4	29.7	39.5
	Avg	**390.2**	**196.3**	**26.7**	**34.7**	**27.6**	**29.7**	**39.4**	**49.5**	**31.5**	**40.1**

Contd...

Table 4.7–*Contd...*

Location	Days	SPM	RSPM	SO$_2$				NOx			
				A	B	C	Mean	A	B	C	Mean
Alambagh	I	381.3	202.6	20.2	33.7	21.0	24.9	34.3	41.1	35.9	37.1
	II	318.8	174.6	19.5	32.1	17.3	22.9	33.3	34.3	29.5	32.3
	Avg	**350.1**	**188.6**	**19.8**	**32.9**	**19.2**	**23.9**	**33.8**	**37.7**	**32.7**	**34.8**
Aminabad	I	378.7	180.6	16.3	19.7	15.4	17.1	19.9	26.9	14.1	20.3
	II	315.7	166.5	15.2	24.0	14.5	17.8	24.7	29.3	11.7	21.9
	Avg	**347.2**	**173.6**	**15.7**	**21.8**	**14.9**	**17.5**	**22.3**	**28.1**	**12.9**	**21.1**
Chowk	I	344.6	205.4	16.4	31.7	19.5	22.5	23.3	46.4	29.7	33.2
	II	327.3	164.1	17.7	25.3	23.1	22.0	28.7	33.9	28.2	30.3
	Avg	**335.9**	**184.7**	**17.1**	**28.5**	**21.3**	**22.3**	**26.0**	**40.1**	**28.9**	**33.2**
Amausi	I	302.3	145.3	13.8	13.6	10.4	12.0	14.1	22.2	15.9	17.4
	II	331.8	168.7	14.1	18.3	15.7	16.1	17.22	22.4	19.2	19.6
	Avg	**317.1**	**157.0**	**13.9**	**15.9**	**13.1**	**14.4**	**15.7**	**22.3**	**17.5**	**18.5**

A: 06:00–14:00 hr; B: 14:00–22:00 hr; C: 22:00–06:0 hr

Source: ITRC 2007.

Table 4.8: Average Concentration (μg/m³) of SPM, RSPM, SO$_2$ and NOx

Area	Location	SPM	RSPM	SO$_2$	NOx
Residential	Aliganj	338.7	172.7	18.4	27.2
	Vikas nagar	312.0	178.0	22.6	23.1
	Indiranagar	348.4	185.0	23.2	30.0
	Gomti nagar	336.4	164.9	20.6	23.5
	Average	**333.9**	**175.2**	**21.2**	**25.9**
Commercial	Hussainganj	419.7	220.7	31.7	37.7
	Charbagh	390.2	196.3	29.7	40.1
	Alambagh	350.1	188.6	23.9	34.8
	Aminabad	347.2	173.6	17.5	21.1
	Chowk	335.9	184.7	22.3	33.2
	Average	**431.8**	**192.8**	**25.0**	**33.4**
Industrial	Amausi	317.1	157.0	14.4	18.5

Source: ITRC 2007.

Vehicles are the prime source of air pollution in urban areas and these air pollutants from vehicles received more attention than ever before with the continuous increase of vehicle demand world widely in recent decades (USEPA, 1991a,1991b; Kenneth, 1994; Larsolov, 1994; Jorgensen, 1996; Bradley *et al.,* 1999; Singer and Harley, 2000; Ye *et al.,* 2000; Charron and Harrison, 2003; Schifter *et al.,* 2003). Gases and particulate emissions from automobiles add air pollution to atmosphere drastically. The major air pollutants emitting from vehicular traffic are oxides of nitrogen (NOx), carbon monoxide (CO), hydrocarbons (HCs), lead (Pb) and sulphur dioxide (SO2) and particulate matter (PM). Due to discharge of excessive amounts of pollutants from vehicular traffic, the environmental air quality is deteriorating continuously. According to the estimation of United Nations, 4.9 billion inhabitants out of 8.1 billion will be living in cities by 2030 (UNCSD, 2001). This makes air pollution in relation to human health a prime policy concern.

References

Bradley, K.S., Stedman, D.H., Bishop, G.A. 1999. A global inventory of carbon monoxide emissions from motor vehicles, *Chemosphere: Global Change Science,* Vol.1, pp. 65–72.

Charron, A. and Harrison, R.M. 2003. Primary particle formation from vehicle emissions during exhaust dilution in the roadside atmosphere, *Atmospheric Environment,* 37(29): 4109–4119.

http://adc.bmj.com/cgi/content

http://en.wikipedia.org/wiki/Sex

http://www.ehponline.org/docs/2007

Jorgensen, K. 1996. Emissions from light and medium goods vehicle in Denmark, *The Science of the Total Environment,* 189-190(28): 131–138.

Kenneth, T.K. 1994. On-road vehicle emissions: U S studies, *The Science of the Total Environment,* 146-147(23): 209-215.

Larsolov, O. 1994 Motor vehicle air pollution control in Sweden, *The Science of the Total Environment,* 146-147(23): 27-34.

Report presented by ITRC Lucknow, on environmental day 5th June 2007.

Schifter, I., Diaz, L., Vera, M., Guzman, E., and Lopez-Salinas, E. 2003. Impact of sulfur-in-gasoline on motor vehicle emissions in the metropolitan area of Mexico City, *Fuel,* 82(13): 1605–1612.

Singer, B.C. and Harley, R.A. 2000. A fuel-based inventory of motor vehicle exhaust emissions in the Los Angeles area during summer 1997, *Atmospheric Environment,* 34: 1783–1795.

Times of India, p-5, 10 Jan 2011, Lucknow edition.

UNCSD 2001. Protection of the Atmosphere- Report to the Secretary General, E/CN.17/2001/2, Commission for Sustainable Development, New York.

USEPA (1991a.) United States Environmental Protection Agency, Draft 1991 Transportation Air Quality Planning Guidelines, USEPA.

USEPA (1991b) United States Environmental Protection Agency, National Air Policy and Emissions Trends Report, USEPA.

WHO Report 6 October 2006.

2015, **Impact of Global Warming and Climate Change on Human and Plant Health**

Editors: **Dr. Arun Arya and Prof. V.S. Patel**

Published by: **DAYA PUBLISHING HOUSE, NEW DELHI**

Pages 45–52

Chapter 5

Women's Influence in Global Warming

Hitesh Solanki* and Ishita Joshi**

Department of Botany, University School of Sciences
Gujarat University, Ahmedabad, Gujarat, India

ABSTRACT

Historically, women power associated with the takeover of the domain of the masculine. Climate change is no longer debatable; it is an undeniable fact. The time for governments and the international community to act is now. Climate change and environmental policies must be intrinsically linked with gender, as women are often the first to be affected by our changing environment. Deforestation or contamination increased the time women spent looking for fuel wood or safe, clean water and also women's are at risk of water-borne disease. Studies showed that while women are responsible for managing household resources, they typically don't have a say in the use and management of environmental resources integral to their households and communities.

Keywords*: Women, Global warming, Environment.*

5.1 Environment: A Gift of God

Environment is a precious gift from God and it is our responsibility to manage the natural resources with concern. From billions of years, human being is facing challenges to maintain and nurture the environment which results from the increase of consumerism and the unorganized development of the world (Halim *et al.*, 2012).

E-mail: *husolanki@yahoo.com; **ishita_joshi1988@yahoo.co.in

As the world is growing fast with new technologies, the environmental issues keep on increasing. Among the entire, major problem is climate change which is the result of global warming. Global warming has become perhaps the most complex issue in the world and now requires more concern. It is becoming alarming day by day and warnings from the scientific community are becoming higher. The main reason behind this is the ongoing build up of human-related greenhouse gases produced mainly by the burning of fossil fuels and forests degradation (Anonymous, 2012a).

Greenhouse gases (GHGs, such as carbon dioxide, methane, and water vapour) are being released plentiful in the upper atmosphere causing a strengthening of the naturally occurring greenhouse effect, which leads ultimately to a warmer earth. This phenomenon will cause the wide range of radical changes in the global climate, including stronger and more frequent storms and floods, crop failure, droughts, and sea level rise, and ultimate effect will be climate change caused by global warming (Henry, 2000).

The occurred damage due to global warming cannot be recovered wholly but efforts can be made to minimize the reasons behind it. Many people, institutes are working for the same and lots of strategies being made. Strategies for adapting to climate change should be relevant to the entire natural world. Most of the efforts are being made on biodiversity conservation.

In conservation planning, focus should be made to support the flexibility of biological diversity as it may possible that certain genotype, species and ecosystem can no longer be conserved in a particular area or region due to changed condition. Other strategies can also be planned for land use, landscape values and water supplies which can promote human needs and fulfil conservation goals. Incorporated fisheries management could reduce the pressure on some coastal fisheries. Enhanced efforts on sustainable agriculture and rural development could make biodiversity more flexible. Wood-fuel conservation by introducing efficient stoves and biogas and other forms of renewable energy, could reduce pressures on forests and thus protect biodiversity (Anonymous, 2003).

Many of the practices mentioned above are traditionally practiced since many years. However, this knowledge has been lost or denigrated due to lack of knowledge of the importance of indigenous system. But the role of indigenous people and their knowledge cannot be ignored.

5.1.1 Importance of Traditional Knowledge

Indigenous knowledge (IK) is the unique, traditional and local knowledge existing within and developed around the specific conditions of women and men indigenous to a particular geographical area. IK or traditional knowledge of local knowledge differs from the developed knowledge by a community from the knowledge generated through universities, government research centres, and private industry (Warren, 1992). Moreover, it is stored in indigenous people's memories and activities and is reflected in stories, songs, proverbs, dances, myths, folklores, cultural values, beliefs, rituals, local language, community laws, agricultural practices, plant species and animal breeds. It is shared and communicated orally, by specific example through culture (Sellato, 2005).

Traditional knowledge in concern with the relationship between human community and the environment can offer valuable insight into how to address the causes and the consequences of climate change. With increased credit and use of Indigenous ecological knowledge, the literature on the subject has grown progressively over the last few decades (Bonny and Berkes, 2008). Much of the earlier research about Indigenous knowledge has emphasized the empirical aspects of that knowledge system (Bonny and Berks', 2008), as in wildlife harvesting (Moller, Berkes *et al.,* 2004; Pearce and Smit, 2010) and vegetation monitoring (Berkes and Kislalioglu, 1998; LaRochelle and Berkes 2003; Garibaldi and Turner 2004; Ghimire and McKey, 2004). The majority of research on Indigenous knowledge is related to environmental management (Davidson-Hunt and Berkes 2001; Long and Tackle, 2003; Watson and Alessa, 2003; Berkes and Davidson-Hunt 2006; Parlee and Berkes, 2006; Berkes and Berkes, 2007; Marie and Sibelet, 2009). Both scientific and Indigenous knowledge are useful because they extend the area of accessible information (Berkes and Berkes 2009).

The world around us is surrounded with biodiversity. Biodiversity is the sum of organisms that include plants, animals, microorganisms and the ecosystems they live in (CBD, 1998). The local/indigenous peoples have been custodians of diversity so far. Through an understanding of the natural processes they have been able to use and share the natural resources wisely Gift giving and receiving, however, is thwarted by the exchange value system. The existence of sacred groves in many parts of the world is a living example of the spiritual aspect in the conservation of biodiversity. Thus knowledge of biodiversity and its use are intertwined with traditional, cultural, spiritual practices of a large number of peoples (Natarajan, 2012). Moreover, the practice of *Tuwa di Pogiwian* (Sabah, Malaysia) aims for the purpose of fertilization and continuity of the plant. In this practice, forest users are not allowed to take the last fruit from a tree in order to make sure the circulation and propagation of the species fruit (Halim *et al.,* 2012).

Indigenous people play a very important role in biodiversity conservation through their traditional knowledge in various fields like agro biodiversity, medicinal diversity and socio-cultural or spiritual aspects. As women are associated with all of these aspects, their contribution in biodiversity conservation and indirectly in preventive measures of global warming is very significant.

5.2 Women's Role in Climate Change and Biodiversity Management

Women, with their central role in the household in village societies, have always been accountable for the food and nutritional needs of their families, and therefore possess a detailed knowledge about the species and ecosystems surrounding them. In traditional agriculture, women are involved in almost all aspects of farming, from seed selection and planting, to harvesting, weeding, winnowing and storing grain. They are often the local educators, passing on traditional knowledge and technologies, such as the proverbial "grandmother's cures", which may hold the key to many curative plant uses. The international community is beginning to acknowledge the importance of women's roles in biodiversity management, and to revise their strategies to conserve traditional knowledge and traditional cultural expressions by facilitating a greater participation by women (www.wipo.int/women-and-ip/en/programs/tk.htm, 2012).

In Sabah region of Malaysia, people refer to the activity of collecting food materials (for building, handicrafts, medicines etc.) from the forest. Only mature ones should be taken with the amount that is required for the family daily needs. It is an acceptable norm to take only what is needed when collecting foods and materials in the forest. The purpose of this practice is to maintain the forest resources and avoid wasting the natural resources of the woodland. And in this practice women are participating actively as male member of family are busy with other works (Tongkul, 2002).

Women play a very significant role in agricultural practices. They are involved in seed selection, production, harvest, storage, processing, and last but not the least cooking (GRAIN 2000, Owen 1998). Arawakan women of the Guainia-Negro region of the Venezuelan Amazon cultivate more than 70 varieties of bitter manioc (Hoffmann, 2003). In one of the many villages of Liberia, Keller women maintained 112 varieties of rice (Thomason, 1991).

A large number of wild plants, vegetable and fruits are the part of diet of indigenous people.

In Kenya a considerable amount of wild biodiversity is used by women during the rainy season for food, medicine and other products. In Bangladesh stagnation of water is common during the monsoons. Women collect seeds of jack fruit, fry them and keep them ready before the onset of the monsoon, as it may be difficult to prepare elaborate meals due to water logging (Natarajan, 2012). Women are also practicing kitchen garden or home gardens where vegetables, greens and other herbs are cultivated. These home gardens also harbour many indigenous varieties that are very often taken care of by women. In rural Bangladesh for example, women select seeds of vine and gourd species, chiefly indigenous varieties that are to be grown the following year (Wilson, 2003).

Plants have a very important role in indigenous people's life as they are used for medicinal purposes. Samoan women healers use about 100 different plant species (Cox, 1995,2000). In India indigenous women from the then State of Madhya Pradesh use a combination of plants as birth control agents (CSE, 1982).

Women are always found to be more religious and actively participating in all religious ceremonies. For example temples and homes in cities, plants and leaves are used for religious purposes and during festivals. According to Ghate (1998) at least 45 species used in different religious ceremonies from Pune city have been enumerated. These species are both cultivated and available as wild plants. In the past decade we have seen women as victims of development and environmental degradation. But what we increasingly realise is that they play a crucial role in the cultivation and the management of diversity. Their contribution to food production is large. In Africa women produce 80 per cent of the food, in Asia 60 per cent and in Latin America 40 per cent. In addition to food production as discussed above women use a number of species to cater to their daily needs, such as wild and domesticated species trees, shrubs, roots, leaves, bark and animals as food and medicines. Thus they rely on a diverse range of species and preserve the same (Natarajan, 2012).

The well known Chipko movement was staged by women in India to guard the trees in the region as they realised the importance of the forest. In Kenya the Green Belt Movement mobilised by 80,000 women to plant trees. Indeed we can see women resisting the destruction of the environment in countries of both the North and the South as it is they who have seen the connections between ecology, health and survival all along (Natarajan, 2012).

Biodiversity management in South India clearly shows how tradition and culture are carefully applied in cultivation and protection of diversity. In areas where finger millet is grown women are involved in seed selection and storage. Seed selection starts with the celebration of diversity in the form of performing rituals, and takes place continuously through field observation, which they do when they work (Ramprasad, 1999). Natural resources are given respect and treated with dignity. They are also appreciated and shared. These aspects are taught and practiced by women of indigenous societies of Northwest America for sustainable development and to fight consumerism (Turner 1992;Turner and Atleo, 1998).

Women farmers have in fact been largely accountable for the development and adaptation of many plant varieties. In the Kalasin region of northern Thailand, women manage the interface between wild and domesticated species of edible plants. They have not only brought new species of wild plants under cultivation in recent years, they have also spurred their communities to carefully regulate collection rights in the face of increasing commercialization (Easton and Ronald, 2000).

As is clear from the previous pages, women's understanding of local biodiversity tends to be broad, containing many unique insights into local species and ecosystems gained from centuries of practical experience. A study in Sierra Leone found, for example, that women could name 31 uses of trees on fallow land and in forests while men could name only eight (Domoto, 1994). In a sample participatory study, women hill farmers in Dehra Dun, India provided the researchers with no less than 145 species of forest plants that they knew and used (Shiva and Dank, 1992).

The efforts of urban women are as important as the knowledge and practice of indigenous women. According to Dr. Sarah Otterstrom, Executive Director of Paso Pacifica, women are leading the crusade against climate change and other important environmental issues that confront the planet at the Clinton Global Initiative. In Nicaragua, women are leading reforestation efforts and have planted over 100,000 native trees. Their work has offset more than 150,000 tons of greenhouse gases and helps to protect watersheds that are crucial to the health of their communities (http://news.yahoo.com/women-lead-fight-against-climate-change-140818855.html, 2011)

References

Anonymous 2003. How will global warming affect my world? A simplified guide to the IPCC's "Climate Change 2001: Impacts, Adaptation and Vulnerability". Published by the United Nations Environment Programme, pp. 10.

Anonymous 2012a. Global Warming and Climate Change. *The Newyork times*. On-line, http://topics.nytimes.com/top/news/science/topics/globalwarming/index.html. <Date accessed online: 25May, 2012>

Anonymous 2012b World intellectual property organization. Women and Traditional Knowledge.On-linewww.wipo.int/women-and-ip/en/programs/tk.htm, <Date accessed online: 25May, 2012>

Berkes F and Davidson IJ, Hunt 2006. Biodiversity, traditional management systems, and cultural landscapes: examples from the boreal forest of Canada. *International Social Science Journal.* 58(187): 35.

Berkes F and Berkes MK 2009. Ecological complexity, fuzzy logic, and holism, In indigenous knowledge. *Futures.* 41(1): 6-12.

Berkes F and Berkes MK 2007. Collaborative integrated management in Canada's north: The role of local and traditional knowledge and community based monitoring. *Coastal Management.* 35(1): 143-162.

Berkes F and Kislalioglu M 1998. Exploring the basic ecological unit: Ecosystem-like concepts in traditional societies. *Ecosystems.* 1(5): 409-415.

Bonny E and Berkes F 2008. Communicating traditional environmental knowledge: Addressing the diversity of knowledge, audiences and media types. *Polar Record.* 44(230): 243-253.

CBD (Convention on Biological Diversity) 1998. Convention on biological diversity, Text and annexes, ICAO. Canada.

Cox PA 1995. Shaman as scientist: indigenous knowledge systems in pharmacological research and conservation. In K. Hostettmann *et al.* (eds.). *Phytochemistry of plants used in traditional medicine*, Clarendon Press, Oxford

Cox PA 2000, Will tribal knowledge survive the millennium? *Science.* 287: 44-45.

CSE (Centre for Science and Environment) 1982 The state of India's environment 1982. A Citizens report. CSE, New Delhi.

Davidson-Hunt IJ and Berkes F 2001. Changing resource management paradigms, traditional ecological knowledge, and non-timber forest products. Forest Communities in the Third Millennium: Linking Research Business and Policy toward. A Sustainable Non-Timber Forest Product Sector. 217: 78-92.

Dixon C 1990. Rural development in the third world, Routledge, London.

Domoto D.T. 1994. Rehabilitation of the End-stage Renal Disease patient: Are the right questions being asked? *Am J Kidney Dis.* 23(3):467-468.

Easton P and Ronald M 2000. Seeds of life: Women and agricultural biodiversity in Africa.

IK Notes No. 23. Africa region's knowledge and learning center. 4pp.

Farnsworth NR 1988. "Screening plants for new medicines," In Biodiversity Eds. E.O. Wilson and Frances M Peter, National Academy Press, Washington, D.C.

Garibaldi A and Turner N 2004. Cultural keystone species: Implications for Ecological conservation and restoration. *Ecology and Society*, 9:3.

Ghate V.S. 1998. Plants in patra-pooja: Notes on their identity and utilization. *Ethnobotany.* 10(1-2): 6-15.

Ghimire SK and McKey D 2004. Heterogeneity in ethno ecological knowledge and management of medicinal plants in the Himalayas of Nepal: Implications for conservation. *Ecology and Society.* 9:3.

Anonymous 2000. Potato: A fragile gift from the Andes. GRAIN, September 2000.

Halim A AB, Othman N, Ismail SR, Jawan JA, and Ibrahim NN 2012. Indigenous Knowledge and Biodiversity Conservation in Sabah, Malaysia. *International Journal of Social Science and Humanity.* 2(2): 159-163.

Henry AD 2000 Public Perceptions of Global Warming. *Human Ecology Review.* 7(1): 25-30.

Hoffmann S 2003 Arawakan women and the erosion of traditional food production in Amazon Venezuela. In Women and Plants, Gender relations in biodiversity management and conservation. Ed. Patricia L. Howard. Zed Books, London and New York. http://news.yahoo.com/women-lead-fight-against-climate-change-140818855.html (2011). <data accessed online 25 May, 2012>

LaRochelle S and Berkes F 2003. Traditional ecological knowledge and Practice for Edible wild plants: Biodiversity use by the Raramuri in the Sierra Tarahumara, Mexico. *International Journal of Sustainable Development and World Ecology.* 10(4): 361-375.

Limson J 2002. The rape of the pelargoniums. in Science in Africa- Africa's first on-line science magazine. http://www.scienceinafrica.co.za/2002/june/pelarg.htm. <Data accessed online: 25 May, 2012>

Long J and Tecle A 2003. Cultural foundations for ecological restoration On the white mountain Apache reservation. *Conservation Ecology.* **8**(1).

Marie CN and Sibelet N 2009. Taking into account local practices and indigenous knowledge in an emergency conservation context in Madagascar.

Moller H and Berkes F 2004. Combining science and traditional ecological knowledge: Monitoring Populations for co-management. *Ecology and Society.* **9**(3).

Natarajan B 2012. Biodiversity and Traditional Knowledge: Perspectives for a Gift Economy. Online, www.gift-economy.com/athanor/athanor_008.html. <data accessed online: 25 May, 2012>

Owen, AL 2008. Grades, gender, and encouragement: A regression discontinuity analysis. MPRA Paper 11586, University Library of Munich, Germany.

Parlee B and Berkes F 2006. Indigenous knowledge of ecological variability and commons management: A case study on berry harvesting from Northern Canada. *Human Ecology.* 34(4): 515-528.

Pearce T and Smit B 2010. Inuit vulnerability and adaptive capacity to climate change in Ulukhaktok, Northwest Territories, Canada. *Polar Record*. 46(237): 157-177.

Ramprasad V. 1999. Women and biodiversity conservation. COMPAS Newsletter-October.

Sellato B. 2005. Forests for food, Forest for trade: between sustainability and extractivism, in Histories of the Borneo Environment. R. L. Wadley, Ed. Leiden: Royal Netherlands Institute of Southeast Asian and Caribbean Studies.

Shiva, V. and Dank E. I. 1992. Women and Biological Diversity: Lessons from the Indian Himalaya. In Cooper, D. Vellve, R. and Hobbelink, H. (eds) 1992. Growing Diversity: Genetic Resources and Local Food Security. Intermediate Technology Publications, London. pp44-50

Thomason GC 1991. Libera's seeds of knowledge, Cultural Survival Quarterly, Summer 1991.

Tongkul F 2002. Traditional System of Indigenous Peoples of Sabah, Malaysia. Penapang: Pecos Trust.

Turner NJ 1992. The earth's blanket: traditional aboriginal attitudes towards nature. *Canadian Biodiversity*. 2(4):.5-7.

Turner NJ and Atleo ER 1998. Pacific North American first peoples and the environment, In Traditional and modern approaches to the environment on the Pacific Rim, tensions and values, Ed. H. Coward. Centre for studies in religion and society, State University of New York Press, *Albany*: 105-124.

Warren MD 1992. Indigenous knowledge, biodiversity conservation and development: keynote address, in *Proc. International Conference on Conservation of Biodiversity*, Nairobi, Kenya. pp 15-30.

Watson A and Alessa L 2003. "The relationship between traditional ecological \ knowledge, evolving cultures, and wilderness protection in the circumpolar north." *Conservation Ecology* 8(1):12-24.

Wilson M 2003 Exchange, patriarchy and status: Women's home gardens in Bangladesh, In Traditional and modern approaches to the environment on the Pacific Rim, tensions and values. Ed. H. Coward. Centre for studies in religion and society, State University of New York Press, Albany: 105-124.

2015, Impact of Global Warming and Climate Change on
Human and Plant Health
Editors: Dr. Arun Arya and Prof. V.S. Patel
Published by: DAYA PUBLISHING HOUSE, NEW DELHI

Pages *53–57*

Chapter 6

Asian Desertification an Impending Danger: Lessons from China

S.K. Basu[1*], M. Dutta[2], X. Chin[3] and S. Datta Banik[4]

*[1]Department of Biological Sciences, University of Lethbridge,
Lethbridge, AB, Canada TIK 3M4*
*[2]Department of Microbiology, Gurudas College, University of Calcutta,
Kolkata – 700 054, W.B., India*
*[4]Jefe del Departamento de Ecologia Humana,
Centro de Investigacion y de Estudios Avanzados (CINVESTAV),
del Instituto Politecnico Nacional (IPN), Carretera Antigua a Progreso Km. 6,
A.P. 73, Cordemex 97310, Merida, Yucatan, Mexico*

ABSTRACT

Desertification is one of the most discussed issues of nowadays, affecting almost all the continents over the globe in some form or other. Environmental factors like uncertainty of rainfall, prevalent draughts and unpredictable climate changes as well as anthropogenic factors such as unrestricted felling of trees, overgrazing, urbanization collectively plays a major role behind desertification. This brief review highlights the problems and challenges associated with desertification in an Asian context, predominantly citing examples from China.

***Keywords**: Desertification, Environment, Anthropogenic, Erosion, Desert, Land resources.*

* *Corresponding author.* E-mail: saikat.basu@uleth.ca

6.1 Introduction

Desertification is considered to be the result of a series of natural and anthropogenic processes leading to gradual environmental degradation and loss of the biological or economic productivity of specific region or locality (Warren 2002). Environmental causes such as degradation of the vegetative cover, reduction in organic matter content, adverse changes in soil physical properties, chemical degradation, and water and wind erosions also account for desertification (FAO/UNEP, 1984). Climatic factors such as rainfall and draught also plays an important role (Ramstein *et al.,* 1997). Anthropogenic factors such as over exploitation of land resource is responsible for constituting a fragile ecosystem and faulty irrigation practices have further enhanced the process of desertification in several instances (FAO/UNEP, 1984). In Asia, the problem of desertification seems to have been spreading in China due to low annual precipitation causing steppe grazing land to become increasingly arid. Sandy desertification in China is predominantly caused by wind erosion as consequence of geomorphological changes (Zhu and Chen, 1994).

6.2 The Desertification Process in China

China is most seriously affected by desertification among Asian countries and is also best reported in primary literature sources. Desertification in interior China is a reflection of climate changes in the Northern Hemisphere (Manabe and Broccoli, 1990). Historically, frequent tribal wars and recurrent land transformation practices from crop to pasture triggered desertification in China is distant past (Bao *et al.,* 1984). The environmental settings conducive to desertification in China are predominantly identified as sandy, loose surface sediment deposits combined with unpredictable and long draught prone windy seasons (Wang and Zhu, 2003). Desertification has modified the existing biogeochemical cycles, negatively altering the vital soil nutrient qualities (Galloway *et al.,* 2003). Desertification in sandy drylands can occur by either direct encroachment of mobile sand dunes upon grazing land, or by deposition of drifting sand over grasses, both under the action of wind.

The effects of desertification also include diminished vegetative cover, reduced productivity, increased soil erosion and invasion of exotic species rapidly replacing the vulnerable native species (Wang *et al.,* 2004). Among other causes of rapid desertification are over cultivation, overgrazing, deforestation and salinization (Wang and Zhu, 2003). In addition to the above factors, excessive gathering of fuel wood and plants for medicinal purposes, mining and construction of transportation routes also have contributed to severe desertification (Sheehy, 1992). Immature policy drafting with no long term planning were also responsible for rapid desertification in drought-prone sandy regions in China between the mid-1960s to mid-1970s (Zhu and Cui, 1996). In the northern Shaanxi Province across the Great Wall of China, wind erosion has depleted soil layers, degraded land quality, reduced land productivity and accelerated the process of desertification (Liu and Jiang, 1996). The Karst rocky desertification in southwestern China is caused by fragile ecosystem, pressure of the local demography, irrational land uses and unchecked pollution (Guan *et al.,* 2000). The Horqin Sandy area has also suffered severe desertification due to anthropogenic activities like excessive biomass extraction for firewood, forest encroachment for agricultural purposes and overgrazing by existing cattle and livestock (Zhao *et al.*, 2005). Windy weather and low vegetation cover in winter and spring along with plenty of loose sand are the fundamental factors that set the ground for aeolian desertification in Hulunbir grasslands (Zhu *et al.*, 1989). It has been reported that desertification in the Mu Us sandy land has been initiated primarily as a consequence of climatic fluctuations during the Ice Age, tectonic activities and more recently by anthropogenic activities (Dong *et al.*, 1988).

6.3 Controlling Desertification

Recently some regional and national level initiatives have been adopted to combat desertification. Afforestation practices such as planting native trees and grasses are often considered appropriate measure to combat desertification. In the severely impacted farming regions, the principle measure has been transformation of croplands severely encroached and impacted by desert into woodlands or grasslands (Zhu and Liu, 1984; Li, 1995). The grazing land is divided into several sub plots and then enclosed so that grazing livestock can only access the pasture after it has a chance to recover organically. Efforts are on way to change current farming practices and to popularize grain-grass intercropping and introduction of fodder grasses into the cropland farming system to reduce pressure of grazing (Li, 1995).

Crop rotations for grain, forage and other cash crops needs to be followed to enhance soil fertility, to develop proper animal husbandry practices and to establish stable agro-ecological system so that the regions affected by desertification can support agro-industry under long term planning (Zhu and Liu, 1984). Other notable efforts include establishing artificial vegetation and restoring natural vegetation to control the shifting sand; establishing large sand-break forest belts to prevent sand encroachment on oases, on roads and in the residential areas. Protective forests have been established in strategic areas to prevent cropland and rangeland from desertification (Li, 1980).

Protecting natural vegetation to prevent reactivation of fixed and semi-fixed sand dunes and sandy rangelands is another commonly used approach for desertification control. Engineering measures include establishing artificial sand fences and spraying chemical binders to stabilize shifting sands have been routinely practiced with limited success (Li, 1980). It is essential to slow down wind velocity through increased surface roughness for rehabilitating desertification. Engineering works such as straw checkerboards can effectively reduce wind velocity and minimize the amount of sand transported, even though their optimal width is still debatable. Common land reclamation strategies in China include windbreaks, irrigation with silt-laden river water, and dune stabilization using straw checkerboards and planted xerophytes. However, technologies are old and the geographical region too big to be effectively managed under current practices.

6.4 Conclusions

The multi-disciplinary approach of rehabilitation and prevention of desertification adapted in China based on ecological principles and economic return has achieved much more success in rehabilitating the deteriorated ecosystem than the biological means. Huge efforts are made China to control desertification since UN held its' first conference on combating desertification 20 years ago. With a combination of rehabilitation and preventive measures following basic ecological principles, desertification can be harnessed and land impacted by desertification be reclaimed for productive uses. New policies are being framed at national and international level to control desertification and rehabilitate the desert impacted areas. Community programmes are being organized to make the mass aware of human factors that play a role in desertification and to prevent them in indulging such activities. Thus it can be said more awareness is being developed against desertification and its harmful effects in past few decades, although much needs to be done in future.

References

Bao, Y., Wen, Z., Dong, S., Zhang, J., and Yang, T. 1984. Inventorying desertification types and its prevention in northern Shaanxi province. *J Desert Res.* 4(2): 35-44.

Dong, G., Shen, J., Jin, J., and Gao, S. 1988. The conceptualization of desertification and desertization. *Arid Land Geog.* 11: 58-61.

FAO/UNEP 1984. Metodolog´ýa Provisional Para la Evaluaci´on y la Representaci´on Cartogr´afica de la Desertizaci´on. Rome: FAO, UNEP.

Guan, W.B., Zeng, W.B., and Jiang, F.Q. 2000. Ecological studies on the relationship between the process of desertification and vegetation dynamics in the west of northeast China: community diversity and desertification process. *Acta Ecol.* Sinica 20: 93-98.

Galloway J.N., Aber, J.D., Erisman J.W., Seitzinger, S.P., Howarth, R.W., Cowling, E.B., and Cosby, B.J. 2003. The nitrogen cascade. *BioSci.* 53(4): 341-356.

Li, M.G. (1980). Principles and measures of shifting sand stabilization on both sides of railway. Research on Shifting Sand Control. The Institution of Desert Research, Chinese Academy of Sciences Ningxia People's Publishing House. pp. 27-48.

Li, Y.T. 1995. Several problems on grassland ecological construction in China. Desertification and its control. China Executive Committee Secretariat on UN convention to combat desertification. China Forestry Sci Press. pp. 51-52.

Liu, X., and Jiang, X. 1996. Extraction of desertification information from the optical and digital processing of multiple-date remotely sensed images. *Arid Land Geog.* 19(3): 1-6.

Manabe, S., and Broccoli, A.J. 1990. Mountains and arid climates of middle latitudes. *Science* 247: 192-195.

Ramstein, G., Fluteau, F., Besse, J.,and Joussaume, S. 1997. Effect of orogeny, plate motion and land-sea distribution on Eurasian climate change over the past 30 million years. *Nature.* 386: 788-795.

Sheehy, D.P. 1992. A perspective on desertification of grazingland ecosystems in North China. *Ambio.* 21: 303-307.

Wang, Y., and Dong, G. 1994. Sand sea history of the Taklimakan for the past 30,000 years. *Geografiska Annaler.* 76A(3): 131-141.

Wang, T., and Zhu, Z.D. 2003. Study on aeolian desertification in China: Definition of aeolian desertification and its connotation. *J Desert* Res. 23(3): 209-214.

Wang, T., Wu, W., and Xue, X. 2004. Spatial-temporal changes of aeolian desertified land during last five decades in northern China. *Acta Geog Sinica*. 59(2): 203-212.

Warren, A. 2002. Land degradation is contextual. Land Degrad. *Develop*. 13: 449-459.

Zhao, H. L., Zhao, X. L., Zhou, R. L., Zhang, T. H., and Drakeb, S. 2005.Desertification processes due to heavy grazing in sandy rangeland, Inner Mongolia. *J Arid Environ*. 62: 309-319.

Zhu, Z., and Cui, S. 1996. Distribution patterns of dessertified land and assessment of its control measures in China. *China Environ Sci*. 16: 328-334.

2015, Impact of Global Warming and Climate Change on
      Human and Plant Health   *Pages 58–71*
Editors: **Dr. Arun Arya and Prof. V.S. Patel**
Published by: **DAYA PUBLISHING HOUSE, NEW DELHI**

Chapter 7

Climate Change: Global Warming and Human Health

V. S. Patel*

*President – IAAPC Consulting Engineer and Visiting Professor,
The Maharaja Sayajirao University of Baroda, Vadodara, India*

ABSTRACT

Evidence is gathering that human activities are changing the climate. This **'climate change'** could have a huge impact on our present and future lives including eco system. The following factors make the change in climate.

☆ Human Influence   ☆ Global Warming

☆ Global Carbon Dioxide   ☆ Methane levels

☆ Greeenhouse effect   ☆ A Rise in Global Sea Level

☆ A change in vegetation zones   ☆ An increase in disease levels

☆ A change in ecosystem   ☆ Human Health effect

☆ Ozone depletion   ☆ Loss of biodiversity,

☆ Air Pollution   ☆ Global warming is " a natural cycle"

The factors above affect the climate naturally. However, we cannot forget the influence of humans on our climate. We have been affecting the climate since we appeared on this earth millions of years ago. In those times, the affect on the climate was small. Trees were cut down to provide wood for fires. Trees take in carbon dioxide and produce oxygen. A reduction in trees will therefore have increased the amount of carbon dioxide in the atmosphere.

* *Corresponding author.* E-mail: vpatel_usa@hotmail.com

The Industrial Revolution, starting at the end of the 19th Century, has had a huge effect on climate. The invention of the motor engine and the increased burning of fossil fuels have increased the amount of carbon dioxide in the atmosphere. The number of trees being cut down has also increased, meaning that the extra carbon dioxide produced cannot be changed into oxygen.

Evidence is gathering that human activities are changing the climate. This **'climate change'** could have a huge impact on our lives and eco system. There is growing concern amongst many natural scientists that human interventions are altering the capacity of ecosystems to provide their *goods* (*e.g.* freshwater, food, pharmaceutical products, etc.) and *services* (*e.g.* purification of air, water, soil, sequestration of pollutants, etc.).

Ecosystem disruption can impact on health in a variety of ways and through complex pathways. These are moreover modified by a local population's current vulnerability and their future capacity to implement adaptation measures. The links between ecosystem change and human health are seen most clearly among impoverished communities, who lack the "buffers" that more affluent communities can afford. Global warming is "a natural cycle." Global warming comes and goes in 2,500 year cycles which may have more to do with cosmic rays than fossil fuel emissions,

Some scientists believe that the human contribution to carbon dioxide in the atmosphere, however small, is of a critical amount that could nonetheless upset Earth's environmental balance. "Today, we are simply near a peak in the current cycle that started about 2,500 years ago,"

Keywords: *Global warming, Climate Change, Human health, Air pollution, Migration.*

7.1 Introduction

Evidence is gathering that human activities are changing the climate. This **'climate change'** could have a huge impact on our lives. The following factors make the change in climate.

- ☆ Human Influence
- ☆ Global Warming
- ☆ Global Carbon Dioxide
- ☆ Methane levels
- ☆ Greeenhouse effect
- ☆ A Rise in Global Sea Level
- ☆ A change in vegetation zones
- ☆ A change in ecosystem
- ☆ Human Health effect
- ☆ Ozone depletion
- ☆ Loss of biodiversity,
- ☆ Air Pollution
- ☆ Global warming is " a natural cycle"

7.2 Human Influence

The factors mentioned above affect the climate naturally. However, we cannot forget the influence of humans on our climate. We have been affecting the climate since we appeared on this earth millions of years ago. In those times, the affect on the climate was small. Trees were cut down to provide wood for fires. Trees take in carbon dioxide and produce oxygen. A reduction in trees will therefore have increased the amount of carbon dioxide in the atmosphere.

The Industrial Revolution, starting at the end of the 19th Century, has had a huge effect on climate. The invention of the motor engine and the increased burning of fossil fuels have increased the amount of carbon dioxide in the atmosphere. The number of trees being cut down has also increased, meaning that the extra carbon dioxide produced cannot be changed into oxygen.

Evidence is gathering that human activities are changing the climate. This **'climate change'** could have a huge impact on our lives. We, human are surrounded by our nation boundary. We cannot migrate to the other part of earth for better climate as birds and animals migrate.

7.3 Global Warming

Global temperatures have risen by 0.6 degrees Celsius in the last 130 years (Figure 7.1). This rise in global temperatures leads to huge impacts on a wide range of climate related factors. The graph below shows the rise of global temperatures since 1860.

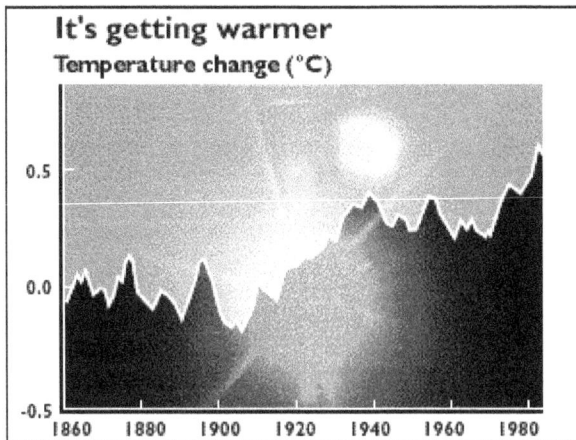

Figure 7.1: Global Temperature Since 1860.
http://news.bbc.co.uk/hi/english/special_report/sci/tech/global_warming

7.4 Global Carbon Dioxide

Levels of carbon dioxide, methane and nitrous oxide gases are rising, mainly as a result of human activities. Carbon dioxide is being dumped in the atmosphere at an alarming rate. Since the Industrial Revolution, humans have been pumping out huge quantities of carbon dioxide, raising carbon dioxide concentrations by 30 per cent

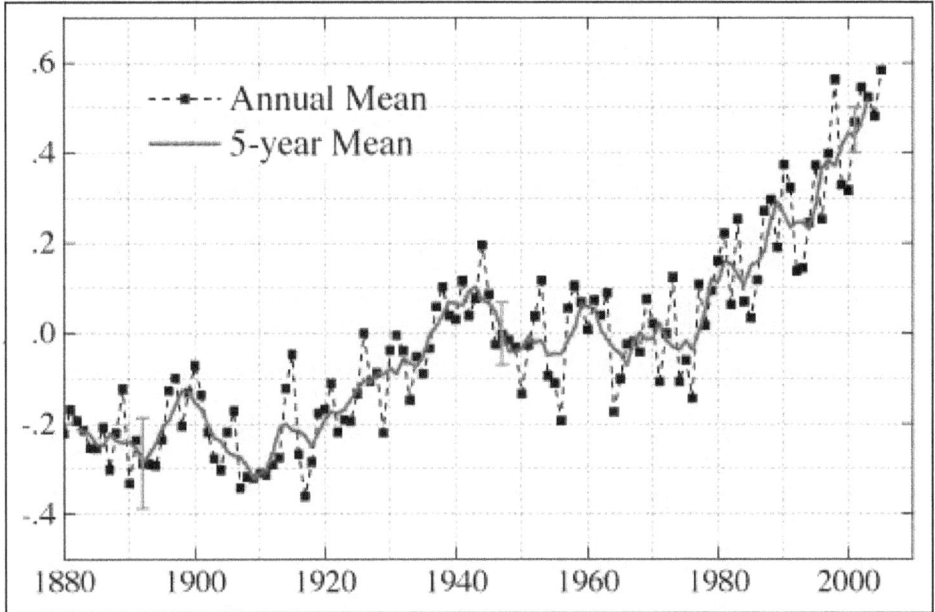

Figure 7.2: Showing Rise in Temperature of Environment.
***Source*: Graph by NASA.**

(Figure 7.3). The burning of fossil fuels is partly responsible for this huge increase. This graph shows carbon dioxide emissions increasing over the last 130 years.

You can see that the rate of increase is huge. How many times greater are the carbon dioxide emissions now than 130 years ago?

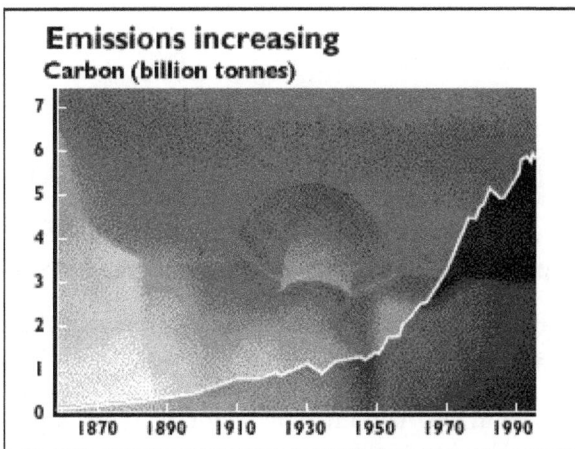

Figure 7.3: Global Carbon Dioxide Emissions Since 1860.
***Source*: Intergovernmental Panel on Climate Change, 1995.**

ONE effect of global warming will be to accelerate the decomposition of soil organic matter, thereby releasing CO_2 to the atmosphere, which will further enhance the warming trend (McNeely *et al.,* 1995, Stern *et al.,* 1973,Kozlowski and Mudd 1975, Vitousek, 1994). Such a feedback mechanism could be quantitatively important, because CO_2 is thought to be responsible for ~55 per cent of the increase in radioactive forcing arising from anthropogenic emissions of gases to the atmosphere (Jenkinson *et al.,* 1991), and there is about twice as much carbon in the top meter of soil as in the atmosphere. Here we use the Rothamsted model for the turnover of organic matter in soil to calculate the amount of CO_2 that would be released from the world stock of soil organic matter if temperatures increase as predicted, the annual return of plant debris to the soil being held constant. If world temperatures rise by 0.03 °C yr^{-1} (the increase considered as most likely by the Intergovernmental Panel on Climate Change (Jenkinson *et al.,* 1991), we estimate that the additional release of CO_2 from soil organic matter over the next 60 years will be 61×10^{15} gC. This is ~19 per cent of the CO_2 that will be released by combustion of fossil fuel during the next 60 years if present use of fuel continues unabated (Jenkinson *et al.,* 1991).

7.5 Global Methane Level and Greenhouse Effect

Methane levels in the atmosphere have increased by 145 per cent since the Industrial Revolution. This increase is a result of gas produced by livestock and paddy fields

There are some uncertainties as to what effects a change in climate might have on the earth. To predict what might happen, we first need to start by understanding how the increase in gases such as carbon dioxide and methane affect our climate. The diagrams below show the effects of these gases on our atmosphere.

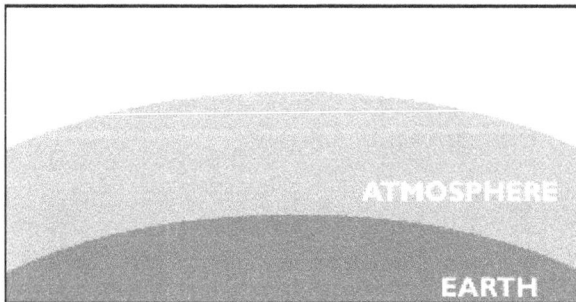

Figure 7.4: Sun and Earth Under Normal Conditions.
http://news.bbc.co.uk/hi/english/special_report/1997

The diagram above left shows the normal conditions of the sun heating the earth. The suns rays hit the earth and some are reflected back into space. However, gases in the atmosphere such as carbon dioxide and methane form a barrier for sunlight. This is shown in the diagram above right. The suns rays hit the earth, but when they are reflected back out into space, they are trapped in the atmosphere. The sun rays cannot escape from the earth's atmosphere, and the earth heats up. This is called the **'Greenhouse Effect'.** Below are some predictions of what might happen to the earth if subjected to a change in climate. The changes may be:

A Rise in Global Sea Levels

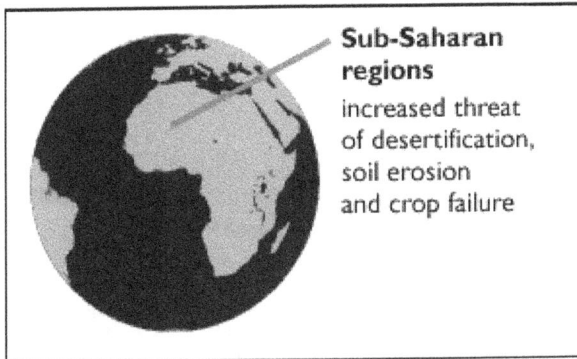

Figure 7.5: The Earth Subjected to a Change in Climate.
http://news.bbc.co.uk/hi/english/

7.6 A Rise in Global Sea Levels

We already know that a global temperature rise of 2.5 degrees Celsius will have a detrimental effect on the earth. One impact of climate change would be a rise in global sea levels. If the heat from the sun cannot escape through the earth's atmosphere then the ice at the north and south poles could melt. This could have a huge effect on the low lying areas of the world. Do you live in a low-lying area, maybe near the sea or in London?

In 1998, 46 million people live in areas at risk of flooding (BBC News, 1997). This amount could increase rapidly if sea levels rose. Scientists estimate that a sea rise of only 50 centimeters would increase the number of people at risk to 92 million. A sea level rise of 1 meter would put 118 million people at risk. Scientists believe that there will be a sea level rise of 50 centimeters over the next 40 to 100 years (BBC News, 1997). The amount of land that would be lost in the Nile Delta and other Delta if the world's oceans rose by 1.5 meters. You can see the huge effect it would have on the area. The rise in sea level would also make millions of people homeless and ruin valuable farmland.

7.7 A Change in Vegetation Zones

A change in climate would have an effect on the world's vegetation zones. We would see a change in the boundaries between grassland, forest and shrub land. This change in vegetation zones could cause famine in arid areas such as Africa that depends on a certain type of crop. The change in vegetation would cause mass movement of people away from arid regions. This could cause huge over-crowding in towns and cities.

The range of pests could also change if the vegetation changed. This could bring about an increase in disease levels. Scientists believe that if the temperature increased by 3-5 degrees Celsius, the number of people potentially exposed to malaria (caught from the mosquito) could increase from 45 per cent to 60 per cent of the world's population. This could lead to an extra 50-80 million cases of malaria a year (BBC News, 1997).

7.8 A Change in Ecosystems

Ecosystems could be affected by a change in temperature. It has been predicted that an increase in temperature would affect species composition. Scientists also believe that up to two thirds of the world's forests would undergo major changes. Scientists believe that deserts would become hotter, and desertification would extend and become harder to reverse.

7.9 Climate Change and Human Health

Climate change is a significant and emerging threat to public health, and changes the way we must look at protecting vulnerable populations

The most recent report of the Intergovernmental Panel on Climate Change confirmed that there is overwhelming evidence that humans are affecting the global climate, and highlighted a wide range of implications for human health. Climate variability and change cause death and disease through natural disasters, such as heat waves, floods and droughts. In addition, many important diseases are highly sensitive to changing temperatures and precipitation. These include common vector-borne diseases such as malaria and dengue; as well as other major killers such as malnutrition and diarrhea. Climate change already contributes to the global burden of disease, and this contribution is expected to grow in the future. The impacts of climate on human health will not be evenly distributed around the world. Developing country populations, particularly in Small Island States, arid and high mountain zones, and in densely populated coastal areas, are considered to be particularly vulnerable. Fortunately, much of the health risk is avoidable through existing health programmes and interventions. Concerted action to strengthen key features of health systems, and to promote healthy development choices, can enhance public health now as well as reduce vulnerability to future climate change. WHO supports member states in protecting public health from the impacts of climate change, and provides the health-sector voice within the overall UN response to this global challenge.

Large-scale and global environmental hazards to human health include climate change, stratospheric ozone depletion, loss of biodiversity, changes in hydrological systems and the supplies of freshwater, land degradation and stresses on food-producing systems.

Appreciation of this scale and type of influence on human health requires a new perspective which focuses on ecosystems and on the recognition that the foundations of long-term good health in populations rely in great part on the continued stability and functioning of the biosphere's life-supporting systems. It also brings an appreciation of the complexity of the systems upon which we depend.

7.10 Ecosystems and Health

There is growing concern amongst many natural scientists that human interventions are altering the capacity of ecosystems to provide their *goods* (*e.g.*

freshwater, food, pharmaceutical products, etc) and *services* (*e.g.* purification of air, water, soil, sequestration of pollutants, etc).

Ecosystem disruption can impact on health in a variety of ways and through complex pathways. These are moreover modified by a local population's current vulnerability and their future capacity to implement adaptation measures.

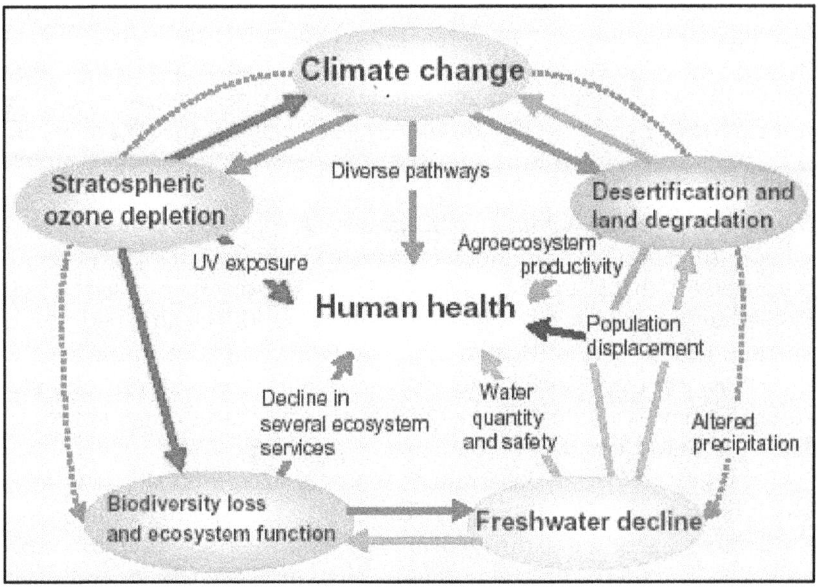

Figure 7.6: Pathways of Ecosystem Disruption and Its Impact on Human Health.

The links between ecosystem change and human health are seen most clearly among impoverished communities, who lack the "buffers" that more affluent communities can afford.

Stratospheric Ozone Depletion, UV Radiation and Health

It has been recognized for several decades that the release of chlorofluorocarbons and other atmospheric pollutants depletes stratospheric ozone, which in turn increases human exposure to ultraviolet radiation, causing skin cancer and cataracts.

The recognition of direct effects on human health effects was a major stimulus to the Montreal Protocol, which acts to reduce emissions of pollutants that weaken the ozone layer. Although this international agreement is proving highly effective in reducing risks in the long term, UV radiation remains a health hazard. The World Health Organization, and partner organizations - through the Intrusion project - have developed and promote the UV Index, a tool to inform and educate the public about sun protection.

7.11 Loss of Biodiversity

Everyone of us depends on biodiversity in some way – the air we breathe, the water we drink, the food we eat, the medicines we use, the work we do – every single day.

Defining Biodiversity

Biodiversity can be divided into four broad categories:

Genetic Diversity

The variety of alleles (different versions of the same gene) within a species. For example, humans have several different alleles for blood type, which are combined in varying ways to produce the many blood types expressed in humans. Other species have similar variances in their DNA, the genetic code that expresses how large they may grow, what colour their fur may be, how fast they may run, or – perhaps most importantly – how effectively their immune systems may respond to a new pathogen.

Species Diversity

The variety of species within a particular area. A species is the most fundamental level of scientific classification among organisms, though the boundaries of this category are often debated among professional biologists and many valid subdivisions (subspecies, races, variants, morphs, etc.) are recognized. The lines between species diversity and genetic diversity cannot be concretely drawn. For example, the right whale (*Balaena glacialis*) has distinct populations in the northern and southern hemispheres. Some scientists believe these are distinct races or genetically unique populations, while others believe them to be separate species entirely (the southern population, if separated from the northern, would be *B. australis*). Measuring species diversity is complicated because species distributions are highly variable from one biome to another, just as the territory covered by each organism varies from species to species. While the ranges of some species, such as the great whales and sea turtles, literally span the world's oceans, the ranges of others may be restricted to a single mountaintop or tiny desert pool. Differing biomes vary in their species diversity, so when measuring the health of a particular area, it must be compared to similar habitats (*i.e.*, rain forest to rain forest, desert to desert) or, preferably, to baseline data recorded in same location at an earlier time.

Cultural Diversity

The variation in behaviours or customs among different groups of the same species. This is most prevalent among humans, but science has recorded distinct cultures in other species, too. For example, humpback whales demonstrate a type of culture in their seasonal songs, as do chimpanzees in their hunting behaviour. Behaviour variations are important because some behaviours may be more favoured than others in the face of sudden evolutionary pressures.

Ecosystem Diversity

The variety of communities of interacting organisms and the environments they occupy. This level of biodiversity is by far the most complex, consisting of the physical (nutrients, soil, water, atmosphere) and the biotic (individual organisms, communities of interacting species) components of the surrounding environment. This level is also complex because the very nature of an ecosystem is broadly encompassing, thus many varied habitats and sub-systems may fall into a larger system that is still in and of itself unique. A good example is the Florida Everglades, a vast wetland ecosystem containing many smaller habitats, such as pine barrens, cypress swamps, and mangrove communities. Eventually, all ecosystems, and by extension all life, on earth are interconnected and interdependent. The earth is enshrouded in a relatively thin mantle, its width extending from the bottom of the ocean to about seven miles up into the atmosphere, of habitat capable of supporting life, called the biosphere.

To the best of our knowledge, nothing lives outside of the biosphere, and – with the sole exception of the sun's rays – everything we need to thrive and prosper must be obtained from this same biosphere. which is also where we must dispose of our waste. Wise and sustainable use of this limited resource is not only what is best for the other organisms who share this space with us, but it is crucial for our very survival as a species.

The physical and biotic components of all ecosystems are continually interwoven through nutrient cycling, the ingestion, absorption, excretion, and decomposition of materials in the environment. For most organisms, the oxygen, nitrogen, and carbon cycles are the most important, with the carbon cycle currently of particular interest to climatologists due to its role in global climate change. Entire ecosystems may interact through hydrologic cycling, the movement of water, which may bring life-giving moisture when clean or may rapidly contaminate a region if polluted.

The fate of the human species is inextricably interwoven with the collective fates of threatened wild spaces, as well as the plants and animals that live in them, around the globe. Healthy habitats are integral to healthy human populations worldwide.

7.12 Air Pollution

Air pollution affects biodiversity on a great scale. The atmosphere, lithosphere, and hydrosphere are negatively affected by pollution (McNeely *et al.,* 1995). Air pollution affects lower life forms more than higher life forms. Plants are generally more affected than animals on land, but not in freshwater. A decline in most species due to pollution is evident except for a minority that increase. It will be focusing on plants and how they are affected by air pollution.

Agricultural crops can be damaged when they are exposed to high concentration of air pollution. The plants will die when they develop yellow colour and there will be reduction in growth of various portions of a plant. Pollution like nitrogen oxides, sulfurdioxide and Ozone cause a direct damage to leaves of a tree.

We can reduce global air pollution on land and climate change. Eager, you have indicated that, "the vehicles are the most important source of atmospheric pollutants as they release carbon monoxide". we can reduce air pollution caused by vehicles by using the cars which use unleaded petrol, this is because leaded petrol are dangerous to both plants and humans. We can also use bicycles or walking, and recycles of newspaper to reduce the pollution on air.

Plants constantly take up atmospheric gases *i.e.* air everyday to sustain their biological processes. Vegetation growing under optimum conditions is most susceptible to air pollution (Stern *et al.,* 1973)]. As air pollution is for the most part man-made, we are the main source of this phenomenon. Pollution can be derived from two kinds of sources namely, stationary and multiple point sources. Stationary point sources include backyard fires (on a small scale) and the burning of a thousand tons of coal each day in coal-fired electrical power plants (on a large scale). Multiple point sources are usually mobile and include automobiles and other vehicles. The vehicles are the most important source of atmospheric pollutants as they release carbon monoxide. This is followed by industrials sources which release sulphur oxides, steam and electric power plants, space heating and lastly refuse burning. Agricultural chemicals also form part of air pollution (Kozlowski and Mudd, 1975)].

The uptake of pollutants depends on the concentration gradient between the ambient air and the absorptive sites within the leaf. It also depends on the conductance of the stomata. The toxic effect of a pollutant may thus be almost directly related to the functioning of the stomata. Stomata openings are related to the physiological activity of the plant in that they regulate gas exchange; correlation exists between the extent of air pollution effects and the degree of opening of the stomata (Stern *et al.,* 1973). Pollutant flow may be restricted by the physical structures of leaves or scavenging by competing chemical reactions. However, as conditions change the ambient dose to which plants are exposed does not necessarily reflect actual cellular exposure. The initial flux of gases to the surface is controlled by boundary layer resistance *i.e.* the amount of gas able to contact the surface. This includes epidermal characteristics and air movement across the leaf.

At slower wind speeds (less that 2m/s), boundary layer thickness decreases as wind speed increases. Thus more pollutant enters the leaf when air is in motion. Pubescence is also important in that leaf hairs provide major areas of impact. Cuticle wax is also important in limiting uptake even if the cuticle is thin. Stomatal resistance is the most critical. Resistance is determined by stomatal number, size, anatomical characteristics for example the degree to which stomata is sunken and the size of the stomatal aperture (Anderson and Threshow, 1991).

Data and Statistics – Effect

Mental health

☆ Depression is ranked as the leading cause of disability worldwide and affects around 120 million people worldwide.

☆ Nearly a million people commit suicide every year, 86 per cent of them live in low- and middle-income countries.

☆ Around 20 per cent of the world's children and adolescents are estimated to have mental disorders or problems.

7.13 Global Warming is 'Just a Natural Cycle'

Some scientists believe that the human contribution to carbon dioxide in the atmosphere, however small, is of a critical amount that could nonetheless upset Earth's environmental balance.

As to why highs and lows follow a 100,000 year cycle, the explanation Essenhigh uses is that the Arctic Ocean acts as a giant temperature regulator, an idea known as the "Arctic Ocean Model." This model first appeared over 30 years ago and is well presented in the 1974 book Weather Machine: How our weather works and why it is changing, by Nigel Calder, a former editor of New Scientist magazine.

7.14 Global Cooling – No It's Warming

Our earth is the planate. Our earth has not got it's own solar energy. Approximately 4.5 billion years ago the earth was nothing more than a giant blob of lava. Since liberated from sun, the earth is gradually cooling. Decades prior, the brilliant Serbian mathematician Milutin Milankovitch had explained how our world warms and cools on roughly 100,000-year cycles due to its slowly changing position relative to the Sun. Milankovitch's theory suggested Earth should be just beginning to head into its next ice age cycle. The surface temperature data gathered by Mitchell seemed to agree; the record showed that Earth experienced a period of cooling (by about 0.3°C) from 1940 through 1970. Of course, Mitchell was only collecting data over a fraction of the Northern Hemisphere – from 20 to 90 degrees North latitude. Still, the result drew public attention and a number of speculative articles about Earth's coming ice age appeared in newspapers and magazines. All other planate all ready cool down and frozen.

Global warming comes and goes in 1,500 - 2500 year cycles which may have more to do with cosmic rays than fossil fuel emissions, according to a new book by two veteran American climate sceptics, Fred Singer and Dennis Avery. If the genuine warming now being seen is caused by human emissions of carbon dioxide, it would have started earlier, according to this book

Mr Avery, who was in London, said: "If this were a CO2 driven warming it should have started in 1940 and risen strongly from there. In fact warming started in 1850 and rose sharply until 1940 then decreased for 35 years."

Mr Avery believes that only half the warming that has happened since 1940 - 0.2 degrees according to his measurements - can be ascribed to man made emissions. The rest he says is natural variability.

"If you factor in the warming from the cyclical trends, it is not very frightening," he said.

Many people blame global warming on carbon dioxide sent into the atmosphere from burning fossil fuels in man-made devices such as automobiles and power plants. Essenhigh believes these people fail to account for the much greater amount of carbon dioxide that enters – and leaves – the atmosphere as part of the natural cycle of water exchange from, and back into, the sea and vegetation.

Compared to man-made sources' emission of about 5 to 6 billion tons per year, the natural sources would then account for more than 95 percent of all atmospheric carbon dioxide, "At 6 billion tons, humans are then responsible for a comparatively small amount - less than 5 percent - of atmospheric carbon dioxide," "And if nature is the source of the rest of the carbon dioxide, then it is difficult to see that man-made carbon dioxide can be driving the rising temperatures. In fact, I don't believe it does."

Some scientists believe that the human contribution to carbon dioxide in the atmosphere, however small, is of a critical amount that could nonetheless upset Earth's environmental balance. But Essenhigh feels that, mathematically, that hypothesis hasn't been adequately substantiated. Here's how Essenhigh sees the global temperature system working: As temperatures rise, the carbon dioxide equilibrium in the water changes, and this releases more carbon dioxide into the atmosphere. According to this scenario, atmospheric carbon dioxide is then an indicator of rising temperatures – not the driving force behind it.

7.15 Five Million Years of Cooling

The last five million years of climate change is shown in the next graph based on work by Lisiecki and Raymo (2005). It shows our planet has a dynamic temperature history, and over the last three million years, we have had a continuous series of ice ages (now about 90,000 years each) and interglacial warm periods (about 10,000 years each). There are 13 (count 'em) ice ages on a 100,000 year cycle (from 1.25 million years ago to the present, and 33 ice ages on a 41,000 year cycle (between 2.6 million and 1.25 million years ago). Since Earth is on a multi-million-year cooling trend, we are currently lucky to be living during an interglacial warm period, but we are at the end of our normal 10,000 year warm interglacial period(Lisiecki and Raymo, 2005).

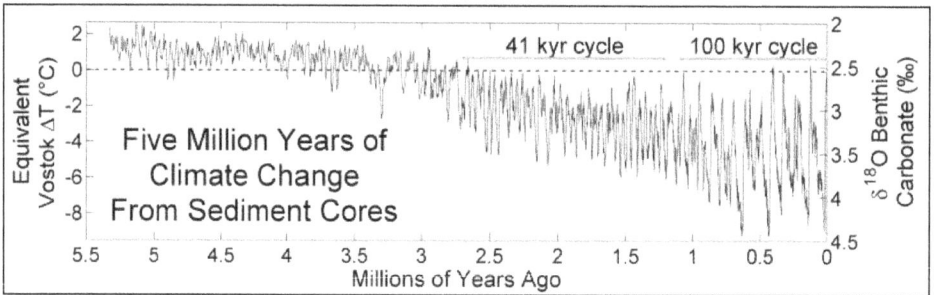

Figure 7.7: The Last Five Million Years of Climate Change.

Long-term, temperatures are now declining (for the last 3,000 years), and we appear to be headed for the next 90,000 year ice age, right on schedule at the end of our current 10,000 year warm period. We have repeated this cycle 46 times in succession over the last 2.6 million years. And in case you are wondering, the previous Antarctic ice cores tell a broadly similar story. The following graph of ice core data from Vostok (vertical scale in degrees C variation from present) shows that Antarctica is also experiencing a long-term (4,000 year) cooling trend mirroring the Greenland GISP2 cooling trend. Though the individual temperature spikes and dips are different than in Greenland, the long-term temperature trend on the planet appears to be down, not up. And since it is so late in our current interglacial period, we could be concerned about global *cooling*.

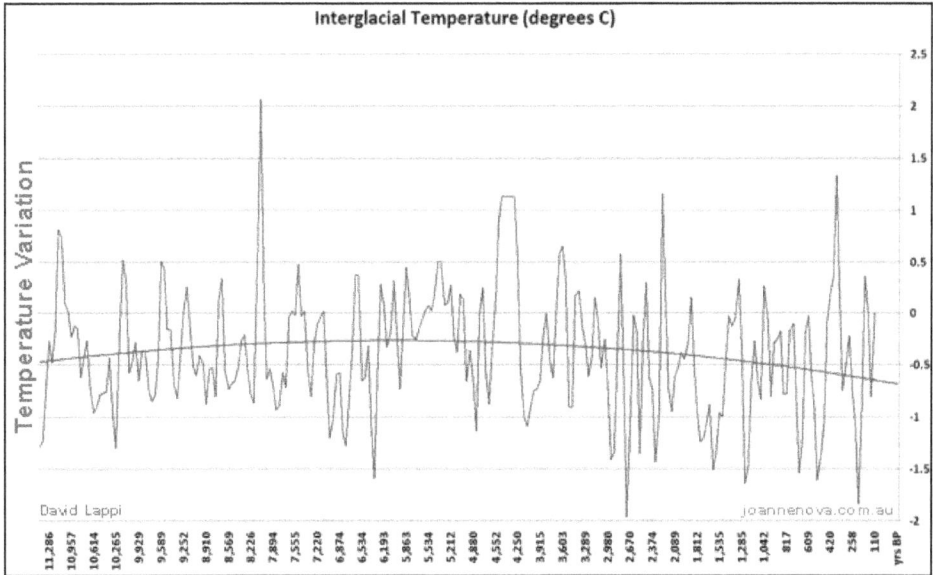

Figure 7.8: Vostok Antarctica, last 12,000 years.

"Today, we are simply near a peak in the current cycle that started about 25,000 years ago," Essenhigh explained.

References

Anderson, F.K., and Threshow, M. 1991. *Plant stress from Air Pollution.* John Wiley and Sons, New York.

Jenkinson, D. S. Adams, D.E. and Wild, A. 1991 Model estimates of CO_2 emissions from soil in response to global warming *Nature* 351, 304 - 306 (23 May 1991); doi:10.1038/351304a0.

Kozlowski, T.T., and Mudd, J.B. 1975. *Responses of Plants to Air Pollution.* Academic Press Inc., New York.

Lisiecki L. E. and Raymo M.E. 2005. *A Pliocene-Pleistocene stack of 57 globally distributed benthic 818O records, Paleoceanography* 20, 1003.

McNeely, J.A., Gadgil, M., Lévêque, C., Padoch, C. and Redford, K. 1995. Human influences on biodiversity. In: Global biodiversity assessment, V.H. Heywood (ed.), Cambridge University Press, Cambridge, pp.

Stern, A.C., Wohlers, H.C., Boubel, R.W., and Lowry, W.P. 1973. *Fundamentals of Air Pollution.* Academic Press, New York.

Vitousek, P. M. 1994. Beyond Global Warming: Ecology and Global Change. *Ecology* 75:1861–1876. http://dx.doi.org/10.2307/1941591.

Zachos, J. Pagani M. Sloan, L. Thomas, E. and Billups, K. 2001 – *Trends, Rhythms, and Aberrations in Global Climate 65 Ma to Present, Science* 292 (5517): 686–693.

2015, **Impact of Global Warming and Climate Change on Human and Plant Health**
Editors: **Dr. Arun Arya and Prof. V.S. Patel**
Published by: **DAYA PUBLISHING HOUSE, NEW DELHI**

Pages 72–77

Chapter 8

A Look at Bird Flu

Sheuli Dasgupta*

Department of Microbiology, Gurudas College,
Kolkata – 700 054, W.B., India

ABSTRACT

The human beings are infected with the Swine flue and influenza viruses. Human flu viruses are those subtypes that occur widely in humans. The activity of the glycoproteins present in the envelope classifies the A type virus into its different sub types. These are: H_1N_1, H_1N_2 and H_3N_2. The ability of viruses to adapt themselves to the changing environment has made it possible to inhabit itself in this vast world for the past millions of years. Ordinarily, swine influenza viruses circulate only among pigs, equine influenza viruses among the Equidae, avian influenza viruses among the birds, and human influenza viruses among humans. The viruses are identified by Revrse Transcriptase Polymerase Chain reaction and ELISA techniques.

Keywords: *Bird flue, H_1N_1, H_1N_2, H_3N_2, DIVA vaccination, Virus, Quarantine.*

8.1 Introduction

Among various other causative agents significant morbidity and mortality is caused by influenza every year which has now acquired the good name as 'epidemic' in temperate zones. Incidence and prevalence of this disease has been irregular with attack rates even upto 50 per cent in some areas. So it was totally unpredictable. A, B, C are the three different types of Influenza virus. They again have their sub types. The

---

* E-mail: sheulidasgupta@yahoo.co.in

envelope protein contains a typical protein denoted as the M protein, which has different structures in the different types of the viruses. Accordingly, classification has been made with respect to the structural properties of the M protein. The A type normally attacks human being and birds also which results in outbreaks of flu. Their genome has been found to contain 8 segmented RNA. Mixing of these segments results in the evolution of the new forms of the viruses. This type is mostly predominant in wild birds which is often isolated from their feces. It also attacks the domestic poultry birds.

We human beings are infected with the human influenza virus. Human flu viruses are those subtypes that occur widely in humans. The activity of the glycoproteins present in the envelope classifies the A type virus into its different sub types. They are:

A. Hem agglutinin activity –H

B. Neuraminidase activity –N.

These two types of proteins that is the H and N types can have different sorts of combinations giving rise to different subtypes of the viruses. In short these viruses ate denoted as H_1N_1. Till date only three known subtypes of human flu viruses are known. They are H_1N_1, H_1N_2, H_3N_2 and it was concluded after extensive research there is genetic similarity of current human flu A viruses with those of the birds. There is a rapid rate of mutation in the viruses so much so that they may even spread among the human beings. This is evoking an alarm in us.

Aetiology of Avian Influenza

Bird flu" is a phrase similar to "swine flu, that it refers to an illness caused by any of many different strains of influenza viruses that have adapted to a specific host. All known viruses that cause influenza in birds belong to the species influenza A virus.

It is A Viral disease of domestic and wild birds characterized by the full range of responses from almost no signs of the disease to very high mortality. The incubation period is also highly variable, and ranges from a few days to a week (3 to 7 days).

☆ Influenza virus A genus of the Orthomyxoviridae family.

☆ They are enveloped, negative stranded RNA viruses.

☆ Influenza A viruses can be divided into 15 Haemagglutinin (H) antigens. 9 Neuraminidase (N) antigens.

☆ Extreme antigenic variability brought about by genetic reassortment in host cells.

8.2 Spread of Bird Flue

Intensive care should be taken while dealing with poultry products and during human consumption because the viruses mainly harbor them and there are all chances of these viruses within us. Although the risk level is low still it should be avoided. Whenever there is outbreak of bird flu statistics shows that there is incidence of human death too. So people should avoid contact with infected birds or contaminated surfaces. Avian influenza A (H5N1) has a unique rate of mutation and also infects other animal species. It causes pathogen city in human being and is fatal too. The infection is carried and propagated through birds, through their feces and consequently healthy birds are also affected in the process. Migratory birds also have been shown to be a prey to this virus.

Scientists have used different virology techniques for virus identification. Effective vaccines are being developed against these viruses. A bird flu vaccine has won federal approval for the first time in 2007 for use in protecting human against the H5N1 influenza virus. The viruses can now be combated for safe and healthy lifestyle.

Swine flu (swine influenza) is a respiratory disease caused by viruses (influenza viruses) that infect the respiratory tract of pigs, resulting in nasal secretions, a barking cough, decreased appetite, and listless behavior. Swine flu produces most of the same symptoms in pigs as human flu produces in people. Swine flu can last about one to two weeks in pigs that survive. Swine influenza virus was first isolated from pigs in 1930 in the U.S. and has been recognized by pork producers and veterinarians to cause infections in pigs worldwide. In a number of instances, people have developed the swine flu infection when they are closely associated with pigs (for example, farmers, pork processors), and likewise, pig populations have occasionally been infected with the human flu infection. In most instances, the cross-species infections (swine virus to man; human flu virus to pigs) have remained in local areas and have not caused national or worldwide infections in either pigs or humans. Unfortunately, this cross-species situation with influenza viruses has had the potential to change. Investigators decided the 2009 swine flu strain, first seen in Mexico, should be termed novel H1N1 flu since it was mainly found infecting people and exhibits two main surface antigens, H1 (hemagglutinin type 1) and N1 (neuraminidase type1). The eight RNA strands from novel H1N1 flu have one strand derived from human flu strains, two from avian (bird) strains, and five from swine strains. Swine flu is transmitted from person to person by inhalation or ingestion of droplets containing virus from people sneezing or coughing; it is not transmitted by eating cooked pork products. The newest swine flu virus that has caused swine flu is influenza A H3N2v (commonly termed H3N2v) that began as an outbreak in 2011. The "v" in the name means the virus is a variant that normally infects only pigs but has begun to infect humans.

Spreading of Avian Influenza

    ☆ MOVEMENT of people, animals, equipment, vehicles

    ☆ CONTACT with neighboring flock

    ☆ CONTACT with insects, rodents, stray animals and pets

    ☆ CONTAMINATED water and feeds

    ☆ INADEQUATE cleaning and disinfection

Most human contractions of the avian flu are a result of either handling dead infected birds or from contact with infected fluids. While most wild birds mainly have only a mild form of the H5N1 strain, once domesticated birds such as chickens or turkeys are infected, it could become much more deadly because the birds are often within close contact of one another. There is currently a large threat of this in Asia with infected poultry due to low hygiene conditions and close quarters. Although it is easy for humans to become infected from birds, it's much more difficult to do so from human to human without close and lasting contact.

Spreading of H5N1 from Asia to Europe is much more likely caused by both legal and illegal poultry trades than dispersing through wild bird migrations, being that in recent studies, there were no secondary rises in infection in Asia when wild birds migrate south again from their breeding grounds. Instead, the infection patterns followed transportation such as railroads, roads, and country borders, suggesting poultry trade as being much more likely. While there have been strains of avian flu to exist in the United States, such as Texas in 2004, they have been extinguished and have not been known to infect humans.

8.3 Avian Influenza : Differential Diagnosis

☆ Infectious Bronchitis

☆ TRT/APV

☆ Newcastle Disease

☆ Respiratory viruses – Mixed infections

8.4 Control Measures

Preventing infection with bird flue normally includes good hygiene and sanitation. Frequent visits to poultry forms should be avoided. Use of protective cloths, gloves and other protective cloths should be done. Quarantine rules should be followed and diseased birds should be buried deep into ground. Precautions should be observed in handling the birds for food. Influenza viruses are susceptible to chemicals like sodium hypochlorite so ionizing radiations can be used. Following precautions should be followed:

a) Biosecurity

b) Quarantine

c) Intensify disinfecting measures

d) Monitoring/Surveillance

e) Stamping Out/Depopulation

f) DIVA Vaccination - only for LPAI and not for HPAI because it might prolong the shedding of the virus

g) Proper Disposal

8.5 Swine Flu (H1N1 and H3N2v influenza virus) Facts

☆ Swine flu is a respiratory disease caused by viruses that infect the respiratory tract of pigs and result in a barking nasal secretions, and listless behavior.

☆ Swine flu viruses may mutate (change) so that they are easily transmissible among humans.

☆ The 2009 swine flu outbreak was due to infection with the so-called H1N1 virus and was first observed in Mexico.

☆ Symptoms of swine flu in humans are similar to most influenza infections: fever (100 °F or greater), cough, nasal secretions_fatigue and headache.

☆ Two antiviral agents, zanamivir (Relenza) and oseltamivir (Tamiflu), have been reported to help prevent or reduce the effects of swine flu if taken within 48 hours of the onset of symptoms.

8.6 How to Stop the Spread of Pandemic Flu Virus ?

Use Good Hygiene Practices

☆ Cover your mouth and nose with a tissue when you cough or sneeze; put the used tissue in a wastebasket and clean your hands.

☆ Clean your hands as soon as possible after coughing, sneezing, or blowing your nose.

☆ Use soap and water and wash your hands for 15 - 20 seconds; or

☆ Use alcohol-based hand wipes or alcohol-based (60-95 per cent alcohol) gel hand sanitizers; rub these on the hands until the liquid or gel dries.

☆ Carry alcohol-based hand wipes or alcohol-based (60-95 per cent alcohol) hand-sanitizing gels with you to clean your hands when you are out in public.

☆ Teach your children to use the hygiene practices, because germs spread from school.

Clean and Disinfect Hard Surfaces and Items in Homes and Schools

☆ Follow label instructions carefully when using disinfectants and cleaners.

☆ Do not mix disinfectants and cleaners unless the labels indicate it is safe to do so. Combining certain products (such as chlorine bleach and ammonia cleaners) can be harmful, resulting in serious injury or death.

☆ Keep hard surfaces like kitchen countertops, tabletops, desktops, and bathroom surfaces clean and disinfected.

☆ Clean the surface with a commercial product that is both a detergent (cleans) and a disinfectant (kills germs). These products can be used when surfaces are not visibly dirty.

☆ Another way to do this is to wash the surface with a general household cleaner (soap or detergent), rinse with water, and follow with a disinfectant. This method should be used for visibly dirty surfaces.

☆ Use disinfectants on surfaces that are touched often. If disinfectants are not available, use a chlorine bleach solution made by adding 1 tablespoon of bleach to a quart (4 cups) of water; use a cloth to apply this to surfaces and let stand for 3 – 5 minutes before rinsing with clean water. (For a larger supply of disinfectant, add ¼ cup of bleach to a gallon [16 cups] of water.)

☆ Wear gloves to protect your hands when working with strong bleach solutions.

☆ Keep surfaces touched by more than one person clean and disinfected. Examples of these surfaces include doorknobs, refrigerator door handles, and microwaves. Clean with a combination detergent and disinfectant product. Or use a cleaner first, rinse the surface thoroughly, and then follow with a disinfectant.

☆ Use sanitizer cloths to wipe electronic

Use Recommended Laundry Practices

☆ Gently gather soiled clothing, bedding, and linens without creating a lot of motion or fluffing; for example, do not shake sheets when removing them from the bed.

☆ Clean your hands after handling soiled laundry items.

☆ Use washing machine cycles, detergents, and laundry additives (like softener) as you normally do; follow label instructions for detergents and additives.

☆ Dry the cleaned laundry items as you normally do, selecting the dryer temperature for the types of fabrics in the load. Line- or air-drying can be used to dry items when machine drying is not indicated.

Use Recommended Waste Disposal Practices

☆ Toss tissues into wastebaskets after they have been used for coughs, sneezes, and blowing your nose.

☆ Place wastebaskets where they are easy to use.

☆ Avoid touching used tissues and other waste when emptying wastebaskets.

☆ Clean your hands after emptying wastebaskets.

References

http://www.mayoclinic.org/diseases-conditions/bird-flu/basics/prevention/con-20030228

www.epa.gov/oppad001/chemregindex.htm.

www.pandemicflu.gov/plan/healthcare/influenzaguidance.html

www.pandemicflu.gov/plan/individual/panfacts.html

Part II
Air Pollution and Climate Change

सुन्दर भविष्य

करे प्रदूषण दूर जहां का,
फैलें फिर खुशियां हर घर में।
जीवन को हम कर दें सुखमय,
शिक्षा को फिर घर—घर लाकर।
जहां हमारा हो अप्रतिम, अद्भुत,
भण्डार सभी जैव विविधता का।
जग अपना हम मंगल कर दें,
और देशहित अपना सब कुछ दें।
न रहे प्रदूषण न हो कोलाहल,
जन—जन में हम अमृत भर दें।
पर्यावरण को जानें समझें,
जीवन को हम फिर सुखमय कर दें।

2015, Impact of Global Warming and Climate Change on
     Human and Plant Health
Editors: Dr. Arun Arya and Prof. V.S. Patel
Published by: DAYA PUBLISHING HOUSE, NEW DELHI

Pages 81–100

Chapter 9

Air Pollution, Climate Change and Mitigation Strategies

Aparna Rathore* and Yogesh T. Jasrai**

Department of Botany, University School of Sciences
Gujarat University, Ahmedabad – 380 009, Gujarat, India

ABSTRACT

Air pollution is the addition of unwanted component of gases and particles into the atmosphere by human activities like burning of fossil fuels, industrial pollution etc. and natural phenomenon like volcanic eruptions, forest fires. All these activities release harmful gaseous and particulate components like CO_2, CH_4, CO, N_2O, SO_2, dust and soot into the atmosphere. The major gaseous component of CO_2, CH_4, N_2O, etc contribute to the phenomenon called as Greenhouse effect, trapping the infrared radiation reflected by the earth thus increasing the average temperature of the earth leading to Global warming. The global warming leads to rising temperatures, melting of ice caps, climatic variability (changes in rainfall pattern, increasing droughts and floods) etc. All these conditions contribute to climate change. All what need to be done to mitigate the climate change is to check the air pollution at industrial, vehicular and house hold level to reduce the greenhouse gas concentration in the atmosphere. This can also be done by stringent rules for pollution check. We can capture the most prominent of the greenhouse gas components by the post-combustion capture and pre-combustion capture, especially of CO_2 from the factories and sequester it in deep geological formations. Switching over to efficient fuels like biofuels and renewable

E-mail: *rathoreaparna@ymail.com; **yjasrai@yahoo.com

sources (solar, tidal, wind, geothermal energy etc.) for energy generation and planting trees, which act as air purifiers will surely be the best method to prevent air pollution and simultaneous climate change.

Keywords: *Air pollution, Greenhouse gases, Global warming, Climate change, Mitigation.*

9.1 Introduction

Atmosphere is a complex system of gases and suspended particles. However, the composition of the atmosphere keeps on changing and hence, its structure is variable in time and space. The atmosphere is mainly composed of gases like nitrogen (N_2) (78.8 per cent), oxygen (O_2) (20.95 per cent), argon (0.93 per cent), and other trace gases like carbon dioxide (CO_2) (387ppm), methane (CH_4) (2ppm), nitrous oxide (N_2O) (0.3ppm), water vapour (variable in ppm) etc. Pollution is the presence of undesirable substance in any segment of the environment, primarily due to human activities and discharging waste products or harmful secondary products, which are harmful to man and other organisms. Air pollution is the addition of unwanted component of gases and particles into the atmosphere by human activities like burning of fossil fuels, industrial pollution etc. and also some natural phenomenon like volcanic eruptions and forest fires. All these activities release harmful gaseous and particulate components like CO_2, CH_4, CO, N_2O, SO_2, dust, soot etc. into the atmosphere hence polluting it. The changes that are occurring now have their origin in the industrial revolution. Long lived gases that are increasing at a substantial rate because of human activities are of particular current interest since they eventually lead to stratospheric ozone depletion, global warming and disturbances in the atmospheric chemistry that many be harmful to the ecosystem.

9.2 Causes of Air Pollution

Air pollution results due to both natural as well as anthropogenic causes.

Natural Causes

Its include the volcanic eruptions during which a lot of ash, dust, smoke and gases such as SO_2, CO_2, etc are released into the atmosphere. The dust from dust storms, wild forest fires and pollens released by the plants, spores of the fungi, bacteria and viruses suspended in the atmosphere also act as pollutants. The death and decomposition of organic matter also emits gases like CO_2, CH_4, H_2S and NO_X into the atmosphere.

Anthropogenic Causes

Its include the activities of the man like deforestation, *jhum* cultivation, modern agricultural practices like use of chemical pesticides and fertilizers, industrial activities, cement factories, thermal and nuclear power plants, waste disposal and biomass burning. The land use changes (like urbanization, industrialization, creation of the SEZ (Special Economic Zone), expansion of agricultural lands etc) leading to the clearing of the forest and increasing number of the vehicles are the major cause of the increasing air pollution. The increasing economic development and a rapidly growing population puts tremendous pressure on the environment and the country's natural resources (http://www.indiaonlinepages.com/population/census2010 accessed on 26.6.2010).

9.3 Types of the Air Pollutants

The air pollutants are basically of two types: **a) gaseous** and; **b) particulate**. The gaseous pollutants include the toxic gases such as sulfur dioxide, carbon monoxide, nitrogen oxides, and chemical vapors released from the automobiles, industries, smoke etc which on reaction in the atmosphere give rise to the smog and acid rain. While, the particulate matter includes the particles like dust, smoke, soot produced due to burning of wood and fossil fuel and also due to volcanic eruptions. The additions of gaseous and particulate material to the atmosphere by the activities of man are called as **pollutants** when their concentration is sufficient to produce harmful effects. Once mixed with the air, some air pollutants persist unaltered and become mixed throughout the atmosphere where they potentially have a global influence.

9.4 Types of Air Pollution

Air pollution can be of various types on the basis of their sources.

9.4.1 Vehicular Air Pollution

The internal combustion engine contributes to the air pollution by exhaust like carbon monoxide, unburnt hydrocarbons, nitrogen oxides, sulphur oxides, lead compounds, smoke, particulates and odour. The vehicular pollution has increased fast throughout the world owing to an increase in the transportation. Baumert *et al.* (2005) in their global survey of the emissions from the transportation have estimated that GHG emissions from the transport sector account for 14 per cent of total GHG emissions, with domestic and international road transport contributing 72 per cent, domestic and international air transport 11 per cent, international marine transport 8 per cent and others 8 per cent.

9.4.2 Industrial Air Pollution

Among the various categories of industries nine prominent groups of industries are considered to be the major pollutant generating industries. The following compilation (Table 9.1) shows the industrial types along with the type of air pollutants released by them;

9.4.3 Air Pollution due to Agricultural Activities

Agriculture originated more than 12,000 years ago and is the largest source of food. With the beginning of industrial revolution there has been modernization in the techniques of agriculture which is contributing to the greenhouse effect with the emissions of the gases like CO_2, CH_4, N_2O, etc. Agriculture contributes to about 60 per cent of the N_2O, 40 per cent of the CH_4 and 1 per cent of CO_2 of the total GHG emissions. The runoff of the chemical fertilizers and pesticides into the water bodies results in eutrophication which depletes the essential nutrients and oxygen leading to the death and decomposition of organisms. The main contributors of the air pollution due to agriculture are application of chemicals, use of tractors, the operation of water pumps, burning of dead plant parts, livestock management, juhm or shifting cultivation by clearing and burning forest. Direct and indirect global emission of GHGs from the agriculture sector is about 17-32 per cent of the total GHG emissions (Figure 9.1). From which the direct GHG emissions are 12 per cent (Bellarby *et al.,* 2008).

Table 9.1: Various Industries with the Pollutants Released by them (Santra, 2005)

Sl.No.	Industries	Pollutants Released
1.	Steel mill	Particulates, smoke, CO, fluoride
2.	Petroleum refinery	SO_2, hydrocarbon smoke, particulates and odour
3.	Acid plant	SO_2, acid mist
4.	Paper mill	SO_2, particulates, odour
5.	Soap and detergent plant	Particulates and odour
6.	Fertilizer plant	Particulates, ammonia, SO_2, NOx, fluoride
7.	Cement plant	Cement dust, SO_2, smoke
8.	Thermal power plant	NOx, SO_2, particulates
9.	Metal smelter	SO_2, NOx, particulates, smoke

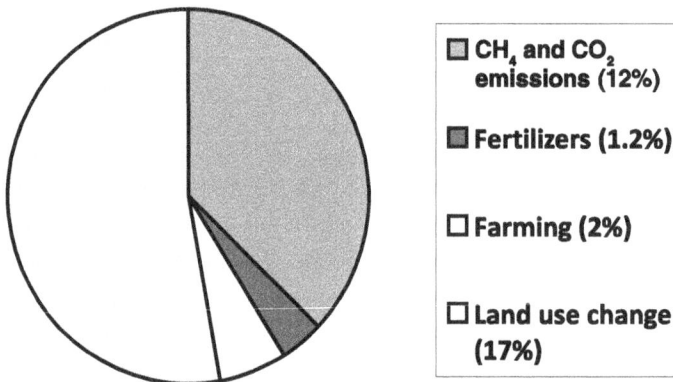

Figure 9.1: Emissions from Agricultural Sector (Bellarby, 2008).

- ☐ CH_4 and CO_2 emissions (12%)
- ■ Fertilizers (1.2%)
- ☐ Farming (2%)
- ☐ Land use change (17%)

9.5 Biological Air Pollution

Biological air pollution results from the natural causes of pollens and spores, wild fires and volcanic eruptions. Climate Change Futures has pointed out an important impact of biological air pollution to be enhanced allergen exposure. Increase in CO_2 levels and warming result in the growth of plants such as ragweed, which produce 61 per cent more pollen than plants grown under normal levels of CO_2. The increase in asthma being observed in both developed and developing countries may be due to a combination of both increasing air pollution and simultaneous pollen allergen exposure (Tibbetts, 2007). Volcanic eruption adds CO_2, CH_4, SO_2, smoke, soot etc to the atmosphere. The correlation between the CO_2 fluxes of the atmosphere–

land and atmosphere–ocean are dependent on the climatic conditions. Bouquets *et al.* (2000) proposed that the amount of CO_2 in the atmosphere varies with the El Niño–Southern Oscillation and as proposed, it was also affected by the climate disturbances arising from the Pinatubo volcanic eruption which occurred in 1991 and due emissions of smoke and SO_2 and simultaneous formation of clouds lead to a fall in the temperature by $1°F$ (Friedlingstein *et al.*, 2006; Bousquet *et al.*, 2000; Jones *et al.*, 2001; Lucht *et al.*, 2002).

The melting permafrost as a result of global warming also adds CH_4 to the atmosphere. The wild fires not only devastate the entire forest but also emit huge amounts of the CO_2, nitrous oxides and other gases into the atmosphere.

9.6 Domestic Air Pollution

The domestic pollution results due to use of the aerosols, CFCs and HFCs releasing electric products and smoking. Pollution also occurs from fuels like coal, fuel wood, kerosene and other biomass fuels used for cooking in urban and rural areas. The report of MoEF 1997 states that the domestic sector contributes to about 7 per cent –8 per cent of the total pollution in Delhi the major contributor of which is the use of inefficient fuels in the poorer households. However in urban areas cleaner and efficient liquefied petroleum gas is more popular and causes less pollution (Singh *et al.*, 2005).

9.7 Greenhouse Gases from Air Pollution

The main natural greenhouse gases are CO_2, CH_4, N_2O, water vapor (H_2O), and ozone (O_3) (Santra 2005). Further human activities in recent decades have also added HFCs (hydro fluorocarbons). Carbon dioxide (77 per cent), nitrous oxide (8 per cent), and methane (14 per cent) are the three main greenhouse gases along with 1.1 per cent of the flours gases that trap infrared radiation and contribute to climate change (IPCC 2007; Leuenberger, 2006). The important greenhouse gases have long lifetimes in the atmosphere and a large fractions of emissions remains there for decades to centuries like CO_2 for around 100 years. Cutting emissions of such long lived gases therefore leads to only slow reductions in their warming effects (Mate *et al.*, 2009). The atmosphere of Venus is very hot because of high concentration of CO_2, whereas Mars is cold because of a lack of CO_2. The following pie chart depicts (Figure 9.2) the global emissions of greenhouse gases with maximum concentration of CO_2 (IPCC 2007).

1. Carbon Dioxide, CO_2

The natural sources of CO_2 is respiration, volcanic eruption, forest fires, soil erosion, earthquake, death and decomposition etc and account for about 19.3 per cent of the total greenhouse gas emission. The anthropogenic sources of CO_2 are industrial emissions, burning of fossil fuel, deforestation, changes in the land use pattern, clearing of forests for agriculture, industrialization, urbanization, transportation and use of inefficient fuel and it contributes to about 57.4 per cent of the total greenhouse gas emissions. Hence the total contribution of CO_2 as a GHG is 77 per cent (IPCC 2007). CO_2 has its absorption spectrum in the infrared region and absorbs maximum amount of the infrared radiations emitted by the earth's surface leading to an increase in the Earth's temperature. The CO_2 levels in the pre-industrial

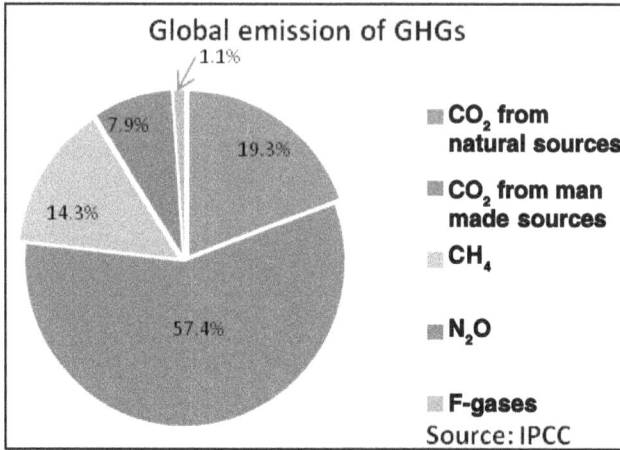

Figure 9.2: Global Emissions of GHGs (IPCC 2007).

time were only 270 ppm which increased to 385 ppm in 2009. So in one million molecules of air, approximately 385 will be CO_2. In the last four hundred thousand years, the amount of CO_2 concentration has never been larger than 290 ppm. The speed with which it is increasing clearly states that human influence is the cause (Ramanathan, 2006).

The general circulation models have predicted that if the atmospheric concentration of carbon dioxide doubles, the mean annual temperature may increase from about 2 to 4.5°C. The rise in temperature at higher latitudes will be greater than at low latitudes and that the temperature changes will be greater in winter than in summer (Gates 1990).

2. Methane, CH_4

The main natural source of CH_4 is the enteric fermentation by the ruminant cattle followed by emissions from marshes, peat bogs and wetlands. The anthropogenic sources of CH_4 emission are rice cultivation, preparation of biogas, landfills, sewage, coal mining, petroleum refineries and burning of fossil fuels and biomass. Methane emissions account for about 14.3 per cent of the total greenhouse gas emissions with global warming potential 25 times greater than CO_2. The average life span of methane is only 10 years, but the final product of disintegration of CH_4 is again CO_2 and water which are good in absorbing more amount of heat and hence enhance the greenhouse effect.

3. Water Vapour

The water vapour component is present in the atmosphere as a result of transpiration and evaporation of the water from water bodies. It has a variable composition in trace amount in the atmosphere depending upon the humidity. It also has absorption spectrum densest in the region of the infrared light so it has a high capacity to absorb the infrared radiations emitted by the earth's surface and increases the global temperature.

4. Nitrous Oxide, N₂O

The natural sources of the nitrous oxides are the volcanic eruptions, wild fires while the anthropogenic causes are deforestation, manure production, land conversion for agriculture, fossil fuel burning including use and production of the fertilizers (1). N_2O emissions account for about 8 per cent of the total GHG emissions (IPCC 2007).

5. Fluoro Gases

Fluorocarbons or F-gases are the quintessential greenhouse gases designed to be stable and durable. They account for only 1.1 per cent of the total GHGs causing global warming. IPCC 2005 calculated that the cumulative building up of these gases in the atmosphere was responsible for at least 17 per cent of the global warming (IPCC 2005, 2007). The F-gases include CFCs, HFCs, PFCs and SF_6. These chemicals are used in the refrigerators and air-conditioning, as solvents, blowing agents in foam, aerosol or propellants and in fire extinguisher. The HFCs have a life time of 14 years (IPCC 2007). Addition of one molecule of CFCs to the atmosphere has the same greenhouse effect as that caused by adding more than ten thousand molecules of CO_2 (Ramanathan, 2006).

Table 9.2: The Sources of the Various Fluoro Gases (Mate *et al.*, 2009)

Fluoro Gases	Sources
HFCs (Hydro fluorocarbons)	Leakage from refrigerator, aerosols, air conditioners
PFCs (Per fluorocarbons)	Aluminium production, semiconductor industry
SF6 (Sulphur hexa fluoride)	Electrical insulation, magnesium smelting

6. Black Carbon

Black carbon is a product of incomplete combustion of fossil fuels, biofuels and biomass. It is a component of soot an important climate forcing aerosol and the second leading cause of global warming after carbon dioxide. It remains in the atmosphere for few days or weeks and contributes to climate change in two ways. Firstly, it warms the atmosphere directly by absorbing solar radiations and converting them to heat. Secondly, it indirectly causes global warming by darkening the surface of ice and snow by getting deposited on them, reducing the albedo and therefore increasing heat absorption and increasing melting. It has revealed that black carbon on snow warms earth three times more than an equal forcing caused by carbon dioxide (Clare, 2009). The Indian Ocean Experiment (INDOEX) in1999, involving more than two hundred scientists observed the brown haze and predicted that the black carbon and other absorbing particles in the brown haze over the Indian Ocean and the Arabian Sea reduces sunlight by about 10–15 percent and enhances atmospheric solar radiating heat by as much as 50–100 percent. This increases the cloud cover, which reflects the solar radiation, adding to the surface-cooling effect; and decreases the size of cloud drops and suppresses precipitation (Ramanathan, 2006).

9.8 Global Warming Potential of the Greenhouse Gases

Global warming potential (GWP) may be defined as a gas's heat-trapping power relative to carbon dioxide over a particular time period. Global warming potential allows the observers to compare the contribution to global warming made by various greenhouse gases that have a varying warming effects and life spans. For example a CH_4 molecule has 25 times the warming potential of a CO_2 molecule (CO_2 having a life span of around 100 years) and some gases are hundreds or thousands of times more powerful (IPCC, 2007). Some of the greenhouse gases along with their global warming potential have been compiled (Table 9.3).

Table 9.3: Global Warming Potential of some Greenhouse Gases (IPCC, 2007)

Sl.No.	Greenhouse Gases	Global Warming Potential
1.	Carbon dioxide, CO_2	1
2.	Methane, CH_4	25
3.	Nitrous oxide, N_2O	298
4.	Hydro fluoro carbons, HFCs	124 – 14,800
5.	Per fluoro carbons, PFCs	7,390 – 12,200
6.	Sulphur hexa fluoride, SF_6	22,800

9.9 Greenhouse Effect and Global Warming

The increasing concentration of the greenhouse gases in the atmosphere allow the shortwave radiation to pass through the atmosphere from which 50 per cent of the radiations are reflected back by the atmosphere, clouds and the earth's surface and the rest 45-50 per cent of the solar radiations are absorbed by the earth's surface. The earth emits the heat in the form of infrared radiations which are absorbed by the greenhouse gases and not allowed to escape into the space, instead they are reflected back to the earth and hence they increase the earth's temperature like a greenhouse in which sun light once trapped cannot escape back. This phenomenon is called the **Greenhouse effect**. The earth's global temperature in the past years has increased due to an increase in the greenhouse gas concentrations, hence causing what is called as **Global warming** (Santra, 2005; Caldeira, 2009).

9.9.1 India and Air Pollution

In a recent study India has been ranked seventh among the top ten worst climate polluters of the world. In order the first ten polluters are Brazil, the US, China, Indonesia, Japan, Mexico, India, Russia, Australia and Peru (www.gits4u.com/envo/envo4.htm). However another study conducted by 80 scientists from 17 institutes confirmed on April 11, 2010 that India is the world's fifth-biggest polluter with more than 3 per cent increase in its greenhouse gas emissions annually between 1994 to 2007.

Researchers at the Indian Institute of Tropical Meteorology (IITM) noted that the emissions of the gaseous pollutants has increased in India over the past 20 years and this has been attributed to the rapid industrialization, urbanization and vehicular growth (www.gits4u.com/envo/envo4.htm).

The data about the levels of NO_x (nitrogen oxide) as acquired from the Global Ozone Monitoring Experiment (GOME) instrument aboard ESA's ERS-2 satellite for a period of 1996-2006 revealed that NOx emissions in India are increasing at a rate of 5.5 per cent per year and the location of the emissions correlated with those of thermal power plant, metropolitan cities, urban and industrial areas.

India ranked 101 in 2005 according to the Environmental Sustainability Index among 146 countries on the basis of its air quality. According to another report of the CPCB formulated NAAQS (National Ambient Air Quality Standards) in September, 2007 revealed 50 cities of India to be most polluted. They include Gobindgarh in Punjab to be the most polluted followed by Ludhiana, Raipur and Lucknow, Faridabad stood 10th, Agra 11th, Ahmedabad 12th, Indore 16th, Delhi 22nd, Kolkata 25th, Mumbai 40th, Hyderabad 44th, Bangalore 46th and Angul (Orissa) was declared the 50th most polluted city of the country (www.gits4u.com/envo/envo4.htm).

Population

The present population of India is around 1.15 billion and stands next only to China. It is estimated that by 2030, the population of India will be largest in the world totaling to about 1.53 billion. Population of India at the time of Independence was only 350 million and since then it has increased more than three times. The present rate of increase in population is 21.34 per cent which was earlier as much as 30-35 per cent (Figure 9.3).

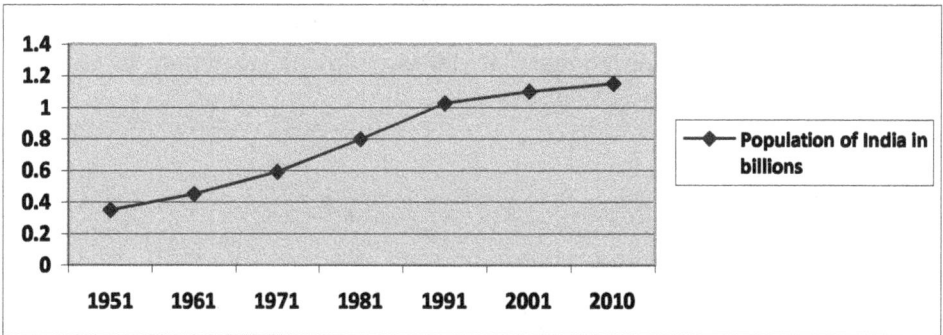

Figure 9.3: The Rapid Increase in the Population of India since Independence.
(www.indiaonlinepages.com/population/census2010).

India is very mysterious in terms of population dynamics as it supports 15 per cent of the global population having only 2.4 per cent of the world's land area. The maximum population (70 per cent) resides in rural areas covering about 550,000 villages (ww.indiaonlinepages.com/population/census2010).

The increasing population pressurizes the already limited resources and further depletes them and in the process further contributes to over exploitation and adds to the pollution level. Hence increasing population is the main root cause of the increasing pollution.

Industries

Industrial pollution has quadrupled in the past 20 years. Environmental problems seem to be further increased owing to India's dependence on the coal for power generation. Three-fourth of the country's electricity generation is from coal which meets half the energy need of India. India has huge reserves of coal which are being used for its economic development. But due to its high ash and carbon content it is degrading the environment. The carbon emissions in India have increased nine times over the past 40 years due to use of coal as the major energy source (www.indiaonlinepages.com/population/census2010 26).

IIT Delhi and Central Pollution Control Board (CPCB) on conducting a comprehensive environmental assessment of industrial areas have shown 10 industrial centers, Ankleshwar and Vapi in Gujarat, Ghaziabad and Singrauli in Uttar Pradesh, Korba in Chhattisgarh, Chandrapur in Maharashtra, Ludhiana in Punjab, Vellore in Tamil Nadu, Bhiwandi in Rajasthan and Angul Talcher in Orissa to be causing maximum environmental pollution (www.gits4u.com/envo/envo4.htm).

Vehicles

In India, the vehicular population is growing at a rate of 5 per cent per annum and today the vehicle population is approximately 40 million with the two-wheelers accounting for a share of 76 per cent of the total vehicular population (SIAM 2010, Singh *et al.*, 2005). 70 per cent of the country's air pollution result due to the vehicular emissions. There has been an eight time increase in the vehicular exhausts over the last 20 years (www.gits4u.com/envo/envo4.htm). According to an estimate the number of personal vehicles has increased from 9.75 lakhs in 1999 to 27.05 lakhs in 2009 which is more than the number of the buses which increased from 2806 in 1999 to 4235 in 2009 in Chennai (Srivatsan 2010).

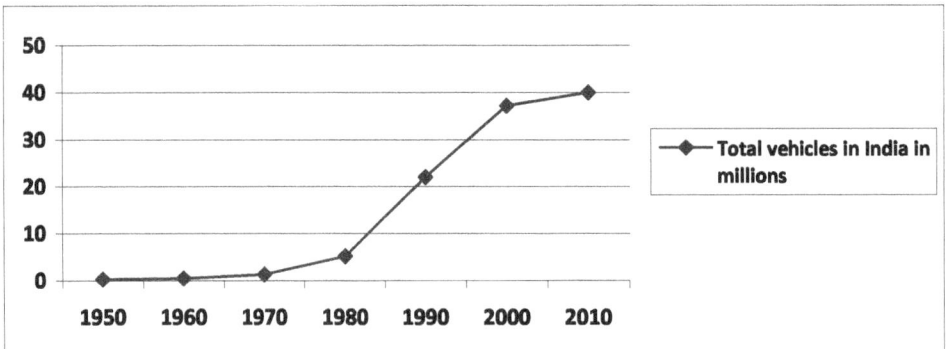

Figure 9.4: The Increase in the Vehicular Population in India since 1950 (Singh *et al.*, 2005).

A study comparing the rates of economic growth and the rates of growth of vehicular pollution and industrial pollution shows that during 1975–1995, the Indian economy grew by 2.5 times, but the industrial pollution load grew by 3.47 times and the vehicle pollution load by 7.5 times (24) this is with no improvement in the pace of pollution control (WWW;/gits4u.com/envo/envo4.htm).

9.10 Impact of Air Pollution

The most important impact of air pollution is that it increases the level of GHGs to exceed the optimum limit and causes enhanced greenhouse effect leading to global warming. Other than this it is hazardous to health of plants and animals. Warmer and more variable weather may cause increase in ground-level ozone which intensify respiratory diseases by damaging lung tissue, reducing lung function and sensitizing the respiratory tract to other irritants. More air conditioning use, due to warmer temperatures, could cause an increase in potentially harmful power plant emissions. Exposure to particulate matter from these and other combustion-related sources can aggravate chronic respiratory and cardiovascular diseases, alter host defenses, damage lung tissue, lead to premature death, and possibly contribute to cancer. Also the exposure to the particulate matters from cement factories and mines cause asbestosis, silicosis and bronchitis. In addition, changes in green plant distribution or pollen production could affect the timing and/or duration of hay fever and other seasonal allergies. The air pollution also impacts the plants and causes chlorosis, necrosis, bleaching and bronzing. The animals are also influenced by the air pollution which results in suffering due to lung infections and other diseases (Tibbetts, 2007). Sulfur dioxide, nitrogen dioxide and ozone cause a decrease in crop yield, acidification of lakes, damage to certain metals and monuments by acid rain. The Taj Mahal, Agra is turning yellow due to the oxidation of the $CaCO_3$ (limestone/calcium carbonate) reacting with other noxious chemicals. It also causes acidification of soil, lakes, pond etc damaging the fisheries and affecting animal and human health. Excess of nitrogen dioxide near the coastal environment causes algal blooms (Singh *et al.*2005).

9.11 Air Pollution and Climate Change

Climate change refers to a statistically significant change in either the mean state of the climate or in its variability (in terms of temperature, atmospheric pressure, precipitation status etc.) persisting for an extending period (typically decades or longer). Climate system is a complex interactive system consisting of atmosphere, land surface, snow and ice, oceans and other water bodies or in other words a climate system is an interactive system comprising of the atmosphere, lithosphere, cry sphere and hydrosphere (Karl and Trenberth, 2006). Air pollutants cause a considerable change in global climate and associated processes like greenhouse effect, ozone depletion, acid precipitation, *El Nino* effects etc. The causes of climate change are almost similar to that of the air pollution. Climate change results due to both; **a) anthropogenic causes** and **b) natural causes**.

a) Anthropogenic Causes

The human activities imparting a negative impact on the existing climate resulting in climate change are included under this category. It involves the following factors:

1. Industrialization

From the beginning of industrial revolution, there has been a constant rise in the pollution level, which is leading to a simultaneous rise in the temperature. The

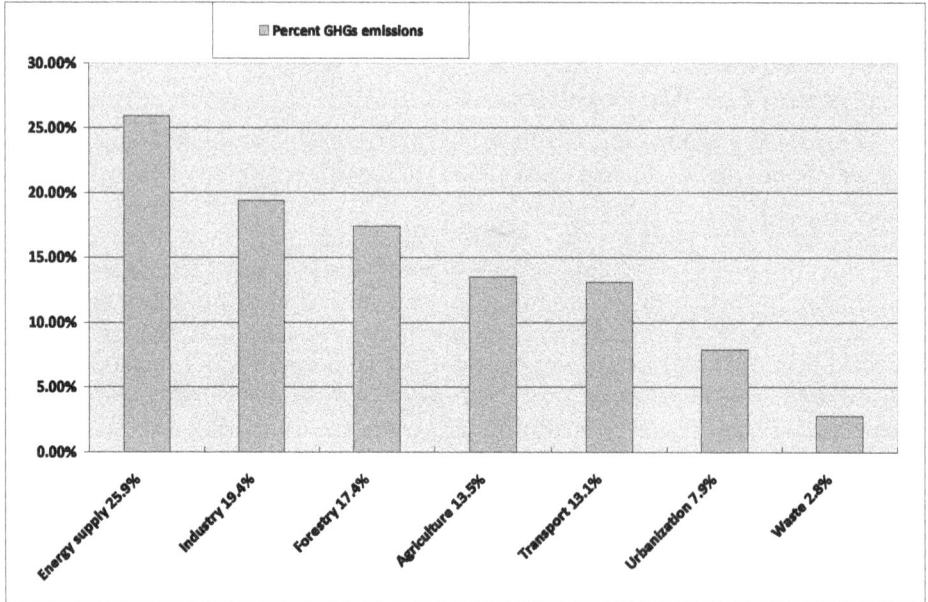

Figure 9.5: The Per cent GHGs Emissions by the Various Sectors (IPCC, 2007).

average global temperature shows a marked increase from 13.8°C in the 1950s to 14.6°C now. This is due to the increase in the greenhouse gases basically carbon dioxide which contributes to 50 per cent of the total global warming. The carbon dioxide concentration has increased from 270ppm in 1800s to 385 in 2009. According to IPCC 2007 the main reason for such a rapid increase is the CO_2 emissions is industrialization which accounts for a total GHG emission of 19.4 per cent (IPCC, 2007).

2. Deforestation

It is the second major cause of carbon dioxide being emitted in the atmosphere and accounts for about 20 per cent of all the climate-changing carbon emissions and 17.4 per cent of the total GHGs emissions (IPCC 2007) released into the atmosphere each year which is more than from all the planes, trains and automobiles (Foerstel, 2009). The ever increasing human population and its increasing demand for food, shelter and many commercial purpose is leading to indiscriminate cutting down of trees. More than 37 million acres of forest are destroyed every year. In Indonesia, deforestation produces 80 percent of the country's annual carbon emissions, placing it among the world's top emitters alongside the United States and China (Foerstel, 2009).

3. Vehicles/Transportation

The number of vehicles has increased a lot in the past few years. The vehicles are found to emit 30 billion tons of carbon dioxide. According to IPCC 2007 the transportation sector contributes to 13.1 per cent of the total GHGs emissions (IPCC 2007).

4. Urbanization

It is increasing at a very fast rate as the population is increasing leading to cutting of trees for space, establishment of cement factories etc. causing the emission of greenhouse gases, hence contributing to global warming. According to the IPCC 2007 urbanization causes emission of about 7.9 per cent of the total GHGs (IPCC 2007).

5. Agricultural Practices

The modernization of the agricultural practices with the increase in the use of chemical nitrogenous fertilizers along with the factors of increasing livestock, manure production and storage causes CH_4 and N_2O to be released, hence causing global warming. The agriculture contributes about 13.5 per cent in the total emission of the GHGs.

6. Energy Supply

This sector is the major contributor of GHGs to the atmosphere contributing to 25.9 per cent of the total emissions. It involves the generation of the electricity from thermal and hydroelectric power plants where the coal and other fossil fuel are burnt to release a lot of energy and cause a lot of pollution (www//gits4u.com/envo/envo4.htm).

b) Natural Causes

The contribution of plants, animals, and humans naturally to global warming come under this category. The plants, animals and humans contribute to carbon dioxide emissions by respiration and also due to their death and decomposition. The deciduous plants on defoliation do not absorb the available CO_2 instead contribute in CO_2 emissions from the respiring branches. A significant part of the earth's climate variability is also caused by changes in the solar emissions, which are due to changes in the sun-earth geometry (Milankovitch cycles). The volcanic eruptions liberate a huge amount of smoke and contribute in increasing the global temperature. Simultaneously, contradictory to the fact that the sulphur oxides so released has a high Aledo and hence cools the Earth by reflecting the solar radiations. The 1991 volcanic eruption of Mount Pinatubo in Philippines had released 20 million tons of SO_2 into the atmosphere cooling the Earth by $1°F$ ($0.33°C$) (Gates, 1990; Kunzig, 2008).

9.12 Simultaneous Impacts of Climate Change

The GHG emissions have increased by 70 per cent from 1970 to 2005. According to the IPCC 2007 report the global temperature has risen by about $0.74°C$ from 1906-2005 which is mainly due to anthropogenic causes and will lead to a rise in the global temperature from 2-4.5°C and of India from 2.7-4.3°C by the end of this century (Anonymous, 2009). Rainfall over India is predicted to increase (Anonymous 2009). The change in hydrological cycle is impacting the monsoon circulation and resulting in extreme events like floods (flash flooding as in Mumbai; July, 2005) and droughts (2002 and 2004).

The ENSO (*El Nino* Southern Oscillation) has direct impact on the Indian monsoon (Sigh *et al.,* 2005). The rise in temperature is leading to melting of the glaciers and ice caps and also thermal expansion of water which has resulted in a sea level rise of 1.8mm/year from 1961-2003, but from 1993-2003 this pace has almost doubled. The sea level in India is predicted to rise by about 88 cm by 2100 (anonymous, 2009). This is resulting into the salinity ingress and also the submergence of the low lying areas and islands. The satellite data shows a 2.7 per cent decline in the Arctic ice content by each decade since 1978. Around 2 m sq. km of the permafrost has declined by the year 2000. The IPCC report also confirm of an increase in the number of the droughts since 1970s. An increase in the incidence of heat waves has also been reported all over the world. The winters have become less severe. The thunderstorms, cyclones, hurricanes etc have also registered an increase in number (Anonymous, 2009).

The Scientists believe that global warming resulting in a rise in the temperature, will lead to cooling of the stratosphere along with the ozone leading to a slow pace of ozone hole repairing. The Ozone layer has been depleting for the past many years and scientists say that presently the ozone hole over Antarctica is three times the size of United States and is still growing (Shah, 2002). A study on the decline in rice yields in India shows GHG and air pollution to be the cause (Cramer, 2006). The increase in the number of droughts and floods will also impact the economy as the agricultural yield will show a decline. The forest area will show a high net primary productivity (NPP) due to the high level of CO_2 in the atmosphere but they will reach maturity soon and will become carbon source rather than carbon sinks. The desertification will increase (Anonymous, 2009). There will be water crises and the human health is also going to suffer due to an increase in the tropical vector borne diseases like malaria, dengue, and chikangunia), rodent borne diseases, water borne diseases like cholera, diarrhea and dysentery (Tibbetts, 2007).

9.13 Controlling Air Pollution

In order to reduce air pollution being caused by the release of pollutants from the factories etc certain measures can be implemented. The measures include; i) the replacement of the burning of fossil fuels by renewable sources like solar energy, tidal energy and geothermal energy for energy generation or by using efficient fuels like biofuels (palms, *Jatropha* etc). ii) Another method to control air pollution could be by a number of mechanical devices that can in a way be helpful in reducing the amount of air pollutants during their emission. The mechanical devices used by the industries emitting pollutants are put under two categories (Santra, 2005):

1. Devices to reduce particulate pollutants (Gravity settling chambers, cyclone collector, dynamic precipitators, spray towers (open spray towers and packed towers, venturi scrubbers, electrostatic precipitators, fabric filters) and

2. Devices to reduce gaseous pollutants (include absorption units like spray towers, packed towers and venturi scrubbers and use absorbent like activated alumina, silica gel, activated carbon etc).

Air pollution control for automobiles can suitably be done by installing catalytic convertors, diesel particulate filter and also changing fuel quality. A carbon levy on fuels is proposed by the president of the Swiss Confederation which would increase the fuel prices and thus increase the dependence on public rather than private transport (Leuenberger, 2006). According to a report of RCEP (2000) increasing the fuel prices has lead to the energy savings as in case of UK where a 10 per cent increase in fuel price lead to decrease in the vehicular consumption of the fuel by 6 per cent (RCEP 2000).

Another strategy to control air pollution would comprise fuel efficiency. Coal is and will remain the main fuel for commercial energy production in India in the near future. To ensure more efficient use of coal an up gradation of the technology involving coal-washing and improvement in combustion technology need to be done. Fuel efficiency of oil is being promoted and it is being conserved by upgrading the equipments, energy audits and installation of the waste heat recovery system. The diesel and coal are being substituted by natural gas. PCRA (Petroleum Conservation Research Association) has been established to increase awareness and to develop fuel-efficient equipments (Anonymous 2009).

The replacement of non-renewable source of energy by the renewable energy will lead to energy efficiency and would reduce emissions of GHGs to 100 per cent. The government of India has taken an initiative in this direction by policies to exploit the huge potential in India's North-East for generation of hydropower. Small hydropower projects of 1,423 MW capacity has been established. The photovoltaic (PV) system using solar energy is being used for rural electrification, railway signalling, microwave repeaters, solar lanterns, home and street-lighting systems, standalone power plants, pumping systems, power to border outposts and TV transmission and reception. Until now 9,20,000 Solar PV systems, with a total capacity of 82 MWp have been established in India. India stands fifth in the world in wind power generation with a total capacity of 1,507 MW. The government has installed around 3.2 million biogas plants, 33 million improved stoves in the rural area, biomass generation plant of 358 MW capacity and gasification system of 42.8 MW capacity. The energy from the urban, municipal and industrial waste has been used to generate 15 MW power (Anonymous 2009).

Bureau of Energy Efficiency (BEE) has been set up to put into operation, conservation measures such as energy standards, labelling of equipments, building energy codes, and energy audits (Anonymous, 2009).

Table 9.4: Comparison of the Efficiency of the Various Fuels (35)

Fuel	Carbon Content (kg C/kg fuel)	Energy Content (kWh/kg fuel)	CO_2 Emissions (kgCO$_2$/kWh)
Gasoline	0.9	12.5	0.27
Diesel	0.86	11.8	0.24
LPG (Liquefied Petroleum Gas)	0.82	12.3	0.24
Natural gas	0.75	12	0.23
Bioenergy	0	0	0

It can be concluded that the emissions of CO_2 from gasoline is more than that from diesel, LPG and natural gas (Table 9.4). Biofuels are responsible for zero CO_2 emissions.

The various metro states of India have converted all the cars, buses, three and two wheelers to CNG (compressed natural gas) vehicles (www.stoveonline.uk.co.in). Introduction of BRTS (Bus Rapid Transport System) in Ahmedabad, Gujarat is another major initiative to combat climate change. For controlling transport emission, a norm called Bharat 2000 coming into force from 1st April 2000 has been implemented for all categories of vehicles manufactured in India. In Delhi 50,000 vehicles switched from petrol and diesel to CNG type (Anonymous, 2009).

The agriculture can contribute in lowering the emissions of methane and nitrous oxide by switching to better cultivation practices and use of fuel efficient water pumps.

The nuclear power plants are also being encouraged so as to cause minimum CO_2 emissions. The nuclear energy will also stand out as one of the source of generating power and mitigation of global warming as it leads to a reduction in emission of about 500 million metric tons of CO_2. But then the only problem will be with the hazardous waste being generated.

The government of India has formulated a number of legislations, policies, and programmes for protecting the environment. Some of them related to air pollution are the Air (Prevention and control of pollution) Act, 1981 and the Environment (Protection) Act, 1986 (Singh *et al.,* 2005). Ambient air quality standards (both short term for 24 hours and long-term for a year) have been put forward for industrial, urban and other sensitive areas with respect to pollutants such as SO_2, NO_x, SPM (Suspended Particulate Matter), RPM (Respiratory Particulate Matter), Pb, CO and NH_3.

A guideline for site selection for the industrial area has also been prescribed. Environmental impact assessment (EIA) of the various industries or other SEZ (Special Economic Zone) to be established is done to study the impacts it may have on the ecosystem and then only its construction is sanctioned.

Pollution prevention technologies are being developed and industries are encouraged to use cleaner technology to reduce emission of pollutants. The union budget 2010-11 for the environment includes the setup of a National Clean Energy Fund (NCEF) for funding research in clean energy technology development. A number of judgments relating to stringent vehicle emission norms, fuel quality, introduction of cleaner fuels, phasing-out of older vehicles, and shifting of hazardous industries have provided a great deal of momentum to the efforts for improvement of air quality (Singh *et al.,* 2005).

Increasing the tree plantation and green cover help improve environmental condition by acting as lungs of the city purifying the air (Oza *et al.,* 2005). Fast growing trees should be promoted and simultaneously guarded (Jasrai, 2005). Tree plantations act as one of the sinks of air pollutants. There are many plants tolerant to many pollutants like SO_2, N_2O, CO and O_3 and should be promoted (Leena *et al.,* 2003).

9.14 Global Warming Mitigation Strategies

Global warming mitigation involves reducing the intensity of radioactive forces so as to reduce the effect of global warming and it can be made possible by two aspects; Geo-engineering and Carbon sequestration. Climate change also can be made to reverse by some new scientific and engineering invention. Geo-engineering are the proposals to manipulate the earth's climate so as to decrease the impact of global warming from the greenhouse gas emission. It comprises of the

a) Sulphur dioxide spraying in the atmosphere to cool the earth;

b) Establishment of artificial trees to suck in excess of CO_2 from the atmosphere; which is the major contributor to global warming;

c) Cloud seeding ships which form clouds with an albedo more than the normal clouds;

d) Iron fertilization of the oceans to increase the population of the algae and phytoplankton in the oceans so as to enhance CO_2 absorption from the atmosphere;

e) Limestone fertilization of the oceans so as to enhance the CO_2 absorption capacity of the oceans;

f) Space mirror in the space so as to reflect some amount of the sun rays. All these methods are very expensive. But still taking into consideration these strategies would surely help in reducing the effect of global warming (Caldeira, 2009; Kunzig 2008).

Another technology comprises of the various methods of carbon sequestration. Major point sources of carbon dioxide include large fossil fuel facilities producing coal-fired power, natural gas, fossil fuel–based hydrogen and synthetic fuel. Carbon dioxide emissions from such sources can be captured and stored in underground geologic formations, helping in lowering the severity of climate change. Carbon Capture and Storage (CCS) technologies are already in use widely in industries producing fertilizers, hydrogen and in natural gas processing (Tibbetts, 2007).

In India also the CCS technologies have taken initiatives. The studies involving capture and simultaneous utilization of the captured CO_2 by NETRA (NTPC Energy Technology Research Alliance) has being undertaken. A development of the project for the capture of 35 tones of CO_2 per month at Simhadri is under process (Anonymous, 2009). The world's largest plant for geological CO_2 sequestration has been established as Otway project in Australia. The oldest CCS plant in the world to store CO_2 on industrial scale is Sleepier gas field in Norway. The carbon capture and storage technology is also being implemented in the Badarpur power plant, Delhi. ONGC, India in collaboration with Norwegian oil and gas has also undertaken a project on CO_2 management. The CO_2 generated from the Hazira Plant, Gujarat is being supplied to the Ankleshwar refinery for the oil extraction (Tibbetts, 2007).

National Action Plan on Climate Change was released in Delhi during 2009, involves eight missions which are; national solar mission, national mission for enhanced energy efficiency, national mission on sustainable habitat, national water mission, national mission for sustaining the Himalayan ecosystem, national mission for a Green India through massive tree plantation programmes, national mission for sustainable agriculture and national mission on strategic knowledge for climate change by establishing a knowledge platform on climate change (Anonymous, 2009). Successful implementation of all these plans would surely help reduce global warming.

9.15 Conclusions

Air pollution and climate change are thus correlated. The more we contribute in the form of GHGs to the atmosphere it leads to an enhanced greenhouse effect and finally aggregates as climate change. An increase in the practices like large scale industrialization, deforestation, urbanization, changing agricultural practices and other human practices in association with the natural processes of volcanic eruptions, wild fires add to air pollution. All these processes contributing to air pollution enhance the level of greenhouse gases in the atmosphere resulting in global warming which simultaneously causes climate change owing to the changes in the variability in the temperature ranges, precipitation levels and increasing number of cyclones, hurricanes, heat spells, cold waves, droughts and floods. Reducing the level of air pollution at the grass root level by cutting off industrial and vehicular emissions will surely bring down the levels of greenhouse gases in the atmosphere and thus make earth a better place to live in. This can be done by reducing emissions, strictly banning the use and generation of all the fluoro gases, switching to clean energy, clean development mechanisms, organic farming, and carbon sequestration at industrial level and by tree plantations. As the greenhouse gases remain in the atmosphere for a very long time the effect of the global warming and climate change will remain prevalent for more than 100 years even if we start the process of reducing the greenhouse gases now.

References

Anonymous 2009. *Climate change*: Road to Copenhagen India's Response Framework, pp: 1-24.

Baumert, K., Herzog,T., Pershing, J. 2005. *Navigating the Numbers*. Greenhouse Gas Data and International Climate Policy. World Resources Institute.

Bellarby,J., Foereid,B., Hastings,A., and Smith,P. 2008. *Cool Farming: Climate impacts of agriculture and mitigation potential,* University of Aberdeen, Greenpeace International, Netherland, pp: 1-44.

Bousquet, P., Peylin, P., Ciais, P., C. Le Quere,C., Friedlingstein, P., and Tans, P.P., 2000. Regional changes in carbon dioxide fluxes of land and oceans since 1980, *Science*, 290: 1342–1346.

Caldera, K. 2009. *Geo-engineering to Shade Earth*. In: 2009 State of the World into a Warming World, Eds: Linda Starke, World watch Institute, Washington DC, pp: 96-98.

Census of India population 2010 http://www.indiaonlinepages.com/population/census2010 accessed on 26.6.2010.

Clare, D. 2009. *Reducing Black Carbon.* In: 2009 State of the World into a Warming World, Eds: Linda Starke, World watch Institute, Washington DC, pp: 56-58.

CO_2 emission of fuels. Forestry Commission: A fuel strategy for England www.stoveonline.uk.co.in accessed on 18-2-2010.

Cramer, W. 2006. Air pollution and climate change both reduce Indian rice harvests. *Proc Natl. Acad Sci*, 103: 19609–19610.

Environmental pollution in India www.gits4u.com/envo/envo4.htm accessed on 30.7.2010.

Foretell, K., 2009. *Climate Change Working Globally to Reduce Deforestation* www.natureconservancy.com accessed on 8.10.2009.

Friedlingstein, P., Cox, P., Betts, R., Bopp, L., Vonbloh, W., Brovkin,V., Cadule, P., Doney, S., Eby, M., Fung, I., Bala, G., John, J., Jones, C., Joos, F., Kato,T., Kawamiya,M., Knorr,W., Lindsay, K., Matthews, H.,D., Raddatz,T., Rayner, P., Reick, C., Roeckner, E., Schnitzler, K. G., Schnur, R., Strassmann,K., Weaver,A.J., Yoshikawa,C., and Zeng, N., 2006. Climate-Carbon cycle feedback analysis: Results from the C4MIP Model inter comparision, *Journal of Climate*, 19: 3337-3353.

Gates, D.M. 1990. Climate change and forests, *Tree Physiology*, 7: 1-5.

IPCC 2005. *Safe-guarding the ozone layer and the global climate system*, Geneva, pp: 135.

IPCC 2007. *Climate Change 2007*: Synthesis Report, Geneva, pp: 4.

IPCC. 2007. *Climate Change 2007*: The Physical Science Basis, Cambridge University Press, Cambridge, UK, pp: 212-213.

Jasrai, Y.T. 2005. *Remedies in our hand for controlling air pollution because of urban traffic.* In: Urban Pollution Issues and Solutions, Eds: A.Arya, S.J.Bedi, Y.T.Jasrai and V.S.Patel, Nidhi Book Centre, Delhi, India, pp: 46-49.

Jones, C.D., and Cox, P.M. 2001. Modelling the volcanic signal in the atmospheric CO_2 record, *Global Biogeochemical Cycles*, 15: 453–466.

Karl, T. R. and Trenberth, K.E. 2006. *What is Climate Change?* In: Climate Change and Biodiversity, Eds: Lovejoy, T. E. and Hannah, L., TERI Press, New Delhi, India, pp: 15-90.

Kunzig, R., 2008. A sunshade for planet Earth. *Scientific American India*, 3: 24-33.

Leena, A., Jasrai, Y.T., and Garge, S.K. 2003. *Green Belt Plant Scavengers for Combating Air Pollution.* In: Air Pollution Development at What Cost? Eds: Y. T. Jasrai and A. Arya, Daya Publishing House, Delhi, India, pp: 32-40.

Leuenberger, M. 2006. *A Global Carbon Levy for Climate Change* Adaptation. Sustainable Development Opinion. IIED, London.

Lucht, W., *et al.,* 2002. Climatic control of the high latitude vegetation greening trend and Pinatubo effect, *Science*, 296: 1687–1689.

Mate, J., Davies, K., and Kanter, D. 2009. *The risk of other greeenhouse gases.* In: 2009 State of the World into a Warming World, Eds: Linda Starke, World watch Institute, Washington DC, pp: 52-55.

Oza, B.D., Sisodia P.S. and Jasrai Y.T. 2005. Counting of Green heads of Vadodara city, *International Journal of Bioscience Reporter,* 3: 148-155.

Ramalingaswami,V., Aggarwal,P., Chhabra,S.K., Desai,P., Ganguly,N.K., Gopalkrishnan, K., Kacker,S.K., Kalra,V., Kamat,S.R., Kochupillai,V., Nag,D., Pande,J.N., Raina,V., Ray,P.K., Saiyed,H., Seth,P.K., Trehan,N., and Wasir,H.S. 1999. Urban Air Pollution. *Current Science,* 77: 334-336.

Ramanathan, V. 2006. *Global Warming,* Bulletin of the American Academy Spring 2006: 36-38.

RCEP, 2000. *The Changing Climate.* 22nd Report of the Royal Commission on Energy

Rhodes, J., and Keith, D. 2008. Biomass with capture: Negative emissions within social and environmental constraints: An editorial comment. *Climate Change,* pp: 321-328.

Santra, S.C. 2005. *Environment Science,* New Central Book Agency (P) Ltd, Kolkata, pp: 169-207.

Shah, A. 2002. *The Ozone Layer and Climate Change* http://www.globalissues.org accessed on 30.7.2010.

SIAM, 2010. Vehicular technology in India, Emission norms www.siamindia.com assessed on 30.7.2010.

Singh, R.P., Prasad, A. K., Chauhan, S.S, and Singh, S. 2005. Impact of growing urbanization and air pollution on the regional climate over India, Country report, *International Association for Urban Climate IAUC Newsletter,* 14: 5-10.

Sreevatsan, A. 2010. 300 per cent rise in vehicle population in 15 years. *The Hindu,* March 2, 2010.

Tibbetts, J. 2007. *Health effects of climate change,* Environmental Health Perspectives, 115: 196-203.

2015, **Impact of Global Warming and Climate Change on** *Pages 101–107*
        Human and Plant Health
Editors: **Dr. Arun Arya and Prof. V.S. Patel**
Published by: **DAYA PUBLISHING HOUSE, NEW DELHI**

Chapter 10
Studies on Ozone and Climate Change

B.V. Kamath*

*Institute of Infrastructure, Technology, Research and Management (IITRAM),
IITRAM, Maninagar (East), Khokhra, Ahmedabad – 380 008, Gujarat, India*

ABSTRACT

The ozone layer in stratosphere is acting as shield against UV radiation and life was not possible on earth surface when this layer was not developed millions of years ago. Chemicals like CFCs have damaged the ozone layer and scientists are suggesting corrective measures to arrest this depletion. Climate change has resulted into decrease of crop yield. Present paper describes some finding on effects of O_3 on crop plants. The tropospheric ozone produced by vehicles can reduce the plant growth and crop yields.

Keywords: Ozone, Stratosphere, Troposphere, Climate change, Plant growth, Health effects.

10.1 Introduction

Human, plant and animal life on earth is protected by a fragile layer of ozone gas (O_3), a naturally occurring form of oxygen, which is highly poisonous. At ground level, ozone contributes to smog and acid rain. But high up in the stratosphere (25-30 km above the earth), ozone forms a screen against the sun's lethal ultraviolet (UV) rays. Without this ozone layer, UV radiations would kill all life on this planet.

Ozone (MW 48) is a three atom allotrope of Oxygen (MW 32). It has sweet odour at low concentration. Depletion of ozone over Antarctica in 1982 resulted into Ozone

---

* E-mail: kamathbv1@rediffmail.com

hole. Efforts are being made to reduce the emission of CFCs and related chemicals. Antarctic expedition in 1987 verified huge losses of ozone, same year Montreal Protocol was signed between 43 countries to reduare CFCs. Researches of Solomon and McIntyre are described. How ozone can affect crop plants at tropospheric level and other products like rubber, fabric, metals and plastic are described. Ozone is not created directly,but is formed when nitrogen oxides and volatile organic compounds mix in sunlight. That is why ozone is mostly found in the summer (Bhatt and Dhamecha, 2009).

10.2 What's Happening to the Ozone Layer?

There is growing scientific evidence that persistent, synthetic (man-made) chemical pollutants are rapidly destroying the ozone layer. A hole has appeared in the layer above the Antarctic and there is a general thinning of ozone around the world, allowing increasing amounts of harmful UV radiation to reach the earth's surface.

A) What Does UV Radiation Do?

A small amount of UV already penetrates the ozone layer, causing human skin cancer, which kills some 12,000 people a year in the United States alone. UV also affects the immune system, our inbuilt resistance to disease. This means that it is easier for cancers to establish themselves and grow, and that we are more vulnerable to diseases such as herpes. In addition, it causes cataracts and dims eyesight. Increased radiation will make us more susceptible to all these health problems.

But UV radiation doesn't only harm human: it debilitates all living things. It damages more than two-thirds of the world's plant species. Higher UV levels are likely to reduce crop yields, which will seriously affect food supplies. Marine life is also threatened. Especially vulnerable are the microscopic plankton that drift on the surface of the sea. These tiny organisms play a vital role in the marine food chain and absorb over half the world's carbon dioxide emissions.

B) How is Ozone Destroyed?

Many chemicals react with ozone to destroy it. They also contribute to the warming up of the climate, known as the "greenhouse effect". The black list includes nitric and nitrous oxides from vehicle exhausts, and carbon dioxide produced by burning fossil fuels such as coal. Other gases such as halons and methyl bromide (used as pesticide) also damage the ozone layer, but the most destructive chemicals by far are a group of chlorine-containing substances called chlorofluorocarbons (CFCs).

CFCs float upwards from the earth's surface, taking about eight years to reach stratosphere. The intense UV radiation they encounter in the stratosphere gradually breaks them down. As the CFCs disintegrate they release chlorine, which reacts with ozone, converting it into ordinary oxygen that offers no protection against UV radiation. Meanwhile the chlorine, which acts simply as a catalyst, is not affected by its contact with the ozone and lives on to destroy thousands more ozone molecules. A single CFC molecule can destroy 100,000 molecules of ozone.

Solomon () began to work with Rolando Garcia to develop a coupled, two-dimensional chemical-dynamical model of the stratosphere and the mesosphere. It quickly led to a better physical understanding of how stratospheric wind moves around such trace chemicals as stratospheric methane and ozone. When the ozone hole was discovered in 1985, Solomon and Garcia were fortunate in having such a good model to examine possible explanations for its mysterious occurrence and found heterogeneous chemistry (in particular, the reaction of hydrochloric acid with chlorine nitrate) as the likely cause. Solomon says this turned out to be a good guess, and it is now widely acknowledged as the key initiating step in producing the ozone hole.

In the mid 1980s, she worked with Jeff Kiehl of the Climate and Global Dynamics Division and began to put a better treatment of radiation into their two-dimensional stratospheric model. In 1986 and 1987, Solomon served as the head project scientist of the National Ozone Expedition at McMurdo Station, Antarctica, where she made some of the first measurements showing that chlorofluorocarbon chemistry indeed is responsible for the ozone hole. Since then, she has combined this dynamical and chemical knowledge to look at the effects of volcanic eruptions on ozone depletion, at gravity waves and mesospheric species, and at a number of other intriguing chemical-dynamical problems.

Solomon is widely known for her crucial role in efforts to determine the cause of the Antarctic ozone hole and showed how chlorofluorocarbons (CFCs) interact in the unique Antarctic environment to cause ozone depletion there. She is currently focusing on research in many fascinating areas, including photochemistry and transport processes in the stratosphere and troposphere; remote sensing of the atmosphere by spectroscopic methods and their interpretation; interpretation of ozone depletion at mid-altitudes and in polar regions; coupling between trace gases and the Earth's climate system; and, stratospheric chemistry, especially observations and interpretation of the chemistry of the Antarctic ozone hole.

10.3 What are CFCs?

CFCS are gases used as (I) propellants in aerosols (ii) insulating materials in refrigerators and (iii) artificial plastic foams for use as food containers, and in furniture and carpets.

When these cheap-to-produce and extremely stable chemicals were first discovered, they were greeted as miracle substances that would revolutionize modern life. Now, however, manufacturers have to look around for alternatives.

10.4 What State is the Ozone Layer in Today?

Every October, an ozone hole opens above Antarctica. Each year, scientists observe that this hole is getting bigger, and that its effects are becoming more obvious. In early 1992, there were reports of blindness in fish, sheep, and rabbits in Southern Chile. Doctors in the area found they were treating significantly increased numbers of patients with allergies and skin and eye complaints.

The northern hemisphere's ozone layer is suffering too. The United States, most of Europe, northern China, and Japan have lost as much as 6 per cent of their protective filter. The United Nations Environment Programme (UNEP) calculates that every 1 per cent loss of ozone results in an additional 50,000 skin cancers and 100,000 cases of blindness from cataracts worldwide.

10.5 Urban Ozone Levels

High ozone concentrations near the surface of the earth are usually associated with urban areas and periods of high temperature and sunlight. Although it first gained notoriety as a pollutant in Los Angeles smog, high levels of ozone are now observed in most urban areas, including most of the eastern United States during the summer months. The daily maximum ozone concentration occurs between noon and 5 P.M. in most central or downtown urban areas. The concentration then decreases during night and often becomes zero at ground level. For locations downwind of these urban centers, peaks in ozone concentrations are observed even during the night.

The highest levels of ozone occur during air pollution episodes. Figure 8.6 shows the ozone concentration during a three-day episode at Montague, Massachusetts, which is characterized by two daily peaks. The first peak occurs in the early afternoon and is due to ozone production locally. A second peak, often larger is evident in the evening due to ozone transported from other locations. As we will see later, ozone production requires a lot of sunlight, so ozone peaks in the late afternoon or evening indicate that ozone produced somewhere else earlier in the day is arriving in the area of concern, transported by the wind.

Measurements like the ones in Figure 8.6 suggest that ozone in the eastern United States is both a local and regional problem. In other words, ozone is both produced from local emissions and transported long distances from cities upwind of an affected area. As a result a blanket of ozone often covers the eastern United States, with concentrations exceeding 80 ppb on geographical scales of over 1000 km. during the "ozone season" from about May to October, these high concentrations of ozone are present for several hours per day, persist for several days during smog episodes and occur in both urban and rural areas.

Despite efforts at the local, regional and federal levels to control ozone during the last 25 years, ambient ozone concentrations in urban, suburban and rural areas of the United States continue to be a major environmental and health concern. Ninety eight countries across the United States did not meet the ozone national air quality standard (maximum one hour average concentration of 120 ppb) in 1990.The nonattainment areas are shown in Figure 8.7 together with the degree of violation of the standard. Areas are characterized by the U.S. Environmental Protection Agency (EPA) as "extreme" if the ozone levels exceed the standard by more than 133 percent, "severe" from 50 to 133 percent above the standard, "serious" from 33 to 50 percent, "moderate" from 15 to 33 and "marginal" from 0 to 15 per cent. It is clear from Figure 8.7 that high ozone concentrations are a widespread problem.

10.6 What are we Doing about it?

Governments have been working to phase out CFCs. Between 1988 and 1992 the world's consumption of the lethal chemicals dropped by 40 per cent. In February 1992, the European Community and the US announced that they would phase out CFCs and other ozone depleting substances (halon, carbon tetrachloride, and methyl chloroform) by 1995.

In addition, industry should recycle CFCs so that the gases can be reused. Some electrical appliance makers are already incorporating the cost of recycling CFCs in the purchase price of their new refrigerators. Aerosol manufacturers have introduced ozone-friendly aerosols, which replace CFCs with alternative propellants. However, natural spray perfumes, hand pumps, and trigger spray products (which use no propellant of any kind) are more likely to provide an environmentally sound alternative to aerosols.

It is possible to replace CFCs by other synthetic chemicals, but it is uncertain how effective these are, and whether or not they might be equally damaging to the environment. For example, solvents like methyl chloroform, carbon tetrachloride and methylene chloride are toxic chemicals in their own right, some of them linked to cancer. They should not, therefore, be used as CFC replacements.

New generation HCFCs (CFCs with a hydrogen atom added) are being produced. These have much lower ozone depletion potential. However, safety tests revealed that HCFC142b, used as a foam-blowing agent and refrigerant, is flammable, while others are poisonous and cause eye deformities in mice

HCFCs 123 and 134a (used principally as foam-blowing agents and aerosol propellants) are thought to be safe to humans, though just how effective they will be remains to be seen. For instance, HCFC134a uses more electricity to maintain the same temperature, so more fossil fuel has to be consumed to produce the necessary electricity. This means that more carbon dioxide is emitted. So helping solve one problem may in fact create others. It is important for the world to develop alternative technologies so that in the future aerosols, refrigerators, and air-conditioners can sell function without using dangerous chemicals. It is also vital that the richer countries in the industrialized world help developing countries in the South to phase out CFCs. The ozone layer, after all, protects the whole planet.

McIntyre and his coworkers have made several notable contributions to atmospheric science research, centered around understanding the fluid dynamics of the Earth's atmosphere, with emphasis on the stratosphere, the layer lying between altitudes of about 10 to 50 kilometres. The stratosphere contains the bulk of the ozone shield that protects the Earth from harmful solar ultraviolet radiation. McIntire's research has helped to explain why the strongest human made ozone depletion occurs in the Southern Hemisphere in the form of the so-called Antarctic ozone hole, even though the chlorofluorocarbons and other chemicals known to cause it enter the atmosphere mainly in the Northern Hemisphere. Part of the answer is that the chlorofluorocarbon molecules go on epic journeys, circumnavigating the globe and visiting both hemispheres many times in the lower atmosphere before eventually arriving in the stratosphere.

10.7 Assessment of Exposure of Ozone on Crops

Ozone has been identified as the most important air pollutant in terms of spatial distribution and impacts on agricultural yields. With the importance of United States agriculture to both domestic and world consumption of food and fiber, significant reduction in their supply would have substantial economic consequences. Various methods have been used to assess the economic impact of ozone, many of which have been simplistic. Reliable assessment procedures should use theoretically justified economic methodologies which consider the effects to both the producer and consumer. These methods usually address price changes due to adjustments in production and the role of the producer input and output substitution strategies. The resulting estimates more accurately assess the true economic impact than other procedures.

Numerous studies have attempted to assess the dollar losses to crop production resulting from ambient ozone. The quality of these various estimates is variable. For example, the earlier economic loss data (U.S. Environmental Protection Agency, 1978) frequently used simplistic traditional approaches which were not theoretically sound. Thus, those previous estimates should be viewed with caution. Most of the recent economic assessments of agricultural losses (since 1978) have focused on regional losses. Crop loss estimates for southern California ranged from 45 (Adams *et al.,* 1982) to approximately 100 million dollars (Leung *et al.,* 1982). These studies used different assessment methodologies and considered the effects on different crops. The economic impact of ozone on corn, wheat, and soybeans for the 'Corn Belt' was estimated at 688 million dollars (Adams and McCarl, 1984), which for Illinois alone the losses were estimated at 55 to 200 million dollars (Mjelde *et al.,* 1984).

Only a few studies have attempted to assess the national economic consequences of ambient ozone. The economic losses have been estimated between approximately 2 (Adams and Crocker, 1982) and 3 billion dollars (Shriner *et al.,* 1982). Although these estimates include more complete dose-response information for more major commodities and better air quality data than previous national estimates (U.S. Environmental Protection Agency, 1978), they should be considered preliminary, because of limited data.

It is apparent that the current dollar estimates of crop damage are useful primarily as indicators of the magnitude of impact. A full accounting of the economic mechanisms underlying agricultural production is required to provide definitive estimates of the extent of agricultural losses. Such accounting should include both annual and perennial crops (agronomic and horticultural) and the associated dynamic adjustments of agricultural production. It must consider the effects on intermediate consumers (such as livestock growers and food processors) and final consumers (both domestic and foreign). The effect of ozone on ornamentals, both physically and economically, has also not been addressed.

References

Bhatt.H and Dhamecha.H.V.(2009). Monitoring and control of urban air pollution in india. In ECO-degradation due to air pollution (eds). Arya.A, Bedi.S.J and Patel.V.S. Scientific Publishers india, Jodhpur pp. 121-133.

Kamath, B.V. and Patel, V.S. (2003). Ozone Depletion-A global Problem, in Air Pollution Development at what Cost (ed. S) Jasrai, Y.T. and Arya A. Daya Publishing House, Delhi, pp. 214-220.

Bjom,L.O.(1976). Light and Life, Hodder and Stoughton. London

Development and the Environment. World Development Report 1992.Oxford University Press.

Ingersoll,A.P (1983). The Atmosphere. In: *Scientific American.* September 1983.

Kaku, M. (1998). Visions. Oxford University Press.

Macalady, D.L. (1998). Perspective in Environment Chemistry. Oxford University Press.

2015, Impact of Global Warming and Climate Change on
Human and Plant Health
Editors: Dr. Arun Arya and Prof. V.S. Patel
Published by: DAYA PUBLISHING HOUSE, NEW DELHI

Chapter 11

Fluxes and Controlling Edaphic Factors of CO_2, CH_4 and N_2O from Soil under *Eucalyptus* Plantation in Gujarat, India

J.I. Nirmal Kumar[1]*, Kanti Patel[1],
Rita N. Kumar[2], Priyakanchini Gupta[1]

*[1]P.G. Department of Environmental Science and Technology,
Institute of Science and Technology for Advanced Studies and Research (ISTAR),
Vallabh Vidyanagar- 388 120, Gujarat, India
[2]Department of Bioscience and Environmental Science,
N.V. Patel College of Pure and Applied Sciences,
Vallabh Vidyanagar - 388 120, Gujarat, India*

ABSTRACT

Greeenhouse Gases like CO_2, CH_4 and N_2O fluxes from soil under a Eucalyptus plantation in central Gujarat, Western India were measured for three month (February to April, 2011) at fifteen days interval using closed static chamber technique and gas chromatography method. The results showed that the soil in our study was a sink of atmospheric CO_2, CH_4 and N_2O. Soil CO_2, CH_4 and N_2O flux varied from -65.27 to 14.6, -0.005 to 0.07 and -0.03 to 0.33 mg m^{-2} h^{-1} respectively. CO_2 emissions were found maximum compared to other gases like CH_4 and N_2O emissions. Variations in soil N_2O emissions could be primarily explained by the differences in litter C:N ratio and soil total N stock. Differences in soil CH_4 uptake could be mostly attributed to the differences in mean soil CO_2

* *Corresponding author.* E-mail: istares2005@yahoo.com

flux and water filled pore space (WFPS). Soil C:N ratio could largely account for variations in soil CO$_2$ emissions. A strong positive relationship existed between CH$_4$ flux and soil temperature. The N$_2$O flux correlated with water filled pore space. The global warming potential of N$_2$O is highest compared to other two principal gases.

Keywords*: Greenhouse gases, Eucalyptus plantation, Global warming potential.*

11.1 Introduction

The enhanced production and reduced consumption of naturally occurring greenhouse gases (GHGs) such as carbon dioxide (CO$_2$), nitrous oxide (N$_2$O) and methane (CH$_4$), are responsible for approximately 90 per cent of the global warming and climate change phenomenon (Solomon *et al.,* 2007). Among them the most important individual greenhouse gas is carbon dioxide but substantial contributions to global warming are also made by methane and nitrous oxide. Soils can store and release considerable quantities of carbon through natural processes including litter deposition, decomposition and root respiration (Drewitt *et al.,* 2002). Although the atmospheric CH$_4$ concentration (1.8 ppmv) is much less than that of CO$_2$ (370 ppmv),however, CH$_4$ is 23 times more effective per molecule as a greenhouse gas than is CO$_2$ in a period of 100 years (Ramaswamy *et al.,* 2001). The CH$_4$ increase accounts for 20 per cent of the increased greenhouse warming potential of the atmosphere. Likewise, N$_2$O is a long half-life gas in atmosphere that is 296 times as effective as CO$_2$ in a period of 100 years as a greenhouse gas and accounts for about 6 per cent of the greenhouse effect (Ramaswamy *et al.,* 2001).

A considerable amount of atmospheric GHG is produced and consumed through soil processes (Tang *et al.,* 2006). Soils provide the largest terrestrial store for carbon (C) as well as the largest source of atmospheric CO$_2$ through autotrophic and heterotrophic respiration (Paul *et al.,* 2002). Soils are also the greatest source (<"60 per cent) of N$_2$O through the microbially mediated processes of nitrification and denitrification (Robertson *et al.,* 2000). Well-drained soils, the only biological sink of atmospheric CH$_4$ through the activity of methanotrophic bacteria (Dalal *et al.,* 2008), are responsible for 6 per cent of global CH$_4$ consumption (Bodelier and Laanbroek 2004). Although N$_2$O and CH$_4$ are added to the atmosphere at much lower concentrations than CO$_2$, but the global warming potentials of N$_2$O and CH$_4$ are approximately 298 and 25 times greater, respectively (IPCC 2007).

Afforestation and reforestation can greatly affect soil GHG fluxes by changing key physical and chemical properties that influence soil nutrient and C cycling and microbial activity (Merino *et al.,* 2004; Kelliher *et al.,* 2006). Tree species are considered to alter soil chemical, physical (*e.g.* moisture and temperature) and biological processes through their root system, crown structure, foliage, leaf structure and litter quality (Ullah *et al.,* 2008). Plantations are becoming a key component of the world's forest resources and playing an important role in the context of overall sustainable forest management. Well-designed, multi-purpose plantations can reduce pressure on natural forests, restore some ecological services provided by natural forests and mitigate climate change through direct C sequestration (Paquette and Messier 2010). In the present study investigated the concentration, fluxes, emissions, global warming potential of important greenhouse gases like CO_2, CH_4 and N_2O and relation between edaphic factors of soil in eucalyptus plantation. Moreover, Kumar and Viyol (2009) examined CH_4 emission in relation to organic carbon, sulphate, phosphate contents of two rice fields of central Gujarat. Dissolved Methane fluctuations in relation to hydrochemical parameters in Tapi estuary, Gulf of Cambay, India investigated by Nirmal Kumar *et al.* (2010).

11.2 Materials and Methods

Study Site

This study was carried out in a *Eucalyptus* plantation (situated on 22°30' 57.14" N, 72°56' 35.69" E) in Anand, Central Gujarat, and Western India. Meteorological data from Anand Agriculture University, Anand showed that maximum and minimum temperature recorded was 42 °C and 26 °C respectively. The mean annual precipitation was 900 mm; rains occur in monsoon season (July to October). The average relative air humidity was about 30 per cent. The total area of plantation is 8093.8 m^2 and is dominated by *Eucalyptus grandis.* Soil under the plantation is sandy loam and litter layer was normally about 2-3 cm; humus layer was about 1-2 cm.

Soil CO_2, CH_4 and N_2O Measurements

Soil CO_2, CH_4 and N_2O fluxes were measured using the static chamber and gas chromatography techniques. Static chamber was established in plot and fabricated by non-reactive materials PVC (Brechet *et al.,* 2009). Chambers were 56 cm-diameter ring and anchored 5 cm into soil. During flux measurement 15-cm-high chamber top was attached to the ring and a fan (about 8 cm in diameter) to ensure good mixing of the air during collection (Mo *et al.,* 2008). Air was sampled from chamber between 11:00 to 3:00 h at each sampling date because of maximum GHGs emission measured. Fluxes of soil CO_2, CH_4 and N_2O were measured 15 days interval during the experiment from February to April, 2011. Gas samples were collected with 100 ml plastic syringes attached to three-way stopcock at 0, 60, 120 and 180 min intervals after chamber closure into glass vials of 30 ml with butyl rubber stoppers which had been evacuated beforehand. The first 100 ml gas was abandoned, because it might contain the gas taken at the latest sampling. N_2O, CH_4 and CO_2 concentrations in the samples were analyzed within 48 h using gas chromatography (Perkin Elmer Auto system Gas Chromatograph)which was equipped with an electron capture detector ECD for N_2O analysis and a flame ionization detector (FID) for CH_4 and CO_2 analysis carried out in Sophisticated Instrumentation Centre for Applied Research and Testing (SICART), Vallabh Vidyanagar, Gujarat, India.

Gas fluxes were calculated from linear regressions of concentrations inside the chambers against the closure time according to the following equation:

$$F = \rho \frac{V}{A} \frac{P}{P_0} \frac{T_0}{T} \frac{dC_t}{dt}$$

where,

F is CH_4 and N_2O gas flux (mg/m²/h)), ñ is gas density at the test temperature (mg/m³), V is chamber volume available (m³), A is bottom area of the chamber (m²), P is atmospheric pressure in the field (hPa), P_0 is atmospheric pressure under standard conditions (hPa), T_0 is absolute air temperature under standard conditions (25°C), T is absolute air temperature in chamber at the time of sampling (°C), C_t is concentration of mixed volume ratios of gases in chamber at time t (10–6).

GHGs emission (t/year) measured by following equation

GHGs (CH_4, CO_2, N_2O) emission (t/year) =Area of land (m²) **×** Average daily CH_4, CO_2, N_2O Emission rate Mg/m²/year **×** Conversion **factor t/mg (10⁻⁹) ×** Molecular/ Atomic ratio (Global Environment Division, 1998).

Global warming potential calculated by GHG (CH_4, CO_2, N_2O) emission multiplied by global warming potential for hundred years of CO_2, CH_4, and N_2O is 1, 21 and 310 respectively.

The Greeenhouse Gas budget provides an estimate of the net budget of CO_2, CH_4, and N_2O based on the Global Warming Potential (GWP) of each gas. GWP is the contribution that a gas makes to the greenhouse effect according to its capacity to absorb radiation and its residence time in the atmosphere. Mean day flux of each gas was estimated and budget of each trip was calculated from the below given formula. It is expressed in terms of 'carbon dioxide equivalents' (CO_2 eq) as follows:

$$CO_{2eq} = (N_2O \times 310) + (CH_4 \times 21) + (CO_2 \times 1)$$

Micro-Environmental Data Measurements

Air temperature at 1.5 m above ground was measured simultaneously. Soil temperature and moisture at 5 cm below soil surface were monitored at each chamber. Soil temperature was measured using a digital thermometer. Soil moisture was measured by gravimetric method. Soil moisture values were converted into water filled pore space (WFPS) by the following formula:

$$WFPS(\%) = \frac{Vol(\%)}{1 - \frac{bd(g\ cm^{-3})}{2.65(g\ cm^{-3})}}$$

where, bd is bulk density, Vol is volumetric water content and 2.65 is density of quartz.

11.3 Soil Sampling and Measurements

Soil samples were collected at different depth 0-10, 11-20 and 21-30 cm and a total of six soil cores collected using an 8 cm diameter stainless steel core in plot and were air dried at room temperature (25 °C), then passed through a 2 mm mesh sieve to remove coarse living roots and gravel and ground with a mill before chemical analysis. Meanwhile pits were sampled to measure bulk density in plot.

Soil was analyzed for total organic C by Walkley and Black method. Total nitrogen (N) was analyzed using the Kjeldahl method. Available phosphorus was estimated by phosphomolybdic blue colorimetric method; sulphate by turbidimetric method and nitrate by colorimetric method using phenol disulphonic acid. Soil pH was measured in a 1 mol L^{-1} KCl solution using a glass electrode. Conductivity measured by potentiometric method. Particle density was estimated by gravimetric method using the standard book by Maiti,(2003).

Statistical analysis. In order to examine the relationship between soil parameters and the measured CO_2, CH_4 and N_2O fluxes, Correlation and linear regression analysis was performed using Sigma Plot 11.0 statistical analysis software.

11.4 Results and Discussion

Climatic Conditions

Soil temperature and WFPS exhibited the monthly variation in plantation. The sampling period in February 2011 was a particularly cool-dry season and in April 2011 was a hot-humid season in this study (Figurs 11.1a-b). The physico-chemical property of the soils (0–30 cm depth) in Eucalyptus plantation was given in Table 11.1.

Table 11.1: Physico-chemical Properties of the Soils (0–30 cm depth) in Eucalyptus Plantation (n=6)

Parameters	Depth		
	0-10	*10-20*	*20-30*
Soil bulk density (g cm^{-3})	0.82±0.01	0.88±0.01	0.91±0.02
Soil conductivity (µ mho/cm)	1.09±0.03	1.15±0.02	1.14±0.04
Soil WFPC (per cent)	21.96±1.03	23.47±1.13	25.84±1.08
Soil Particle Density	2.15±0.08	3.48±0.12	5.07±0.06
Soil pH	7.16±0.02	7.21±0.01	7.11±0.01
Soil organic C (Mg ha^{-1})	44.93±0.54	32.54±0.48	27.22±0.59
Soil total N (Mg ha^{-1})	3.79±0.09	3.21±0.04	2.99±0.08
Soil C:N	11.85±0.5	10.14±0.4	9.10±0.7
NH_4^+-N content (mg kg^{-1})	5.20±0.05	4.84±0.13	4.12±0.18
NO_3^--N content (mg kg^{-1})	2.34±0.03	1.87±0.09	1.51±0.01
Soil total sulphate (mg kg^{-1})	1.25±0.01	1.28±0.07	1.31±0.03
Soil available phosphorus (mg kg^{-1})	0.12±0.02	0.09±0.01	0.08±0.01

GHGs Concentration, Flux and Emission

Soil CO_2 concentration fluctuated from 190 to 869.5 ppmv during the experiment from February to April, 2011. The maximum concentration was found between 12:00 to 1:00 period. Fluctuations in CO_2 concentration was observed in all the subsequent intervals, perhaps it could be due to hot humid temperature (Figure 11.2a). CH_4

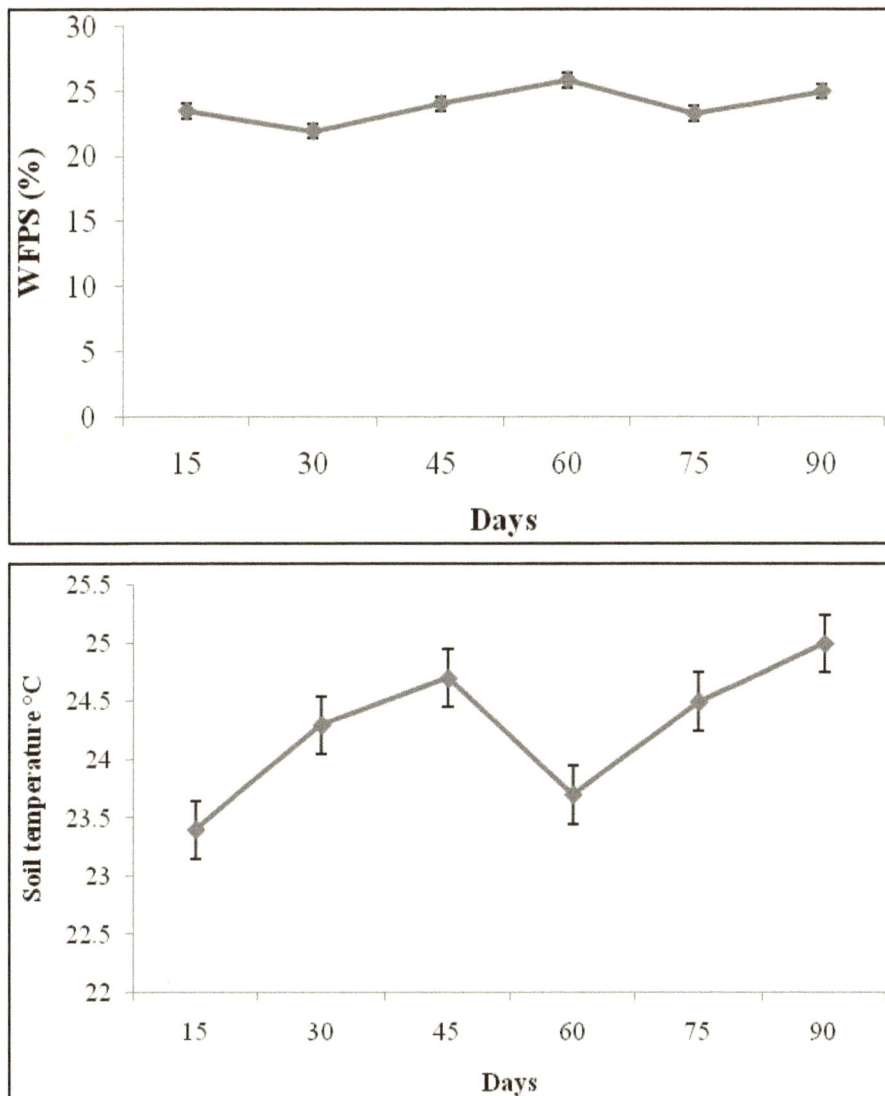

**Figure 11.1: Seasonal Patterns of Soil Water Filled Pore Space,
WFPS (a) and Soil Temperature (b) Measured in the Eucalyptus Plantations.
Error bars indicate standard error (n=6).**

concentration ranged from 0.59 to 2.14 ppmv with the highest value reported in noon period. (Figure 11.2b). N_2O concentration observed between 15.29 to 136.69 ppmv with the maximum during 1:00 to 2:00 period. The fluctuation of GHGs was distinct as time interval increased (Figure 11.2c). Statistical significant differences were set with P values < 0.05 between the times and days interval of GHGs concentration.

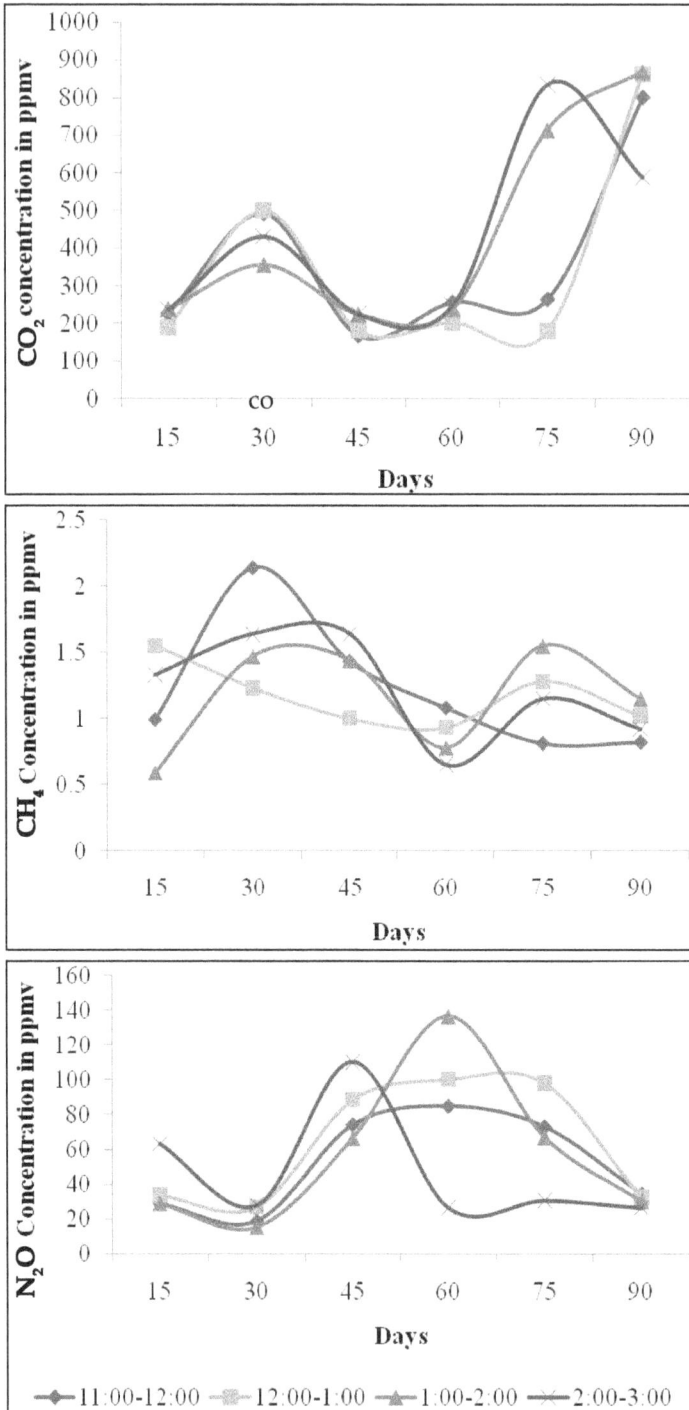

Figure 11.2: GHGs Concentration in Eucalyptus Plantation in Times and Days Interval a. CO_2, b. CH_4, c. N_2O

Soil CO_2, CH_4 and N_2O flux varied from -65.27 to 14.6, -0.005 to 0.07 and -0.03 to 0.33 mg m^{-2}h^{-1} respectively (Figures 11.2 a, b and c). CO_2 flux highest value registered in fifth interval during 2:00 p.m. to 3:00 p.m. (Figure 11.3a). The greater value of methane concentration was registered during the second trip between 2:00 p.m. to 3:00 p.m. and the flux fluctuated diurnally and uptake was recorded maximum in fourth interval (Figure 11.3b). The maximum value of N_2O flux was observed during the third and fourth interval between 2:00 p.m. to 3:00 p.m and flux reduced at 1:00 to 2:00 p.m.in all intervals. The soil is consuming atmospheric N_2O maximum in third interval during 1:00 to 2:00 p.m. (Figure 11.3c). Negative values of GHGs flux indicate gas consumption by the soil while positive results indicated the GHGs release through the soil. Soil CO_2 flux positively related to WFPS where as CH_4 flux positively related to soil temperature besides, N_2O flux positively related to WFPS while CH_4 uptake reduced with increased WFPS in plantation (Figure 11.4). Soil CO_2, CH_4 and N_2O emission in eucalyptus plantation represented in Figure 11.5. CO_2 emissions were found maximum compared to other gases, on the contrary, CH_4 and N_2O emissions were negligible as compared to CO_2 emission (Figure 11.5).

GHG flux correlated with edaphic factor of soil was shown in Table 11.2. Soil C: N ratio negatively correlated with CO_2 and CH_4 flux and positively correlated with N_2O flux. WFPS was positively correlated with CO_2 and CH_4 flux and negatively correlated with N_2O flux which the observations are very well corroborated with the findings of Liu *et al.* (2008) Soil-atmosphere CO_2 exchange.

The soil CO_2 mean emission rate of 19.54 mg C m^{-2} h^{-1} measured in this study is similar to that measured in temperate forests by Wang *et al.* (2006), subtropical forests by Tang *et al.* (2006), and tropical rain forests by Sotta *et al.* (2004). Soil CO_2 annual mean emission was lower than in broadleaf plantations (between 56.38 and 72.15 mg C m⁻2h⁻1), as observed in some studies when compared coniferous with broadleaf forest/plantations by Livesley *et al.* (2009). Soil CO_2 emission, as the result of soil respiration generates mainly from autotrophic (root) and heterotrophic (microbial) activity (Janssens *et al.*, 2001). Much of the spatial variations in soil respiration obtained across the topographic gradient was explained by differences in soil water content, bulk density, root biomass and soil organic matter by Epron *et al.* (2006). According to the above-mentioned studies, we input the relevant variables into a multiple linear model to assess the importance of different factors for influencing the variations in annual mean soil CO_2 emissions in the plantation with different tree species in this study. The differences in the magnitude of mean soil CO_2 emissions among the plantations could best be explained by differences in litter C:N ratio, while the other input parameters were rejected by the multiple linear regression model because of low significance (Table 11.3). Hättenschwiler *et al.* (2005) emphasized that soil C:N ratio, as a good indicator of substrate quality, was an important factor regulating microbial activity and thus influencing litter decomposition. The Soil C:N ratio in the *Eucaluptus* plantation was higher in 0-10 cm depth (Table 11.1), indicating that high microbial activity but heterotrophic respiration may be lower in the 11-20 and 21-30 cm depth. The temporal variations in soil CO_2 emissions in all plantations coincided with those in soil temperature and moisture (Figures 11.2 and 11.3), indicating that soil temperature and moisture exert the significant effects on the temporal variations

Table 11.2: Correlation between GHG Flux and Edaphic Factor of Soil in Eucalyptus Plantation

	1	2	3	4	5	6	7	8	9	10	11	12	13	14	15
Soil bulk density	1	0.169	0.869*	0.896*	-0.091	-0.867*	-0.672	-0.933*	-0.784	-0.667	0.235	-0.649	0.043*	0.174	-0.371
Soil conductivity		1	0.121	-0.011	0.790*	0.092	-0.829*	-0.030	0.161	-0.767*	-0.608	-0.439	-0.090	0.942*	-0.483
Soil WFPC			1	0.972*	-0.340	-0.861*	-0.570	-0.835*	-0.951*	-0.634*	0.219	-0.338	0.120*	0.256*	-0.123*
Soil Particle Density				1	-0.447	-0.871*	-0.477	-0.829*	-0.973*	-0.566	0.424	-0.381	-0.018	0.075	-0.138
Soil pH					1	0.319	-0.542	0.120	0.608	-0.361	-0.761	-0.369	0.112	0.674	-0.479
Soil organic C						1	0.373	0.951*	0.847*	0.330	-0.297	0.193	-0.380*	-0.038*	0.372*
Soil total N							1	0.530	0.305	0.958*	0.380*	0.730	0.040*	-0.794*	0.454*
Soil C:N								1	0.759*	0.457	-0.120	0.419	-0.387*	-0.122*	0.351*
NH$_4^+$-N content									1	0.415	-0.467	0.185	-0.047	0.024	-0.007
NO$_3$–N content										1	0.216	0.747	0.252	-0.720	0.300
Soil total sulphate											1	0.023	-0.497	-0.695	0.001
Soil available phosphorus												1	0.418*	-0.239	0.212*
CO$_2$ Flux													1	0.162*	-0.089
CH$_4$ Flux														1	-0.391
N$_2$O Flux															1

* $P < 0.05$.

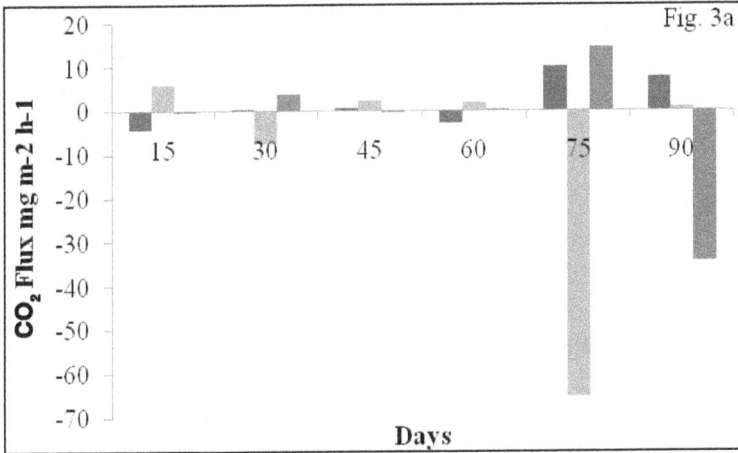

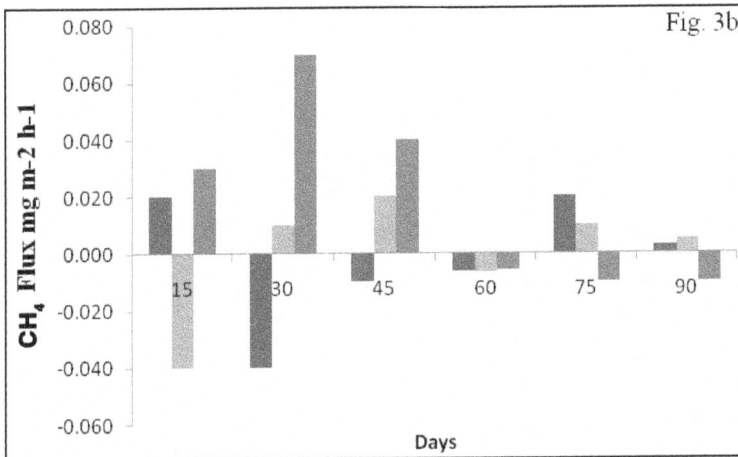

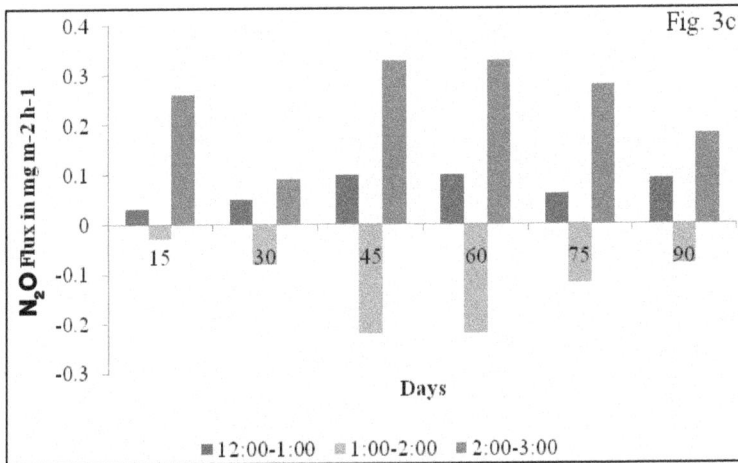

Figure 11.3: GHGs Flux in Eucalyptus Plantation in Times and Days Interval a. CO$_2$, b. CH$_4$, c. N$_2$O.

Figure 11.4: Relationships between Soil N$_2$O, CH$_4$, and CO$_2$ Fluxes, and Soil Temperature and Soil Water Filled Pore Space (WFPS) in the Eucalyptus Plantation.

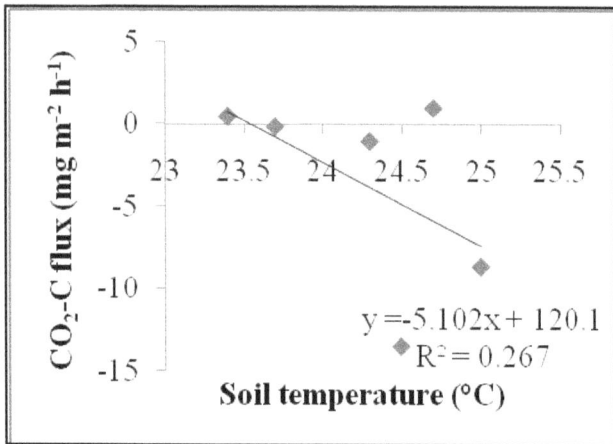

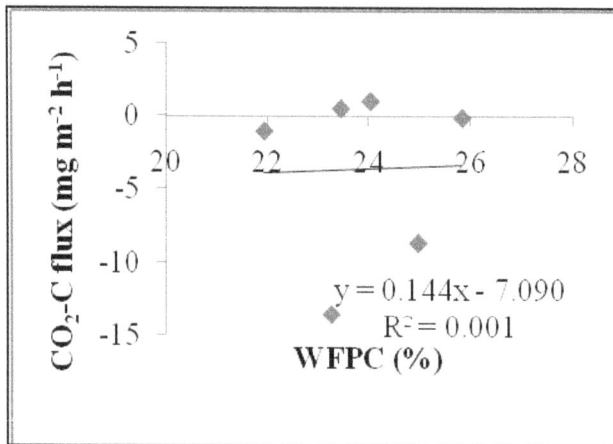

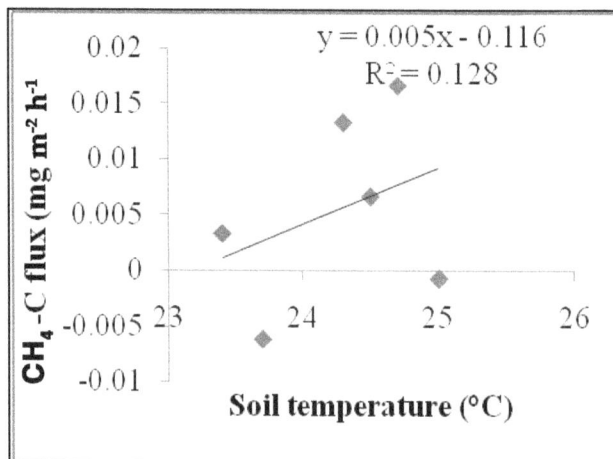

$$y = -5.102x + 120.1$$
$$R^2 = 0.267$$

Soil temperature (°C)

$$y = 0.144x - 7.090$$
$$R^2 = 0.001$$

WFPC (%)

$$y = 0.005x - 0.116$$
$$R^2 = 0.128$$

Soil temperature (°C)

Contd...

Figure 11.4– *Contd...*

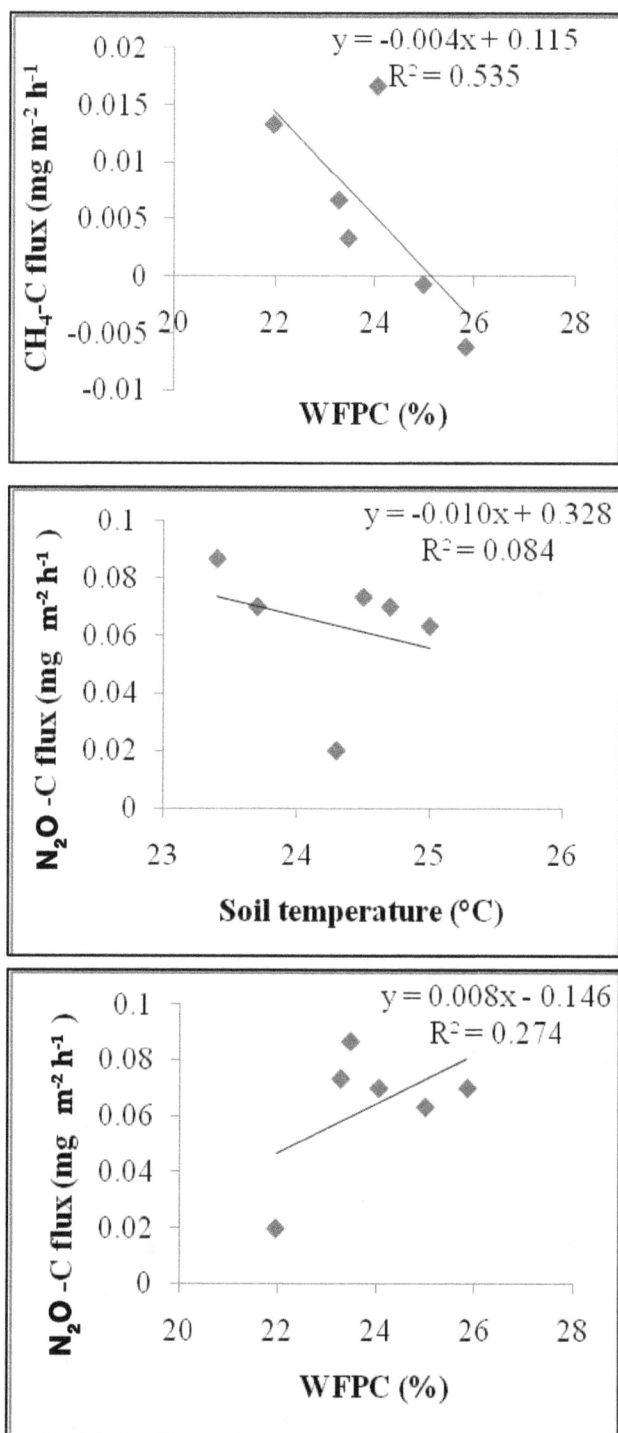

$y = -0.004x + 0.115$

$R^2 = 0.535$

CH₄-C flux (mg m⁻² h⁻¹)

WFPC (%)

$y = -0.010x + 0.328$

$R^2 = 0.084$

N₂O -C flux (mg m⁻² h⁻¹)

Soil temperature (°C)

$y = 0.008x - 0.146$

$R^2 = 0.274$

N₂O -C flux (mg m⁻² h⁻¹)

WFPC (%)

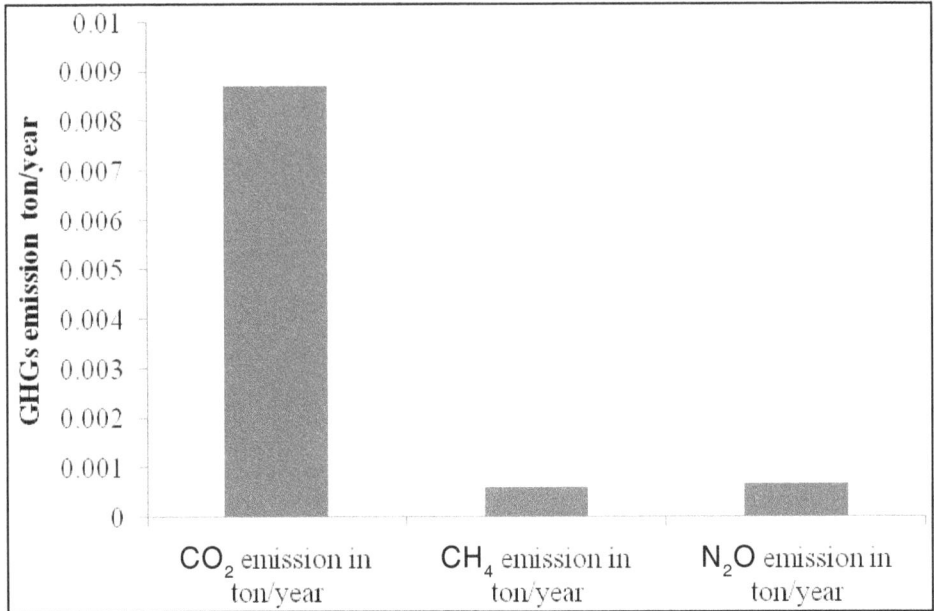

Figure 11.5: GHGs Emission (ton/year) in Eucalyptus Plantation.

in soil CO_2 emissions in the subtropical plantations where the previous studies in subtropical India had supported our results (Bhatiya *et al.,* 2004).

11.5 Soil-Atmosphere : CH_4 Exchange

CH_4 measurements indicated a consistent net soil consumption of CH_4 (*i.e.* negative CH_4 flux) in plantation (Figure 11.3b). The highest soil CH_4 uptake rate of -40.12 µg C $m^{-2}h^{-1}$ measured in plantation is similar to that measured in other forests (Verchot *et al.,* 2000; Borken and Beese 2006; Tang *et al.,* 2006; Fest *et al.,* 2009), but less than that measured in more productive natural forest systems (Merino *et al.,* 2004; Werner *et al.,* 2007). Soil-atmosphere CH_4 exchange is the result of simultaneously occurring production and consumption processes in soils, and is thus controlled by CH_4- producing methanogens operating at anaerobic conditions and CH_4-consuming methanotrophs that depend on oxygen as a terminal electron acceptor (Topp and Pattey 1997). Activity and population sizes of these microbes are dependent on a multitude of soil factors, like soil temperature, moisture, pH, substrate availability, and aeration of soil profile (Verchot *et al.,* 2000; Merino *et al.,* 2004; Reay and Nedwell 2004; Werner *et al.,* 2007). Thus, the relevant parameters were performed in a multiple linear regression analysis to access the importance of different factors for explaining the variations in soil CH_4 uptake in the plantation. Using mean CO_2 efflux and mean WFPS, where we were able to observed variations in soil CH_4 fluxes in plantation.

The high soil respiration rates can create anaerobic microbe sites as O_2 is consumed, resulting in CH_4 production in soil (Verchot *et al.,* 2000). Therefore, soils should consume less CH_4 when CO_2 production by root and high microbial respiration. Furthermore, soil CH_4 uptake should decrease due to the increased anaerobiosis resulting from elevated soil moisture (Werner *et al.*, 2007). Under high soil WFPS, anoxic conditions enhance and CH_4 diffusion to the methanotrophs in the subsurface is restricted (Ball *et al.*, 1997). Tree species might affect CH_4 uptake by changing soil chemical and physical properties and/or by changing in the population and diversity of methanotrophic bacteria (Borken and Beese, 2006). Our results suggested that tree species could influence soil CH_4 uptake through the variations in soil CO_2 efflux and moisture in the plantation (Table 11.3). Tree species composition includes diversity and density has influence on soil CO_2 efflux through abiotic and biotic factors (Borken and Beese 2005). Tree species also can affect soil moisture as a result of canopy structure and canopy interactions with the atmosphere (Borken and Beese 2006). The temporal variations in soil CH_4 fluxes displayed dependency on soil WFPS (Figures 11.2 and 11.3). This is similar with other studies in tropical and temperate forests, where soil CH_4 uptake rates were negatively related to soil moisture (Castro *et al.*, 2000; Verchot *et al.*, 2000) whereas CH_4 uptake is dominated by aeration of the soil profile (Khalil and Baggs, 2005).

Table 11.3: Results of Multiple Linear Regression Analysis of Biogeochemical Parameters and Annual Mean GHG Flux in Eucalyptus Plantation (n=6)

Parameters	*Models*
	CO_2 flux (mg C m⁻² h⁻¹) (Y_1)
Soil C:N ratio (X_1)	$Y_1 = -40.726 + 3.969 X_1$, $R^2 = 0.551$, $P < 0.001$
	CH_4 Flux (mg C m⁻² h⁻¹) (Y_2)
Mean CO_2 flux (mg C m⁻² h⁻¹) (X_2)	$Y_2 = 0.0091 X_2 + 0.171 X_3 - 4.04$, $R_2 = 0.826$, $P < 0.01$
Mean WFPS (per cent) (X_3)	
	N_2O (mg N m⁻² h⁻¹) (Y_3)
Soil C:N ratio (X_4)	$Y_3 = -0.0096 X_4 + 0.051 X_5 + 0.01$. $R^2 = 0.466$, $P < 0.001$
Soil total N (Mg ha⁻¹) (X_5)	

Soil-atmosphere N₂O Exchange

The mean soil N_2O emission of 1.52 mg N m⁻²h⁻¹ measured in plantation which agrees with the estimates from other forest studies (Castaldi *et al.*, 2006; Livesley *et al.*, 2009), but is less than that measured in some moist tropical or boreal forests (Hall *et al.*, 2004; Werner *et al.*, 2007). Zhang *et al.* (2008) showed the atmospheric deposition rates of N were high due to the rapid expansion of industrial and agricultural activities in southern India and the abundant N inputs strongly increased the reactive N in the soil and thus the production of N_2O. However, our study area is in the undeveloped area in Western India (Wang and Tang 2007) and results revealed the effects of the tropical tree species on soil N_2O emissions. The general soil N_2O emission potential is predominantly controlled by soil pH (Stevens *et al.*, 1997), soil moisture (Merino *et al.*, 2004), soil C and N stocks (Li *et al.*, 2005), soil inorganic N contents (Merino *et al.*, 2004) and C:N ratio of litter and soil (Werner *et al.*, 2007). The importance of different biogeochemical parameters for explaining the variations in soil N_2O emissions in the plantation was assessed with a multiple linear regression analysis. Differences in the magnitude of soil N_2O emission in the plantation could be explained primarily by differences in soil C:N ratio (negatively correlated), and furthermore by differences in soil total N stock (positively correlated) (Table 11.3). N_2O flux in forest soils has been shown to correlate with gross nitrification rates (Ambus *et al.,* 2006) and to soil C:N ratio, as that greatly determines soil nitrification activity (Erickson *et al.,* 2002). The importance of soil N status as a scalar to explain differences in soil N_2O emissions among the plantations is in agreement with previous observations by Regina *et al.* (1996) in boreal soils and by Zhang *et al.* (2008a) in subtropical forests. The temporal variations in soil N_2O emissions were attributed to those in soil temperature and moisture (Figures 11.2 and 11.3). Similar results were reported in other subtropical forests (Tang *et al.,* 2006; Liu *et al.,* 2008). It has been reported that N_2O production by nitrification and denitrification can increase strongly with the increasing soil temperature and moisture in temperate forests (Borken and Beese, 2006) and in subtropical forests (Tang *et al.,* 2006; Liu *et al.,* 2008).

11.6 Global Warming Potential (GWP)

The overall balance between the net exchange of CO_2, CH_4, and N_2O constitutes the net global warming potential (GWP) of any terrestrial ecosystem. Storage of atmospheric CO_2 into stable organic carbon pools in the soil can sequester CO_2. The results reveal that global warming potential of N_2O is highest compared to other two principal gases. However, the negligible portion of GWP was encountered for methane (Figure 11.6). Adviento *et al.* (2007) also found similar results of global warming potential.

GHG Budget

Greeenhouse gas budget was calculated in terms of CO_2 equivalent. Total budget of the study time was found 129.9 CO_2 equivalents. Maximum budget was found in

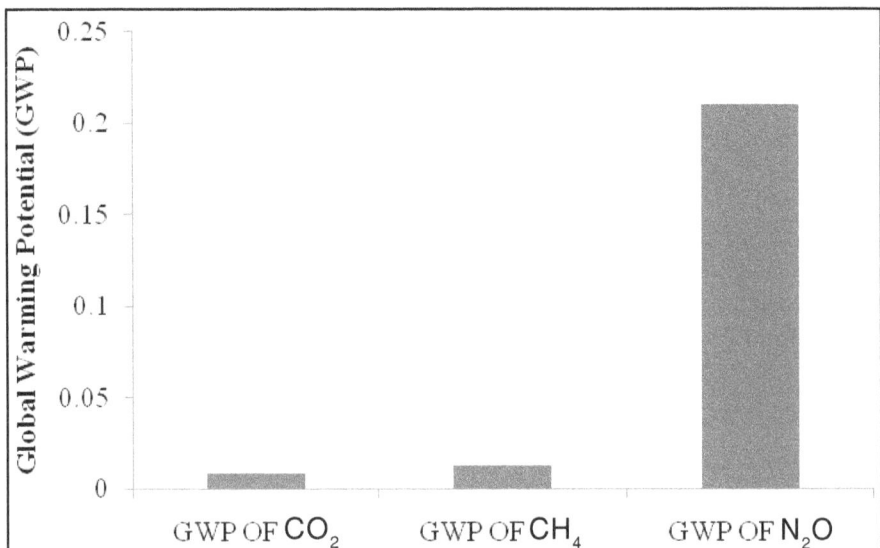

Figure 11.6: Global Warming Potential (GWP) in Eucalyptus Plantation.

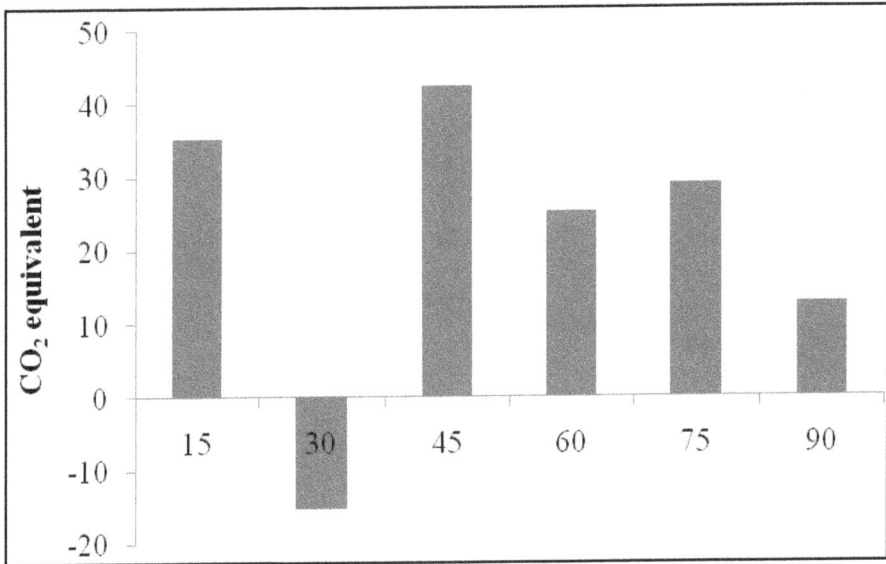

Figure 11.7: GHGs Budget in Eucalyptus Plantation.

the third interval and minimum budget was found in second interval, which was very negligible (Figure 11.7). Among all the three gases, N$_2$O is the largest contributor to the global atmospheric greeenhouse gas budget mainly via microbial process of nitrification and denitrification.

11.7 Conclusion

In this study we estimated diurnal CO$_2$, CH$_4$ and N$_2$O fluxes through simple linear regression between gases and WEPS and temperature. The study confirmed that soil temperature was an important factor influencing soil CH$_4$ flux, while WEPS was identified as a key factor regulating N$_2$O emission.

Acknowledgements

Authors are highly thankful to Ministry of Environment and forests (MoEF), New Delhi India for financial assistance and grateful to Sophisticated Instrumentation Center for Advanced Research and Testing (SICART), Vallabh Vidyanagar, Gujarat for analysis of gas samples.

References

Adviento-Borbe, M.A.A., M.L., Haddix, D.L., Binder, D.T. Walters and A., Dobermann (2007). Soil greenhouse gas fluxes and global warming potential in four high-yielding maize systems. Global Change Biology13, 1972–1988, doi:10.1111/j.1365486.2007.01421.

Ambus P, Zechmeister-Boltenstern S, Butterbach-Bahl K (2006). Sources of nitrous oxide emitted from European forest soils. *Biogeosciences* 3:135–145

Ball BC, Dobbie KE, Parker JP, and Smith K.A. (1997). The influence of gas transport and porosity on methane oxidation in soils. *J Geophys Res-Atmos* 102:23301–23308

Bhatiya A., Pathak, H, Agrawal P.K.(2004). Inventory of methane and nitrous oxide emissions from agricultural soils of India and their global warming potential. *Current Science*, 87 (3): 317-324.

Bodelier PLE, Låånbroek H.J. (2004). Nitrogen as regulatory factor of methane oxidation in soils and sediments. *FEMS Microbiol Ecol* 47:265–277. doi:10.1016/S0168-6496(03). 00304-0

Borken W, and Beese F. (2005). Soil carbon dioxide efflux in pure and mixed stands of oak and beech following removal of organic horizons. *Can J For Res* 35:2756–2764. doi:10.1139/x05-192

Borken W, and Beese F. (2006). Methane and nitrous oxide fluxes of soils in pure and mixed stands of European beech and Norway spruce. *Eur J Soil Sci* 57:617–625. doi:10.1111/j.1365-2389.2005.00752.x

Bréchet L, Ponton S, Roy J, Freycon V, Coûteaux M, Bonal D, Epron D (2009). Do tree species characteristics influence soil respiration in tropical forests? A test based on 16 tree species planted in monospecific plots. *Plant Soil* 319:235– 246. doi:10.1007/s11104-008-9866-z

Castaldi S, Ermice A, Strumia S (2006). Fluxes of N_2O and CH_4 from soils of savannas and seasonally-dry ecosystems. *J Biogeogr* 33:401–415. doi:10.1111/j.1365-2699.2005.01447.x

Castro MS, Gholz HL, Clark KL, Steudler PA (2000). Effects of forest harvesting on soil methane fluxes in Florida slash pine plantations. *Can J For Res* 30:1534–1542. doi:10.1139/cjfr-30-10-1534

Dalal R, Allen D, Livesley S, Richards G (2008). Magnitude and biophysical regulators of methane emission and consumption in the Australian agricultural, forest, and submerged land-escapes: a review. *Plant Soil* 309:89–103. doi:10.1007/s11104-007-9446-7

Drewitt, G.B., Black, T.A., Nesic, Z., Humphreys, E.R., Jork, E.M., Swanson, R., Ethier, G.J., Griffis, T., Morgenstern, K. (2002). Measuring forest floor CO_2 fluxes in a Douglas-fir forest. *Agric. For. Meteorol.*, 110, 299–317.

Epron D, Bosc A, Bonal D, Freycon V (2006). Spatial variation of soil respiration across a topographic gradient in a tropical rainforest in French Guiana. *J Trop Ecol* 22:565– 574. doi:10.1017/S0266467406003415

Erickson H, Davidson EA, Keller M (2002). Former land-use and tree species affect nitrogen oxide emissions from a tropical dry forest. *Oecologia* 130:297–308. doi:10.1007/s004420100801

Fest BJ, Livesley SJ, Drösler M, Gorsel E, Arndt SK (2009). Soil-atmosphere greenhouse gas exchange in a cool, temperate Eucalyptus delegatensis forest in south-eastern Australia. *Agr Forest Meteorol* 149:393–406. doi:10.1016/j.agrformet.2008.09.007

Global Environment Division, Greenhouse Gas Assessment Handbook, A Practical Guidance Document for the Assessment of Project level Greenhouse Gas Emissions, 1998.

Hall SJ, Asner GP, Kitayama K (2004). Substrate, climate, and land use controls over soil N dynamics and N-oxide emissions in Borneo. *Biogeochemistry* 70:27–58. doi:10.1023/B:BIOG.0000049335.68897.87

Hättenschwiler S, Tiunov AV, Scheu S (2005). Biodiversity and litter decomposition in terrestrial ecosystems. *Annu Rev Ecol Evol Syst* 36:191–218. doi:10.1146/annurev. ecolsys.36.112904.151932

IPCC (2007). Climate change 2007: The scientific basis. Cambridge University Press, Cambridge, UK, Contribution of Working Group I to the Fourth Assessment Report of the Intergovernmental Panel on Climate Change.

Janssens IA, Lankreijer H, Matteucci G (2001). Productivity overshadows temperature in determining soil and ecosystem respiration across European forests. *Global Change Biol* 7:269–278. doi:10.1046/j.1365-2486.2001.00412.x

Kelliher FM, Clark H, Zheng L, Newton PCD, Parsons AJ, Rys G (2006). A comment on scaling methane emissions from vegetation and grazing ruminants in New Zealand. *Funct Plant Biol* 33:613–615. doi:10.1071/FP06088

Khalil MI, Baggs EM (2005). CH_4 oxidation and N_2O emissions at varied soil water-filled pore spaces and headspace CH_4 concentrations. *Soil Biol Biochem* 37:1785–1794. doi:10.1016/j.soilbio.2005.02.012

Li CS, Frolking S, Butterbach-Bahl K (2005). Carbon Sequestration in Arable Soils is Likely to Increase Nitrous Oxide Emissions, Offsetting Reductions in Climate Radiative Forcing. *Clim Change* 72:321–338. doi:10.1007/s10584-005-6791-5

Liu H, Zhao P, Lu P, Wang YS, Lin YB, and XQ R. (2008). Greenhouse gas fluxes from soils of different land-use types in a hilly area of South China. *Agr Ecosyst Environ* 124:125–135. doi:10.1016/j.agee.2007.09.002

Livesley SJ, Kiese R, Miehle P, Weston CJ, Butterbach-Bahl K, and Arndt S.K. (2009). Soil-atmosphere exchange of greenhouse gases in a *Eucalyptus marginata* woodland, a clover-grass pasture, and *Pinus radiata* and *Eucalyptus globules* plantations. *Global Change Biol* 15:425–440. doi:10.1111/j.1365-2486.2008.01759.x

Maiti, S.K. 2003. *Handbook of Methods in Environmental Studies Vol.2: Air, Noise, Soil, and Overburden Analysis.* ABD Publisher, Jaipur, India, ISBN: 81-85771-58-8.

Merino A, Perez-Batallon P, Macias F (2004). Responses of soil organic matter and greenhouse gas fluxes to soil management and land use changes in a humid temperate region of southern Europe. *Soil Biol Biochem* 36:917–925. doi:10.1016/j.soilbio.2004.02.006

Mo JM, Zhang W, Zhu WX, Gundersen P, Fang YT, Li DJ, Wang H (2008). Nitrogen addition reduces soil respiration in a mature tropical forest in southern China. *Global Change Biol* 14:403–412. doi:10.1111/j.1365- 2486.2007.01503.x

Nirmal Kumar, J. I., Kumar, R. N., and Viyol, S. (2010). Dissolved Methane Fluctuations in Relation to Hydro-chemical Parameters in Tapi Estuary, Gulf of Cambay, India. *International Journal of Environmental Research* 4 (4): 893-900.

Nirmal Kumar,J.I and Viyol S. (2009). Short term diurnal and Temporal measurement of Methane emission in relation to organic carbon, sulphate, phosphate contents of two rice fields of central Gujarat, India. *Paddy and Water Environment*.7 : 11-16

Paquette A, Messier C (2010). The role of plantations in managing the world's forests in the Anthropocene. *Front Ecol Environ* 8:27–34. doi:10.1890/080116

Paul KI, Polglase PJ, Nyakuengama JG, Khanna PK (2002). Change in soil carbon following afforestation. *For Ecol Manage* 168:241–257. doi:10.1016/S0378-1127(01)00740-X

Reay DS, and Nedwell DB (2004). Methane oxidation in temperate soils: effects of inorganic N. *Soil Biol Biochem* 36:2059– 2065. doi:10.1016/j.soilbio.2004.06.002

Regina K, Nykanen H, Silvola J, and Martikainen PJ (1996). Fluxes of nitrous oxide from boreal peatlands as affected by peatland type, water table level and nitrification potential. *Biogeochemistry* 35:401–418. doi:10.1007/BF02183033

Solomon S, Qin D, and Manning M. (2007). Technical Summary. In: Climate Change 2007: The Physical Science Basis. Contribution of Working Group I to the Fourth Assessment Report of the Inter-governmental Panel on Climate Change. Cambridge University Press, Cambridge, UK

Sotta ED, Meir P, Malhi Y, Nobre AD, Hodnett M, and Grace J. (2004). Soil CO_2 efflux in a tropical forest in the central Amazon. *Global Change Biol* 10:601–617. doi:10.1111/j.1529-8817.2003.00761.x

Stevens RJ, Laughlin RJ, Burns LC, Arah JRM, and Hood R.C. (1997). Measuring the contributions of nitrification and denitrification to the flux of nitrous oxide from soil. *Soil Biol Biochem* 29:139–151. doi:10.1016/S0038-0717(96)00303-3

Tang XL, Liu SG, Zhou GY, Zhang DQ, and Zhou CY (2006). Soil atmospheric exchange of CO_2, CH_4, and N_2O in three subtropical forest ecosystems in southern China. *Global Change Biol* 12:546–560

Topp E, and Pattey E (1997). Soils as sources and sinks for atmospheric methane. *Can J Soil Sci* 77:167–178

Ullah S, Frasier R, King L, Picotte-Anderson NP, Moore TR (2008). Potential fluxes of N_2O and CH_4 from soils of three forest types in Eastern Canada. *Soil Biol Biochem* 40:986–994. doi:10.1016/j.soilbio.2007.11.019

Verchot LV, Davidson EA, Cattanio JH, and Ackerman I.L. (2000). Land-use change and biogeochemical controls of methane fluxes in soils of eastern *Amazonia. Ecosystems* 3:41–56. doi:10.1007/s100210000009

Wang CK, Yang JY, and Zhang Q.Z. (2006). Soil respiration in six temperate forests in China. *Global Change Biol* 12:2103– 2114. doi:10.1111/j.1365-2486.2006.01234.x

Wang XZ, and Tang Y.H. (2007). Route selection in undeveloped regions –research of Guangxi. *Productivity Research* 24:48–50

Werner C, Kiese R, Butterbach-Bahl K. (2007). Soil-atmosphere exchange of N$_2$O, CH$_4$, and CO$_2$ and controlling environmental factors for tropical rain forest sites in western Kenya. *J Geophys Res-Atmos* 112: DO3308. doi: 10.1029/2006JD007388

Zhang W, Mo JM, Yu GR, Fang YT, Li DJ, Lu XK, and Wang H. (2008). Emissions of nitrous oxide from three tropical forests in Southern China in response to simulated nitrogen deposition. *Plant Soil* 306:221–236.

2015, Impact of Global Warming and Climate Change on
Human and Plant Health
Editors: Dr. Arun Arya and Prof. V.S. Patel
Published by: DAYA PUBLISHING HOUSE, NEW DELHI

Pages 128–140

Chapter 12

Pollution Status of Chandlodia Lake Located in Ahmedabad, Gujarat

Hitesh A. Solanki, Pradeep U. Verma*
and Deepika K. Chandawat

Ecology Lab
Department of Botany, School of Sciences,
Gujarat University, Ahmedabad - 380 009, Gujarat, India

ABSTRACT

Pollution is viewed as the release of substances and energy as waste product of human activities which result in harmful changes within the natural environment. In the present study water of Chandlodia lake was analyzed for various physico-chemcial parameters. The study was carried out for a period of one year. Monthly data's were collected and were represented seasonally along with standard error. Different parameters studied were Temperature, Electrical conductivity, Turbidity, Total dissolve solids, pH, Alkalinity, Total Hardness, Calcium, Magnesium, Dissolved Oxygen, Biochemcial oxygen demand, Chloride, Sodium, Nitrate and Phosphate. Analysis of physical and chemical parameters were carried out by using the method suggested by APHA (1985), Kumar and Rabindranath (1998) and Trivedy and Goel (1984). The result obtained during study was compared with WHO (1971) and BIS (1991) standards. Therefore from the above result it was concluded that water of Chandlodia lake shows very high level of pollution.

Keywords: Pollution, Physicochemical characters, Chandlodia lake, Ahmadabad, Gujarat.

---

* Corresponding author. E-mail: drpradeepverma0205@gmail.com

12.1 Introduction

Pollutants are defined as the substances which cause pollution. They can be physical or chemical. Each lake system is unique, and its dynamics can be understood only to a limited degree based on information from other lakes. Just as a physician would not diagnose an individual's medical condition or prescribe treatment without a personal medical examination, a limnologist or hydrologist cannot accurately assess a lake system or suggest a management strategy without data and analysis from that particular lake and its environment. The study was carried out to check the pollution status of Chandlodia lake. Chandlia lake is located in Ahemdabad.

Figure 12.1: Chandlodia Lake.

Chandlodia lake is an artificial lake, constructed by AUDA. Chandlodia lake is also located in the western part of Ahmedabad city. The half portions of the lake become dry during the summer season. The people in the surrounding region use to throw waste in the lake and the cattle also use to take bath in the lake water. The lake covers an area of 22,940 m². And its exact geographical location is 23°05′00.05″ N Latitude and 72°33′09.43″ E Longitude.

12.2 Materials and Methods

The present study was carried out for Chandlodia Lake, located in Ahemdabad city. In the present study the sampling was done during morning hour. The water samples were collected in the polyethylene bottles. The closed bottle was dipped in the lake at the depth of 0.5 to 0.7 m, and then a bottle was opened inside and was closed again to bring it out at the surface. The samples were collected from five different points and were mixed together to prepare an integrated sample. From the time of sample collection to the time of actually analyses, many physical and chemical reactions would change the quality of the water sample; therefore to minimize this change the sample were preserved soon after the collection. The water samples were preserved by adding chemical preservatives and by lowering the temperature. The water temperature, pH, DO, EC and TDS were analyzed immediately on the spot after the collection, whereas the analyses of remaining parameters were done in the laboratory.

The study was carried for a period of 1 year (March 2009 to February 2010). Monthly data was collected, but results were represented season wise. Four month make one season [March to June summer season, July to October monsoon season, and November to February winter season].The collected water samples were brought to the laboratory and relevant analysis was performed. pH was determined electrometrically using digital pH meter, electrical conductivity was measured by conductivity meter, dissolved oxygen is measured by DO meter, total dissolve solid was measured by using TDS meter and similarly turbidity is measured by Nepthalo turbidity meter. Alkalinity, chloride, TDS, calcium, magnesium, total hardness, nitrate and phosphate were determined by method suggested by APHA (1985), Kumar and Rabindranath (1998) and Trivedy and Goel (1984). Estimation of sodium was done by Flame Photometric method. The mean value of the monthly data was calculated as season wise and standard error was also calculated by using following formula

Standard Deviation

$$\sigma = \sqrt{\frac{\sum (X_i - m)^2}{n-1}}$$

Standard Error

$$\sigma_x = \frac{\sigma}{\sqrt{n}}$$

12.3 Results and Discussion

Temperature

The temperature plays a crucial role in physico-chemical and biological behavior of aquatic system (Dwivedi and Pandey, 2002). Whereas according to Singh and Mathur (2005) temperature is one of the most important factors in the aquatic environment. The temperature of Chandlodia lake ranges between 16±1.47 to 29 ±1.58. The maximum temperature was recorded during summer season and minimum was recorded during winter season. Generally water temperature correspond with air temperature indicating that the samples collected from shallow zone has a direct relevance with air temperature, shallow water reacts quickly with changes in atmospheric temperature. This type of observations were made by Welch (1952); Joshi and Singh (2001); Ghose and Basu (1968); Young (1975); Sehgal (1980) and Jayanti (1994) for the different water bodies studied by them.

Table 12.1: Physico-chemical Parameters of Chandlodia Lake

Sl.No.	Parameters	2009–2010		
		Summer Mean + S.E.	Monsoon Mean + S.E.	Winter Mean + S.E.
1.	Temperature in ºC	29 ±1.58	21 ±1.29	16 ±1.47
2.	Electrical conductivity in mhos/cm	2.78 ±0.07	3.24 ±0.26	2.94 ±0.07
3.	Turbidity in NTU	16 ±1.47	22 ±0.91	18 ±1.58
4.	Total Dissolve Solid in ppm	1014 ±37	1264 ±39	963 ±31.5
5.	pH	8.7 ±0.17	9.1 ±0.18	8.9 ±0.15
6.	Alkalinity in ppm	192 ±3.65	214 ±6.87	208 ±5.6
7.	Total Hardness in ppm	312 ±5.89	328 ±8.29	304 ±6.58
8.	Calcium in ppm	76 ±3.92	82 ±4.32	66 ±3.92
9.	Magnesium in ppm	30 ±2.58	31 ±1.73	34 ±2.12
10.	Dissolved Oxygen in ppm	2.14 ±0.07	5.08 ±0.07	3.28 ±0.22
11.	Biochemical Oxygen Demand in ppm	1.96 ±0.08	3.38 ±0.3	2.20 ±0.21
12.	Chloride in ppm	122 ±4.97	93 ±4.65	88 ±4.97
13.	Sodium in ppm	44 ±3.65	65 ±2.89	38 ±4.97
14.	Nitrate in ppm	5.8 ±0.18	7.4 ±0.22	6.7 ±0.32
15.	Phosphate in ppm	1.66 ±0.12	1.92 ±0.11	1.24 ±0.05

S.E. = Standard error.

12.4 Electrical Conductivity

Electrical conductivity in the water is due to salt present in water and current produced by them. It measures the electric current which is proportional to mineral matter present in water. A high level of conductivity reflects on the pollution status as well as trophic levels of the aquatic body (Ahluwalia, 1999). Conductivity of water depends upon the concentration of ions and its nutrient status and variation in dissolve solid content. Electrical conductivity recorded in Chandlodia lake ranges between 2.78 ±0.07 to 3.24 ±0.12. The high value of conductivity was recorded during monsoon season were as low value was recorded during summer season.

The water during the summer decrease as a result some of the aquatic plant got destroyed and very few plants remain in the water. Therefore aquatic plants are in very rare amount in monsoon season, thus electrical conductivity is more in monsoon because water is free from vegetation and aquatic life therefore all the ion are accumulated in water. The decomposition of plants and animals ion are released back in water after summer Vora *et al.* (1998); Ahluwalia, (1999) and (Solanki and Pandit, 2006).

12.5 Turbidity

Turbidity is the measure of the light scattered by suspended particles. The substances not present in the form of solution cause it. According to Das and

Shrivastava (2003) clay, slit, organic matter, phytoplankton and other microscopic organisms cause turbidity in pond water. Light penetration is also highly affected by turbidity. Turbidity in Chandlodia lake recorded ranges between 16 ±1.47 to 22 ±0.91. The maximum turbidity in water was recorded during monsoon season and minimum turbidity was recorded during summer season.

High turbidity in lake water during monsoon season is due to addition of sand, clay, slit, dung and various other pollutant along with rain water from the surrounding area into the lake. Similar results were also observed by Saxena *et al.* (1966); Ansari and Prakash (2000) and Solanki (2001). Dagaonkar and Saksena (1992) and Garg *et al.* (2006) have also reported high turbidity during monsoon season this may be due to inflow of storm water from the surrounding area.

12.6 Total Dissolved Solids

Total dissolved solids denote mainly the various kinds of mineral present in the water. However if some organic substances are also present, as more often in the polluted waters, they may also contribute to the dissolved solid, Dissolved solid do not contain any gas and colloids. In natural water dissolved solids are composed mainly of carbonates, bicarbonates, chlorides, sulphates, phosphates and nitrate of calcium, magnesium, sodium, potassium, iron and manganese etc. The amount opf total dissolve solid in Chandlodia lake ranges between 963 ±31.5 to 1264 ±39. The maximum amount of total dissolve solid was recorded during monsoon season and minimum was recorded during winter.

The high value of TDS during monsoon may be due to addition of domestic wastewater, garbage and sewage etc in the natural surface water body. Indeed, high concentration of TDS enriches the nutrient status of water body which were resulted into eutrophication of aquatic ecosystem Similar result was observed by Singh and Mathur (2005) and Swarnlatha and Rao (1998).

12.7 pH

pH measure the concentration of hydrogen ion in water. It is the measurement of acidity or alkalinity. Verma *et al.* (1978) and Sharma *et al.* (1981) have reported that generally in India many small confined water pockets particularly, are alkaline in nature.The pH value ranges between 8.7 ±0.17 to 9.1 ±0.18. The maximum pH was recorded during monsoon and and minimum pH was recorded during summer season.

According to Prescott and Vinyard (1965); George (1962) and McCombie (1953), high pH values were found to promote the growth of algae and results in blooms. Nandan and Patel (1992); Moitra and Bhattacharya (1965) and Verma and Mohanty (1995) also observed that high pH values promote the growth of algae and results in heavy bloom of phytoplankton. Wani and Subla (1990) reported that the pH value above 8 in natural water are produced by photosynthetic rate that demand more CO_2 than quantities furnished by respiration and decomposition. The pH of water also depends on the relative quantities of calcium, carbonate and bicarbonate.

12.8 Alkalinity

Alkalinity in natural water is due to free hydroxyl ion and hydrolysis of salts formed by weak acid and strong bases and also due to salt containing carbonates and bicarbonates silicate and phosphate along with hydroxyl ion in the Free states. The change in alkalinity depend on carbonates and bicarbonates, which in term depend upon release of CO_2. Change in carbonates and bicarbonates also depend upon release of CO_2 through respiration of living organisms. The amount of total alkalinity in Chandlodia lake ranges between 192 ±3.65 to 214 ±6.87. The minimum value of alkalinity was recorded during summer season and maximum value was recorded during monsoon season

The addition of large amount of sewage waste and organic pollutant in the lake also effect photosynthesis rate, which also result in death of plants and living organism. The degradation of plants, living organism and organic waste might also be one of the reasons for increase in a carbonate and bicarbonate, resulting an increase in alkalinity value. Similar results were also observed by Chaurasia and Pandey (2007); Abbasi *et al.,* 1999 and Jain *et al.,* 1997. Padma and Periakali (1999) reported that increase in alkalinity during monsoon was due to input of freshwater and dissolution of calcium carbonate ions in the water column.

12.9 Total Hardness

Water hardness is the traditional measure of the capacity of water to react with soap, hard water requiring a considerable amount of soap to produce lather. Hardness of water is not a specific constituent but is a variable and complex mixture of cations and anions. The total hardness recorded in the water of Chandlodia lake ranges between 304 ±6.58 to 328 ±8.29. The maximum amount of total hardness in the water of Chandlodia lake was recorded during monsoon season and minimum amount was recorded during winter season.

Patel and Sinha (1998) noted that total hardness is mainly due to calcium magnesium and eutrophication. The high value of hardness during monsoon may be due to presence of high content of calcium and magnesium in addition to sulphate and nitrate in the sewage waste added during monsoon. Angadi *et al.* (2005) also observed similar result in Papnash pond of Karnataka. Tripathi and Pandey (1990) observed a higher value of total hardness in one of the lake and stated that it may be due to polluted water.

12.10 Calcium

Ansari and Prakash (2000) observed that the calcium is an important nutrient for aquatic, organism Calcium is commonly present in all water bodies. The amount of calcium in the water of Chandlodia lake ranges between 66 ±3.92 to 82 ±4.32. The maximum amount of calcium in the water of Chandlodia lake was recorded during monsoon season and minimum amount was recorded during winter season.

Calcium is present in water naturally, but the addition of sewage waste might also be responsible for the increase in amount of calcium. Udhayakumar *et al.* (2006) and Angadi *et al.* (2005) also observed similar result in their studies of water bodies. The decrease may be due to calcium being absorbed by living organisms.

12.11 Magnesium

Magnesium is found in various salt and minerals, frequently in association with iron compound. Magnesium is vital micronutrient for both plant and animal. Magnesium is often associated with calcium in all kind of water, but it concentration remain generally lower than the calcium(Venkatasubramani and Meenambal, 2007). Magnesium is essential for chlorophyll growth and act as a limiting factor for the growth of phytoplankton (Dagaonkar and Saksena, 1992). The amount of magnesium recorded in the water of Chandlodia lake ranges between 30 ±2.58 to 34 ±2.12. The maximum amount of magnesium in the water was recorded during winter season and minimum amount was recorded during summer season

Decrease in level of magnesium reduces the phytoplankton population. Govindan and Devika (1991) have suggested that the considerable amount of magnesium influence water quality. Magnesium is essential for chlorophyll bearing plant. Magnesium enters into combination with anions other than CO_2 in lakes such as chloride and sulphate (Jhingran, 1975). Various sub-processes like bating, picking, tanning, dyeing and fat liquoring causes water pollution (Bolton and Klein, 1971). Magnesium is vital micronutrient for both plant and animal.

12.12 Dissolved Oxygen

The oxygen in water can be dissolved from air or is produced from the photosynthetic organism like algae and aquatic plants oxygen is poorly soluble gas in water and it solubility depend on the temperature of water and its partial pressure. The solubility of oxygen also decreases with increasing salinity of water. Vijayaraghavan (1971) has established a direct relationship between photosynthesis and dissolved oxygen. Measurement of dissolved oxygen is a primary parameter in all pollution studies. The amount of dissolved oxygen recorded in the water of Chandlodia lake ranges between 2.14 ±0.07 to 5.08 ±0.07. The minimum amount of dissolved oxygen in the water of Chandlodia lake was recorded during winter season whereas maximum amount was recorded during monsoon season

The high temperature and addition of sewage and other waste might be responsible for low value of DO. Woodward (1984) and Mathuthu *et al.* (1993) made similar observation in their study of water bodies. According to Kataria *et al.* (2006) depletion of dissolve oxygen in water is due to high temperature and increased microbial activity. Dissolve oxygen with high value observed during monsoon may be as a result of the increased solubility of oxygen at lower temperature (Prasannakumari *et al.,* 2003).

12.13 Biochemical Oxygen Demand

BOD refers the oxygen used by the microorganism in the aerobic oxidation of organic matter. Therefore with the increase in the amount of organic matter in the water the BOD increases. The BOD value in Chandlodia lake ranges between 1.96 ±0.08 to 3.38 ±0.3. The minimum demand of oxygen in the water was recorded during summer season, whereas maximum demand was recorded during monsoon season

The higher value of BOD during monsoon was due to input of organic wastes and enchanced bacterial activity. Kumar and Gupta (2002) made similar observation in certain freshwater ecosystem of Santal Pargana, (Jharkhand). The reason of high BOD in monsoon might also be due to presence of several microbes in water bodies, which accelerate their metabolic activities with the increase in concentration of organic matter in the form of municipal and domestic waste discharge into water bodies and so the demand of oxygen increased Kumar and Sharma (2005) made similar observation.

12.14 Chloride

The greater source of chlorides in lake water is disposal of sewage and industrial waste. Human body release very high quantity of chlorides through urine and fasces. The chloride concentration was used as an important parameter for detection of contamination by sewage. Prior to development of bacteriological and other test like BOD and COD.The amount of chloride recorded in the water of Chandlodia lake ranges between 88 ±4.97 to 122 ±4.97. The minimum amount of chloride in the lake water was recorded during winter season and the maximum amount was recorded during summer season.

The higher concentration of chloride during summer month may be associated with frequently run-off loaded with contaminated water from the surrounding. Sunder (1988) and Kumar (1995) also observed the same pattern. The high chloride concentration of the lake water may be due to high rate of evaporation (Prasad *et al.,* 1985) or due to organic waste of animal origin (Purohit and Saxena, 1990). Many worker like Laxminarayana (1965); Singh (1965); Verma *et al.* (1978); Billore (1981); Venkateswarlu (1969) and Jana (1973) reported an increase in chloride content of water during summer seasons.

12.15 Sodium

Sodium is a natural constituent of raw water, but its concentration is increased by pollutional sources such as rock salt, precipitation runoff, soapy solution and detergent. The amount of sodium recorded in the water of Chandlodia lake ranges between 38 ±4.97 to 65 ±2.89. The minimum amount of sodium in the water of Chandlodia lake was recorded during winter and maximum amount was recorded during monsoon season

The high level of sodium during monsoon may be attributed to the rain water as it carries the salt dissolved from the surrounding area. Sahai and Sinha (1969) made similar observation. The addition of wastewater containing soap solution and detergent from the surrounding slummy area are also responsible for the increase in sodium level in the water bodies.

12.16 Nitrate

Nitrates are contributed to freshwater through discharge of sewage and industrial wastes and run off from agricultural fields. Some groundwaters naturally have high nitrate concentration.The amount of nitrate recorded in the water of Chandlodia lake ranges between 5.8 ±0.18 to 7.4 ±0.22. The minmum amount of nitrate in the water of Chandlodia lake was recorded during summer season, whereas maximum amount of nitrate in the water was recorded during monsoon season.

The high nitrate concentration during monsoon might be due to influx nitrogen rich flood water that bring about large amount of contaminated sewage water. The monsoon season was the period with the highest nitrate-nitrogen concentration which is known to support the formation of blooms. Similar observation was made by Blomqvist *et al.* (1994); Anderson *et al.* (1998) and Zimba *et al.* (2001). Nitrate content was higher in monsoon season, which can be attributed to the nitrate leached from the surrounding area. But lower concentration in summer was due to utilization by plankton and aquatic plants. Similar result was observed by Kannan (1978).

12.17 Phosphate

Excess amount of phosphate may cause eutrophication leading to extensive algal growth called algal blooms. Total phosphates in water include both organic and inorganic phosphates. Organic phosphates are part of living and dead plants and animal; over 85 per cent of total phosphorous is usually found in organic form. The amount of phosphate recorded in the water of Chnadlodia lake ranges between 1.24 ±0.05 to 1.92 ±0.11. The minimum amount of phosphate recorded in the water of lake was during winter season and the maximum amount was recorded during monsoon season.

The maximum value of phosphate during monsoon may be attributed to surface runoff during rainy season receiving huge quantity of domestic sewage, cattle dung and detergents from the surrounding catchment area. Catchment area activities are enriching phosphate in the lake (Tamot and Sharma, 2006). The lower value of phosphate in summer month may be due to more uptake of phosphate for luxuriant growth of macrophytes.

References

Abbasi SA, Khan FJ, Sentilevelan K, and Shabuden A. (1999). *Indian J. Env. Hlth*. **14(3)**: 176-183

Ahluwalia AA (1999). Limnological Study of wetlands under Sardar Sarovar command area. *Ph.D. Thesis*. Gujarat University, Ahmedabad.

Anderson DM, Cembella AD and Hallegraeff GM (1998). *Physiological Ecology of Harmful algal blooms*. 1ˢᵗ Edn., Springer-Verlag, Berlin, **pp**: 647-648.

Angadi SB, Shiddamaliayya N and Patil PC (2005). Limnological study of papnash pond, Bidar (Karnataka). *J. Env. Biol.*, **26:** 213-216.

Ansari KK and Prakash S (2000). Limnological studies on Tulsidas Tal of Tarai region of Balrampur in relation to fisheries. *Poll. Res.* **19(4):** 651-655.

APHA AWWA (1985). Standard Methods for the examination of water and wastewater. Washington DC 18th Edition.

Billore DK (1981). Ecological studies of Pichhola lake, *Ph.D. Thesis*, Univ. of Udaipur.

BIS (1982). Standard tolerance limits for bathing water. Bureau of Indian Standards. IS. 2296.

Blomqvist P, Petterson A and Hyenstrand (1994). Ammonium –nitrogen: A key regulatory factor causing dominance of non nitrogen fixing cyanobacteria in aquatic systems. *Arch. Hydrobiol.*, **132:** 141-164.

Bolton RL and Klein L (1971). Sewage treatment basic principlesand trends. Butterworths, London. **pp**: 159-160.

Chaurasia M and Pandey GC (2007). Study of physico- chemical characteristic of some water pond of Ayodhya –Faizabad. *Indian J. of Environmental Protection* **27 (11)** 1019-1023.

Dagaonkar and Saksena DN (1992). Physicochemical and biological characterization of a temple tank, Kaila Sagar, Gwalior, Madhya Pradesh. *J. Hydrobiol.* **8 (1):** 11-19.

Das AK and Shrivastva NP (2003). Ecology of Sarni Reservoir (M. P.) in the context of Fisheries. *Poll Res.* **22(4):** 533 – 539.

Dwivedi BK and Pandey GC (2002). Physico-chemical factors and algal diversity of two ponds in Faizabad, India. *Poll.Res.* **21(3):** 361-370.

Garg RK, Rao RJ, Saksena DN (2006). Assesment of physicochemical water quality of Harsi Reservoir, District Gwalior, Madhya Pradesh. *J. Ecophysiol. Occupat. Health* **6:** 33-40.

George MG (1962). Diurnal variation in two shallow ponds in Delhi, India. *Hydrobiol.*, **3**: 265.

Ghose BB and Basu AK (1968). Observation on estuarine pollution of Hoogly by the effluents from a chemical factory complexat Reshase, West Bengal. *Env. Health.* **10:** 209-218.

Govindan VS and Devika R (1991). Studies on Heavy metal profiles of Adyar river and waste stabilization pond. *J. Ecotoxicol. Environ. Monit.*, **1(1):** 53-58.

Jain CK, Bhatica KKS and Vijay T (1997). Groundwater quality in coastal region of Andraapradesh. *Indian J. of Environmental Health*. **39 (3):** 182-190.

Jana BB (1973). Seasonal periodicity of plankton in freshwater ponds in West Bengal, India. *Hydrobiologia.* **58**: 127-143.

Jayanti M (1994). A comprehensive study of three contrasting lentic system in the content of Aquaculture, *Ph.D. Thesis,* Bharathidasan University, Tiruchirappalli.

Jhingran VG (1975). *Fish and Fisheries of India.* Hindustan Publ. Corp. (India) Delhi, **pp**: 954.

Joshi PC and Singh (2001). A. Analysis of certain physicochemical parameters and plankton of freshwater hill stream at Nanda devi biosphere reserve. *Uttar Pradesh J. Zool.*, **21:** 177-179.

Kannan V (1978). The limnology of Sathiar: A freshwater impoundment. *Ph.D. Thesis*, Madurai Kamraj University, Madurai.

Kataria H C, Singh A and Pandey SC (2006). Studies on water quality of Dahod Dam, India. *Poll. Res.* **25(3):** 553 – 556.

Kumar A (1995). Observation on the diel variations in abiotic and biotiuc components of the river Mayurrakshi (Santal Pargana). Bihar. *Indian.J. Ecol*. **22 (1):** 39-43.

Kumar SM and Ravindranath S (1998). "Water Studies – Methods for monitoring water quality". Published by Center for Environment Education (CEE), Banglore, Karnataka, India, **pp**: 191.

Laxminarayana JS (1965). Studies on the phytoplankton of the river Ganges, Varanasi, India, Part I, *Hydrobiologia*, **25:** 119-137.

Mathuthu AS, ZaranYika FM, Jannalagadde SB (1993). Monitoring of water quality in upper Mukavisi river in Harare, Zimbabwe. *Environ. Int.* **19(1):** 51-6.

McCombie AM (1953). Factors Influencing the growth of phytoplankton Canada. *J. Fish Res. Ed.*, **10:** 253-282.

Moitra SK and Bhattacharya BK (1965). Some hydrological factors affecting plankton production in fish pond in Kalyani, West Bengal, India. *Icthyalogia* **4 (1 and 2):** 8 – 12.

Nandan SN and Patel RJ (1992). *Ecological studies of algae in aquatic ecology*. Ashis Publishing House. New Delhi. **pp**. 69-99.

Padma S and Periakali (1999). Physico-chemical and geochemical studies in Pulicat lake, east coast of India. *Indian J.Mar. Sci.*, **28**: 434-437.

Patel NK and Sinha BK (1998). Study of the pollution load in the pond of Burla area near Hirakund dam at Orissa. *J. Env. Poll*. **5:** 157-160.

Prasad BN, Jaitly YC and Singh Y (1985). Periodicity and interrelationships of physic-chemical factors in ponds. In: A.D. Adoni (eds.) *Proc. Nat. Symp. Pure and Apply. Gmnol. Bull. Bot. Soc. Sagar.*, **32:** 1-11.

Prasannakumari AA, Ganaga Devi T, and Sukesh Kumar CP (2003). Surface water quality of river Neyyar- Thiruvananthapuram, Kerala, India. *Poll Res.* **22(4)**: 515 – 525.

Prescott GW and Vinyard WC (1965). Ecology of Alaskan Freshwater algae, V. Limnology and flora of malikpur lake. *Trans. Amer. Micros. Soc.* **84(4):** 427-478.

Purohit SS and Saxena MM (1990). In *water, Life and Pollution*. Agro Botanical Publishers (India), Bikaner, India.

Sahai R and Sinha AB (1969). Investigation on bio-ecology of inlands water of Gorakhpur (UP), India. I. Limnology of Ramgarh Lake. *Hydrobiol*. **34(3)**: 143-447.

Saxena KL, Chakraborty RN, Khan AQ and Chattopadhya SN (1966). Pollution studies of the river Ganga near Kanpur. *Indian J. Environ. Hlth*. **8:** 270-285.

Sehgal HS (1980). Limnology of lake Sruinsar, Jammu with reference to zooplankton and fisheries prospectus. *Ph. D. Thesis*, University of Jammu.

Sharma KD, Lal N and Pathak PD (1981). Water quality of sewage drains entering Yamuna at Agra. *Indian J. Environ Hlth.*, **23:** 118-122.

Singh RN (1965). Limnological relations of Indian inland waters with special reference to water blooms. *Proc. Int. Assoc. Limnol.* **12:** 831-883.

Singh RP and Mathur P (2005). Investigation of variations in physicochemical characteristics of a freshwater reservoir of Ajmer city, Rajasthan, *Ind. J. Environ. Science,* **9:** 57-61.

Solanki and Pandit B R (2006). Trophic status of lentic waters of ponds water of Vadodara, Gujarat state, India. *Int. J. of "Bioscience Reporter"* **4 (1):** 191 – 198.

Solanki HA (2001). Study on pollution of soils and water reservoirs near industrial areas of Baroda. *Ph.D Thesis* submitted to Bhavnagar University, Bhavnagar.

Sunder S (1988). Mounting the water quality in a stretch of river Jhelum. Kashmir. In Book" *Ecol and Poll. of Indian rivers* " Ashish Publishing House. New Delhi. **pp:** 1312-161

Swarnalatha N and Narasingrao A (1998). Ecological studies of Banjara lake with reference to water pollution. *J. Environ. Biol.* **19 (2):** 179-186.

Tomat S and Sharma P (2006). Physico chemical status of Upper lake (Bhopal, India). Water quality with special reference to phosphate and nitrate concentration and their impact on lake ecosystem. *Asian.J. Exp.Sci.* **20 (2):** 289-296.

Tripathi AK and Pandey SN (1990). *Water Pollution.* Ashish Publishing House, New Delhi, **pp:** 312.

Trivedy RK and Goel PK (1984). In: *Chemical and Biological Methods for Water Pollution Studies.* Published by Environmental Publication, Karad, Maharashtra (India).

UdhayaKumar J, Natarajan D, Srinivasan K, Mohansundari C and Balasurami M (2006). Physico-chemical and Bacteriological Analysis of water from Namakkal and Erode Districts, Tamil Nadu, India. *Poll Res.* **25(3):** 495-498.

Venkatasubramani R and Meenambal T (2007). Study of subsurface water quality in Mattupalayam Taluk of Coimbatore district Tamil Nadu. *Nat. Environ. Poll. Tech.* **6:** 307-310

Venkateshwarlu V (1969). An ecological study of the algae of the river Mossi, Hyderabad (India) with special reference to water pollution. II factors influencing the distribution ogf algae. *Hydrobiol.* **33:** 352-363.

Vijayaraghavan S (1971). Seasonal variation in pri. Productivity in three tropical ponds. *Hydrobiol.* **38:** 395-408.

Wani IA and Subla (1990). Physico-chemical features of two shallow Himalayan lakes. *Bull Eviron. Sci.,* **8:** 33-49.

Welch PS (1952). *Limonology,* 2nd Ed., McGraw Hill Book Co., N.Y.**pp:** 536.

WHO (1971). *International Standards for drinking water,* 3rd Ed. Geneva, World Health Organization.

Woodward GM (1984). Pollution control in Hamber estuary. *Water pollution control.* **83 (1):** 82-90

Young JO (1975). Seasonal and diurnal changes in the water temperature of a temperate pond (England) and tropical pond (Kenya). *Hydrobiol.* **47:** 513-526

Zimba PV, Khoo L, Gaunt PS, Brittain S and Carmichael (2001). Confirmation of cat fish, *Ictalurus punciatus* Ralfinesque, mortality from microcystis toxins. *J. Fish Dis.*, **24:** 41-47.

2015, Impact of Global Warming and Climate Change on
    Human and Plant Health
Editors: **Dr. Arun Arya and Prof. V.S. Patel**
Published by: **DAYA PUBLISHING HOUSE, NEW DELHI**

*Pages **141–146***

Chapter 13

Changes in Chlorophyll Content in *Cassia occidentalis* L. in Response to Automobile Pollution

Shiv Kumari*, Ashwini Kumar and Ila Prakash

Botany Department, D.N. (P.G.) College Meerut (U.P.) India

ABSTRACT

The earth is the only planet known in the entire universe capable of supporting life. This is due to its unique environment. Any undesirable change in the environment, which may be due to addition of unwanted substances results in atmospheric pollution and disturbs the normal functioning of the ecosystem. Present study deals with the observation made on *Cassia occidentalis* on different roadsides having heavy pollution load in comparison to control. The studies were made on *Cassia occidentalis* taken from Garh Road, Delhi Road, Railway Road, and University Road. Reduction in chlorophyll contents were found to be depends on pollution concentration. An increase significant reduction was observed at highly polluted sites due to high concentration of automobile pollution, chlorophyll a, chlorophyll b and total chlorophyll content were reduced.

Keywords: *Chlorophyll, Cassia occidentalis, Automobile pollution, Merrut city.*

13.1 Introduction

All natural ecosystems maintain balance between their diverse components. The race for rapid development has resulted in unscrupulous exploitation of natural

---

* *Corresponding author.* E-mail: drshivasharma85@gmail.com

resources. This has disturbed the delicate ecological balance between living and nonliving components of the biosphere. Hermens *et al.* (2009) has reported the effect of pollution on vegetation. Losses incurred in chlorophyll a were relatively higher than chlorophyll b in SO_2 exposed leaves of *Euphorbia hirta* (Gupta and Ghouse, 1987). Function of pollution abatement is the best performed by the pollution tolerant species (Das and Prasad, 2010). Increased SO_2 concentration for longer duration results in considerable decrease in total chlorophyll content (Rath *et al.,* 1994.,Prakash *et al.*1997). Rajput and Agarwal (2004) also observed total chlorophyll content at polluted sites. Chapla and Kamalkar (2004) reported that ozone inhibit the production of necessary enzymes required for chlorophyll synthesis. Wath *et al.* (2006) observed that the plants along roadside with heavy traffic and markets are affected by vehicular emissions which cause a significant decrease in total Chlorophyll. Similar findings were observed in *Oryza sativa* by Prakash *et al.* (2008).

13.2 Materials and Methods

Fresh leaves of *Cassia occidentalis* were collected from different sites of Meerut city. Chlorophyll content was obtained by using Arnon's method (1949). For this purpose 100 mg of fresh leaf tissue was homogenized in 80 per cent acetone with a pinch of sodium bicarbonate. After centrifugation at 5000 rpm for 5 min., the supernatant was collected and the final volume was made up to 10 ml with acetone. The absorbance was measured at 663 nm and 645 nm on a systronic spectrophotometer using 80 per cent acetone as blank. Chlorophyll a, b and total chlorophyll were calculated by using the following formulae:

$$\text{Chl a (mg/g f.wt)} = [12.7(A_{663}) - 2.69(A_{645})] \times \frac{V}{1000\ XW}$$

$$\text{Chl b (mg/g f.wt)} = [22.9(A_{645}) - 4.68(A_{663})] \times \frac{V}{1000 \times W}$$

$$\text{Total (mg/g f.wt)} = [22.2(A_{645}) + 8.02(A_{663})] \times \frac{V}{1000 \times W}$$

where,

A = Absorbance at specific wavelength

V = Final volume (ml) of chlorophyll extract with 80 per cent acetone

W = Weight (g) of leaf tissue.

All the data were subjected to statistical analysis to find out Critical Difference at (CD) 5 per cent and 1 per cent level (Fisher 1951), is superscripted with single star (*) and double star (**) respectively.

13.3 Observations

A reduction was observed in total chlorophyll content in *C. occidentalis* at all polluted sites and it was maximum at Delhi road. The decrease in the value of chlorophyll a was found to be higher than chlorophyll b. In *C. occidentalis* reduction percentage in Chlorophyll a/b content were recorded 7.9 per cent/4.0 per cent in mg/gm fresh weight of leaves at University road and 36.3 per cent/38.2 per cent in mg/gm fresh weight of leaves at Delhi road (Table 13.1 and Figure 13.1).

Total chlorophyll was also recorded a gradual decrease with the increase in automobile pollution. In *C. occidentalis* the total chlorophyll reduction percentage was 39.9 per cent at Delhi road and maximum 4.3 per cent at University road (Table 13.1 and Figure 13.1)

13.4 Discussion

Maximum reduction in pigment concentrations was observed at Delhi road, moderate at Garh road and Railway road and less at University road. Air pollutants are known to cause significant reduction in chlorophyll pigments (Katz and Shore; 1955, Agrawal *et al.,* 1991). A significant reduction in total chlorophyll and protein was observed with reduced leaf area (Wath *et al.,* 2006).The chlorophyll contents in leaves of the plants on polluted sites showed a significant reduction. The inhibition of vital physiological processes like photosynthesis, chlorophyll metabolism and enzymatic activities ultimately led to the reduced plant growth.

However, chlorophyll a was found to be more susceptible than chlorophyll b. Sensitivity of chlorophyll a hampers the plant growth as it plays significant role in the process of photosynthesis. Reduced activity of chlorophyll molecule is associated with deficiency of nitrogen and Mg^{+2} ions in plants. As both these ions are involved in structure and synthesis of chlorophyll, their deficiency leads to the reduction in chlorophyll. This results in a decline of photosynthetic activity. These results are in

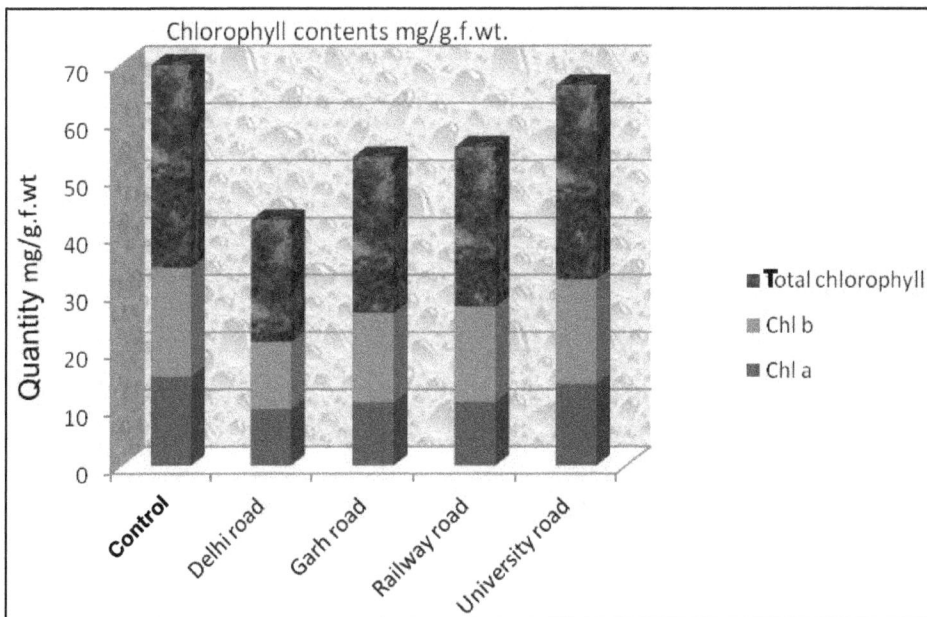

Figure 13.1: Chlorophyll Contents of *C. occidentalis* on Four Roads.

Table 13.1: Chlorophyll a, Chlorophyll b and Total Chlorophyll (mg/g/f/wt) in *Cassia occidentalis* L. at different Study Sites in Meerut City

Cassia occidentalis	Chlorophyll Content in different Study Sites						
	Control	Delhi Road	Garh Road	Railway Road	University Road	CD 5%	CD 1%
Chlorophyll a (mg/g.f.wt)	15.5038± 1.1528	9.8677*± 4.4675	10.9962*± 5.6898	11.0695*± 6.0363	14.272± 1.2905	3.366	8.023
Chlorophyll b (mg/g.f.wt)	18.9860± 0.4510	11.7237*± 2.8477	15.6332**± 2.0244	16.6423*±1.9260	18.2223± 1.6125	1.316	3.318
Total Chlorophyll (mg/g.f.wt)	35.4676± 0.4395	21.2897**± 9.4560	27.1923**± 5.0617	27.9430**± 6.3667	33.9211± 2.25180	1.283	3.058

Values are mean ± Standard Error.

Values are statistically significant at *< CD5 per cent and **< CD1 per cent.

accordance with Farooq *et al.* (1985) and Prakash *et al.* (1989). Increasing order of polluted study sites are as follows:

University road < Railway road < Garh road < Delhi road.

References

Agarwal, M., Singh, S.K., Singh, J. and Rao, D.N. 1991. Biomonitoring of air pollution around urban and industrial sites. *J. Environ. Biol.* 209-220.

Arnon, D.I. 1949. Copper enzyme in isolated chloroplasts, polyphenoloxidasa in *Beta vulgaris. Pl. Physiol.* 24: 1-15.

Chapla, J. and Kamalkar, J.A. 2004. Metabolic responses of tropical trees to ozone pollution.*J. Environ. Biol.* 25(3): 287-290.

Das, S. and Prasad A. 2010: Evaluation of expected performance index for some tree and shrub species in and around Rourkela. *Indian J. Environ. Prot.,* 30 (8): 635-642.

Farooq, M., Masood, A. and Beg, M.U. 1985. Effect of acute exposure of sulphur dioxide on the metabolism of *Holoptelea integrifolia* plants. *Environ. Pollut.* 39: 197-205.

Fischer, R.A. 1951. *The Design of Experiments.* Six Eds. Oliver and Boyd, London

Goswami, R 2002. Toxicity of air pollution to plants. A Ph.D. Thesis. Ç.C.S. Univ., Meerut.

Gupta, M.C. and Ghouse, A.K.M. 1987. Effects of coal and smoke pollutants from different sources on growth, chlorophyll content, stem anatomy and cuticular traits of *Euphorbia hirta* L. *Geobios* 14: 221-229.

Harmens, H., Mills, G., Hayes, F., Jones, L., Norries, D., Cooper, D. and the participants of the ICP vegetation Programme Coordination Centre. 2009. Air Pollution and vegetation: ICP vegetation Annual Report 2008/2009. Bangore, *NERC/Centre for Ecology and Hydrology*, pp. 40.

Katz, M. and Shore, V. C. (1955). Air pollution damage to vegetation. *J. Air Pollu. Cont. Asso.* 5: 144-150.

Prakash., Joshi, C. and Chauhan, A. 2008. Performance of locally grown rice plants (*Oryza sativa*) exposed to air pollutants in a rapidly growing industrial area of Haridwar. *Life Science Journal.* 5(3): 57-61.

Prakash, G., Agarwal, S., Kumar, N. and Verma, S.K. 1989. Changes in growth and yield associated with photosynthetic pigments, carbohydrate and phosphorus content in *Lycopersicon esculentum* exposed to sulphur dioxide. *Acta Bot. Indica* 17: 43-48.

Prakash, G., Tomar,Y.S., Sharma,T. K. and Gupta, V. 1997. Evaluation of *Spinacia oleracea* L. var. Pusa All Green on exposure to sulphur dioxide. *Ad. Pl. Sci. Res.* 5-6: 159- 169.

Rajput, M. and Agarwal, M. 2004. Physiological and yield responses of pea plants to ambient air pollution. *Indian J. Plant Physiol.* 9 (1): 9-14.

Rath, S., Padhi, S.K., Kar, M. and Ghouse, P.K. 1994. response of *Zinnia* to sulphur dioxide exposure. *J. Orn Amer. Horti.* 2 (1-2): 42-49.

Wath, N.D., Shukla, P., Tambe, V., and Scarika, B., Ingle S.T. 2006. Biological Monitorining of roadside plants exposed to vehicular pollution in Jalgaon city. *J. Environ. Bio.* 27(2 Supplement): 419-421.

Part III
Agriculture and Horticulture: Impact of Climate Change

क्यों कोई हमको याद करे?

पर्यावरण है बदल रहा
जलवायु यहां की बदली है।
जैव विविधता खतरे में
सागर में बढ़ता पानी है।
बढ़ता प्रदूषण जल—थल में
न कोई कोना बाकी है।
मानव अस्तित्व बचाने को
बस अंतिम घड़ियां बाकी हैं।
बढ़ रहा प्रदूषण नगर—नगर
गावों में पानी जहरीला है।
ग्रीन हाउस गैसों से बढ़ता
ग्लोबल वार्मिंग का खतरा है।
कम होते कोरल समुद्र में
पृथ्वी पर बढ़ते वेक्टर है।
सूक्ष्म जीवों से फिर
हम सबका जीना मुश्किल है।
आओ फिर हम सब
मिलकर विचार करें,
पर्यावरण प्रदूषण को दूर करें।
हो संसाधनों का संरक्षण
जैव विविधता का संवर्धन
जिससे आगे आने वाली पीढ़ियां
युग—युग तक हमको याद करें।

प्रो. अरूण आर्य

2015, Impact of Global Warming and Climate Change on
Human and Plant Health
Editors: Dr. Arun Arya and Prof. V.S. Patel
Published by: DAYA PUBLISHING HOUSE, NEW DELHI

Pages 149–162

Chapter 14

Nitrification from the Stand Point of Agricultural Implications and Environmental Concerns

R.S. Mann[1], S.K. Basu[2]*, M. Asif[3] and X. Chen[4]

[1]PSWC, Rampura, Punjab – 151 103, India
[2]University of Lethbridge, Lethbridge, Alberta, Canada T1K 3M4
[3]University of Alberta, Edmonton, Alberta, Canada T6G 2P5
[4]Yunan Academy of Agricultural Sciences, Kunming, Yunan, China 650205

ABSTRACT

Nitrogen is one of the principal components of atmospheric gases and also an important constituent of major structural and biological compounds such as chlorophyll, nucleic acids, proteins and enzymes to name only a few. In addition, nitrogen is also a major constituent of several inorganic and organic fertilizers as well as organic matters and compost. In short like hydrogen, oxygen and water cycles, nitrogen cycle is one of the pivotal natural bio-geo-chemical cycles that silently operates and regulates both biotic and abiotic components of our existing environment and ecosystems. The recycling of elemental nitrogen in our natural system thus has monumental impacts on the quality of our life, environment and economy from multiple perspectives. This review captures the essence of nitrification process from the stand point of modern agricultural practices with a focus on its impact on our natural environment.

Keywords: Denitrification, Environment, Fertiliser, Oxidation, Nitrogen, Nitrification.

---

* *Corresponding author.* E-mail: saikat.basu@uleth.ca

List of Abbreviations

AMO: Ammonium Mono Oxygenase; GWP: Global Warming Potential; GYI: Grain Yield Increase; HAO: Hydroxyl Amino Oxidoreductase; N: Nitrogen; NAD: Nicotine Amide Dinucleotide; NEA: Nitrifying Enzyme Activity; NUE: Nitrogen Use Efficiency; SOM: Soil Organic Matter

14.1 Introduction

Nitrogen is the most abundant element in the atmosphere with a concentration of 78 per cent (Brady and Weil, 2008). Nitrogen is an important constituent of all the amino acids, proteins and enzymes which are needed for all the major biological processes. Nucleic acids and chlorophyll also contain nitrogen. The dinitrogen molecule is very stable and inert because of the strong triple bond between two nitrogen atoms and thus cannot be used by plants and animals directly (Brady and Weil, 2008). Amongst all the living organisms, only some microbes (bacteria or cyanobacteria) are known to establish a symbiotic relationship with specific groups of higher plants (such as legumes) and are capable of using atmospheric N_2. All the other living organisms including crop plants like cereals have to depend on the nitrogen which is derived from Soil Organic Matter (SOM) mineralization or supplied by nitrogen fertilizers (Subbarao *et al.,* 2006). Nitrogen is directly available to living organism when it is bonded to hydrogen, oxygen or carbon (*e.g.*, NH_4^+, NO_3^- and R-C-NH_2). The continuous N_2 cycle ensures that the balance of N_2 is maintained in the atmosphere (Brady and Weil, 2008).

14.2 Why Nitrification needs to be Studied?

In the current high production agriculture systems in most of the cases it is seen that the fertilizer application is not a limiting factor, which ultimately results in excess application of fertilizers especially the nitrogen (N) fertilizers with a very little concern about what effect it will have on the environment (Subbarao *et al.,* 2006). Currently about 100 Tg (million Mg) of fertilizers (mainly in the form of ammonia/NH_3) are used to drive the agri-productivity towards feeding the ever expanding global population (Wrage *et al.,* 2001). According to one estimate the global production of N fertilizer will increase to 134 Tg by the year 2020 (Vitousek *et al.,* 1997).

As the process of nitrification converts the added nitrogen into NO_3^- which can be readily taken up by plants, but this NO_3^- is also prone to losses through leaching and de-nitrification. So the addition of large quantities of N fertilizer may lead to adverse environmental impacts due to the loss of nitrogen due to leaching subsequent processes like de-nitrification etc. So in this review we will mainly discuss: (i) nitrification and nitrous oxide emissions during nitrification (ii) why it must be studied/why it is important (iii) factors affecting nitrification (iii) methods to control nitrification and (iv) factors affecting nitrous oxide emissions during nitrification.

14.3 Nitrification

Nitrification can be defined as a process in which NH_3 is converted to nitrate through biological oxidation. It has been observed that in order to increase the N_2 use efficiency the nitrification part in the N_2 cycle can be the most useful. The nitrate formed as the process of nitrification can be easily lost through de-nitrification and leaching. This in turn affects the processes by which plants absorb nitrogen and which form N_2 is lost to environment (Subbarao *et al.,* 2006). Nitrification is of two types, as discussed below.

14.4 Autotrophic Nitrification

In the process of nitrification NH_4^+ or NH_3 is oxidized to NO_3^-. It is carried out by autotrophic microorganisms. In the first part of the reaction, the NH_4^+ or NH_3 is converted to NO_2^- by NH_3 oxidizers or primary nitrifiers *e.g.*, *Nitrosomonas europaea*. In the second step, NO_2^- is oxidized to NO_3^- by nitrite oxidizers or secondary nitrifiers *e.g.*, *Nitrobacter winogradskyi* (Wrage *et al.,* 2001). To completely understand the process of nitrification, we have to understand the detailed nitrification pathway. Hydroxylamine (NH_2OH) is the first intermediate which is produced when NH_3 is oxidized; this reaction is catalyzed by ammonia mono oxygenase (AMO). The two electrons needed in this reaction to reduce one of the atomic O_2 to H_2O are derived from the following reaction involving oxidation of NH_2OH to NO_2^-. It is important to study these intermediates produced because the enzymes catalyzing these reactions have a broad range of substrates and they can inhibit there enzymatic activity by competitively or covalently binding to the active site, which in turn inhibits the NH_3 oxidation. Acetylene (C_2H_2) is one of the well known inhibitors of this step (Wrage *et al.,* 2001). The next step in nitrification is the oxidation of hydroxylamine NH_2OH to NO_2^-, which is catalyzed by hydroxylamine oxidoreductase. This enzyme can be inhibited by hydrazine. The NO_2^- thus produced is further oxidized to NO_3^- by the enzyme nitrite oxidoreductase. Chlorate can inhibit this reaction (Wrage *et al.,* 2001).

The process of nitrification provides energy to *Nitrobacter* for the fixation of CO_2. The nitrifying microorganisms are very slow growing and relatively greater substrate is needed to make them grow at faster rates. About 30 g of NH_3 is needed for making 1 g dry matter of *Nitrosomonas* sp. This can be explained by the fact that the microorganisms coming under the family Nictrobacteriaceae are aerobes or obligate autotrophs. The energy which is generated during nitrification is used for the CO_2 fixation. Both NH_3 and NO_2^- are not very efficient energy sources. Hence electrons released during oxidation step of nitrification needs to be directed only to low energy level of the respiratory chain. Therefore, this oxidation step is not linked directly to NAD (Nicotine Amide Dinucleotide) or the first element of respiratory chain. Hence, the energy has to be utilized in driving electrons to high energy level referred to as reverse energy flow (Wrage *et al.,* 2001). Nitrous oxide (N_2O) may also be formed during the process of nitrification. N_2O can be produced by the chemical decomposition of the intermediates such as NH_2OH or NO_2^- which are formed between NH_4^+ and NO_2^- (Wrage *et al.,* 2001). Bremner and Blackmer (1978) reported that when soil is treated with either ammonium sulphate [$(NH_4)_2SO_4$] or with urea, the amount of N_2O released was greater when compared to the soil treated with KNO_3. It was also noticed that the amount of N_2O released increased with the increase in the amount of ammonium N or urea N added. It was observed that most of the N_2O released was produced during nitrification.

14.5 Heterotrophic Nitrification

In addition to the autotrophic nitrifiers some nitrifiers are heterotrophic and therefore, use organic carbon (C) as the source of C and as fuel or energy supply. Although heterotrophic nitrification is common to fungi in soils with low pH, some bacteria are also known to be capable of heterotrophic nitrification. The heterotrophic nitrifiers are different from autotrophic nitrifiers in the sense that they also have the capability to denitrify (Wrage *et al.,* 2001).

When heterotrophic nitrification was studied in *Pseudomonas denitrificans* it was found that the heterotrophic nitrification differs from the autotrophic nitrification with respect to the fact that ammonia monooxygenase enzyme is not inhibited by C_2H_2. Also, the hydroxylamine oxidoreductase (non-haem iron enzyme) found in heterotrophic nitrifiers is a multi-haem enzyme in autotrophic nitrifiers. The end product NO_3^- is produced under aerobic conditions during heterotrophic nitrification which can be used for de-nitrification when the conditions are suitable. It must be noted that the heterotrophic nitrifiers can carry out the process of de-nitrification under aerobic conditions while the conventional de-nitrifiers cannot (Wrage *et al.,* 2001). N_2O is thus generated by heterotrophic nitrifiers as an intermediate in the reduction of NO_2^- to N_2. It was observed that heterotrophic nitrifiers (*e.g., A. faecalis*) produced much more N_2O under aerobic conditions then autotrophic nitrifiers. Factors such as acidic pH, higher oxygen concentration and availability of organic matter are favourable conditions for heterotrophic nitrifiers to generate N_2O in significant amounts (Wrage *et al.,* 2001; Anderson *et al.,* 1993).

14.6 Nitrifier de-nitrification

Nitrifier denitrification takes place during nitrification. During this process, NO_3^- is not formed; rather, the oxidation of NH_3 to NO_2^- is readily followed by reduction of NO_2^- to N_2O and N_2. This is conducted only by autotrophic NH_3-oxidizers (Wrage *et al.,* 2001). This process is different from coupled nitrification-denitrification reaction in that it only involves a singly type of microbe, whereas in the latter two different types of microbes oxidize NH_3 to N_2 (Wrage *et al.,* 2001).

The oxidation of NH_3 to NO_2^- generates from nitrification and the reduction of NO_2^- to N_2 is considered as gentrification (Poth and Focht, 1985). This is because during de-nitrification the N_2 produced by reducing NO_2^- to NO is further reduced to N_2O and finally to N_2 (Poth, 1986). It has been reported that nitrifier de-nitrification is mostly carried out by NH_3 oxidizers. The enzymes involved in nitrifier de-nitrification are same which are involved in nitrification and de-nitrification. Nitrifier de-nitrification is different from heterotrophic nitrification in two ways; first, only autotrophic nitrifiers are involved in nitrifier de-nitrification and second, the enzymes involved in nitrifier de-ntrification are different from those involved in heterotrophic nitrification (Wrage *et al.,* 2001).

The coupled nitrification-de-ntrification is indicated by overlapping boxes between nitrification and de-ntrification. The boxes of nitrifies de-ntrification and nitrification overlap because nitrifies de-ntrification is a pathway of nitrification, but both separate out at NO_2^- where onwards the reactions differ. NO_3^- is an intermediate in coupled nitrification and gentrification whereas NO_3^- is not formed during nitrifies gentrification. We can summarize by saying that nitrifies de-ntrification is carried out by only nitrifies whereas both coupled nitrification-de-nitrification is carried out by both nitrifies and gentrifies (Wrage *et al.*, 2001).

14.7 Why is Nitrification Important?

Nitrification may affect nutrient use efficiency. Ammonium is a caption (NH_4^+) that is held strongly onto the functional groups of soil organic matter and onto the surface of the clay particles which are negatively charged by electrostatic forces. This prevents the leaching losses of N. The soil conversion of NH_4^+ to NO_3^- makes the applied fertilizer prone to losses due to leaching. Sahrawat (1982) revealed that out of the total fertilizer N added; about 90 per cent is applied as NH_4^+ which is subsequently nitrified in about four weeks post-application. Strong and Cooper (1992) reported that the application of NH_4^+-N can be considered equivalent to the application of NO_3^--N in arable soils as nitrification is a very general phenomenon. The NO_3^- is the substrate for gentrification, so the nitrogen present in the soil as NO_3^- can be lost to the atmosphere as N_2 via process of gentrification. As from the definition of de-nitrification we can see that it is the process of reduction of NO_3^- to N_2 which is carried out by heterotrophic microorganisms *e.g.*, *Pseudomonas, Bacillus, Thiobacillus* etc (Wrage *et al.*, 2001). The process of nitrification also provides an opportunity for nitrogen to escape into environment in the form of N_2O, NO and N_2. Thus it can be interpreted that these losses of nitrogen may lower the nitrogen-use efficiency and can affect the environment adversely (Wrage *et al.*, 2001). Slanged and Kerkhoff (1984) reported that if N remains in soil in NH_4^+ form for a longer time then the plants have more opportunity to absorb it and the chances of its leaching are decreased which will be desirable from an agricultural perspective.

14.8 Rates of Nitrification in Intensive Agriculture

It has been observed that in the managed agricultural systems nitrogen is found in the form of NO_3^- whereas in mature ecosystems NH_4^+ is the dominant form. This may be because the rate of nitrification is greater in managed agricultural systems then in natural ecosystems (Subbarao *et al.*, 2006). Intensive agriculture has resulted in the disappearance of the crop rotations of cereals with legumes, and now we find that only a single crop is grown on the same piece of land continuously (Dennis *et al.*, 2002). This happened after World War II when the N fertilizer became available easily and at cheaper prices. The legumes and animal manures were replaced by N fertilizers. The easier availability of excess nitrogen as a substrate may have resulted in increased nitrification and thus concentrations of NO_3^- must have increased in the system. The improvements in the drainage systems and the development of artificial drainage systems resulted in increasing the losses of NO_3^--N by leaching (Dennis *et al.*, 2002). It has been reported that the intensive agriculture systems give rise to the conditions which are favourable for and enhance the rate of nitrification, reduce the nitrogen use efficiency (NUE) and reduce soil organic matter (SOM) (Subbarao *et al.*, 2006).

14.9 Loss of N as Nitrate

The process of nitrification gives rise to many environmental problems/concerns. The major one amongst them is the loss of nitrogen through leaching as NO_3^-. The process of nitrification converts ammonia to NO_3^-, which is lost by leaching and pollutes the groundwater (Giles, 2005). The consumption of water polluted with nitrates can cause harmful disease like methemoglobinemia in infants (also called blue baby syndrome) and nitrate poisoning in animals (Subbarao *et al.,* 2006). According to Malakoff (1998), the amount of NO_3^--N (worth approximately US$750 million) that is leached annually into the Mississippi river is ~15 per cent of total N fertilizer applied to all agricultural systems in USA.

14.10 N_2O Emissions

The anthropogenic altercation of N_2 cycle has severely impacted our immediate local environment. The rate of N_2 entering the land-based N_2 cycle has almost doubled and is still increasing. It has also resulted in the enhanced global concentrations of N_2O, a potent greenhouse gas (Vitousek *et al.,* 2000). The global warming potential (GWP) of N_2O is 296 and 13 folds greater than CO_2 and methane (CH_4) respectively. It can stay in the atmosphere for long time and has a capacity to absorb about 210 times more infra red radiation than CO_2. The concentration of N_2O is accelerating in the atmosphere at an alarming rate of 0.75 per cent per annum (Wrage *et al.,* 2001). About 70 per cent of total anthropogenic emissions of N_2O come from agriculture (Wrage *et al.,* 2004). It is estimated that cultivated lands contribute about 3.5 Tg of N_2O-N annually. The use of synthetic fertilizers alone contributes 1.5 Tg N_2O-N out of the total N_2O emissions from cultivated lands. It has been projected that by the year 2100, total N_2O emissions linked directly to the application of synthetic fertilizer will increase from current rate of 1.5 Tg N yr^{-1} to around 4.2 Tg N yr^{-1} (Kroeze, 1994).

14.11 Benefits of Nitrification

A part from the adverse impacts, nitrification may also have some benefits. In some cases nitrification can be beneficial, *e.g.,* under alkaline conditions, nitrification helps to retain the N by converting NH_4^+ to NO_3^-, where otherwise the ammonia lost by volatilization. The animal and human excreta and other wastes like paper, pulp contain N mainly in ammonium form and for these nitrification acts as a starter and initiates the process of releasing the N from these wastes back into the atmosphere (Kowalchuk and Stephan, 2001; Subbarao *et al.,* 2006).

14.12 Energy Concept for NH_4^+ and NO_3^-

It has been reported that NH_4^+ requires less energy to be assimilated by plants then NO_3^-. NH_4^+ needs only 5 moles of ATP per mole of NH_4^+ while NO_3^- requires 20 moles of ATP per mole of NO_3^- for assimilation. But still it has been observed that the plants which take up N in ammonium form have a lower dry matter production then which take up N in nitrate form. It has been hypothesised that the decrease in yield can be attributed to an inadequate accumulation of nitrate and organic anions which are produced by nitrate reduction (Salasac *et al.,* 1987). The difference in the energy requirements can be exploited to inhibit nitrification. The energy which is saved in the crops which are grown on NH_4^+-N may lead to the greater production of biomass in those crops; however, more research is needed towards this aspect to get some results (Subbarao *et al.*, 2006).

14.13 Factors Influencing Nitrification

Physical Factors

Strong *et al.* (1999) observed the effect of soil texture on nitrogen mineralization. They were of the opinion that soil texture is an important regulator of N transformations in soil. N mineralization and nitrification was significantly influenced by clay and sand when compared to silt. They observed that if the soils with high clay content are kept continuously wet then N mineralization and nitrification is much lesser compared to same soil continuously dried and rewetted. This may be attributed to the fact that the organic matter which is present in the clay soil is located in the small pores and is thus prevented from the attack by the microbes.

14.14 Environmental Factors: Temperature, Moisture, Aeration

The optimum and the maximum temperature for nitrification vary within a wide range depending upon the climatic zones. Malhi and McGill (1982) reported that the rate of nitrification increased rapidly from 4-20 C, being maximum at 20 C. Nitrification showed a great response to temperature within a range of 10-20 C. The greater rate of nitrification at higher temperature was mainly attributed to the increase in the nitrifier activity with increase in temperature. With an increase in temperature the rate of nitrification decreased, the rate recorded at 30 C was only 10-15 per cent of that recorded at 20 C. At 40 C, no NO_3^--N was produced. In another study the response of rate of nitrification was examined over a temperature range of 5-30 C. It was observed that the rate increased throughout the full range of temperatures. No lag phase was observed and also the rate of nitrification was not affected as the residual ammonia concentration decreased (Wild *et al.*, 1971).

14.15 Moisture

Malhi and McGill (1982) reported that nitrification enhanced with higher moisture content. They observed that in three different soil samples, nitrification rates increased with moisture content increasing from -1500 to -33 kPa. A 2.8 fold higher nitrification was observed at -33 kPa, compared to -1500 kPa. In another study by Davidson (1992) to detect sources of NO and N_2O after wetting dry soil, nitrification was reported to be the major source of both emissions when soil water was detected below field capacity. However, under higher soil field capacity, de-nitrification was the major source of N_2O. The authors reported C_2H_2 inhibited N_2O production when soil water was detected below field capacity. In other words, when soil water was above field capacity, de-nitrification could be the possible source of N_2O. Thus when the water content is optimum for nitrification, then the nitrification can be considered as main source of N_2O emissions.

14.16 Aeration

Nitrification is also affected by O_2 and CO_2 concentrations in the soil environment (Yuan *et al.,* 2005). Soil moisture and O_2 levels are inversely related to one other in the soil system. Nitrification occurs in aerated soils being an aerobic process. In one of the experiment, optimum nitrification was observed when soil O_2 concentration was around 20 per cent, which is similar to the atmospheric O_2 concentration (Subbarao *et al.,* 2006).

14.17 Chemical Factors: pH and C:N Ratio pH

Wild *et al.* (1971) concluded that a pH of 8.4 is optimum for nitrification. The study suggested that 90 percent of the maximum rate of nitrification occurred in the pH range of 7.8 to 8.9. Sahrawat (1982) studied eight mineral soils and two histosols under a pH range of 3.4 to 8.6. The soils which had pH greater than 6 had a greater rate of nitrification and released NO_3^- ranging from 98 to 123 µg g^{-1}. When Na_2CO_3 was used to increase the pH of one of the soil samples, it increased the amount of NO_3^- formed; however, the total mineral N recorded in these samples did not vary.

14.18 Substrate Concentration

Malhi and McGill (1982) reported impacts of substrate concentration on nitrification in three Alberta (western Canada Prairies) soil samples. The maximum nitrification rate was recorded at NH_4^+-N concentration of 50-200 µg g^{-1} of soil. It was also reported that nitrification rate increased with increase in NH_4^+-N level from 50-200 µg g^{-1} of soil, but reduced at levels greater than 200 µg g^{-1} of soil. It was found that nitrification rate was about 2.5 folds greater when NH_4^+-N level was 200 µg g^{-1} of soil compared to the rate when the concentration was 50 µg g^{-1} of soil.

14.19 Landscape and Time of Fertilizer Application

The efficacy of the fertilizer can also be impacted by the time and method of fertilizer application. When wheat was grown on low landscapes positions the grain yield, total crop N uptake, grain yield increases (GYI) and crop nitrogen use efficiency (NUE) for wheat was greater when the urea was banded in late fall compared to that in early fall. However in the high landscapes the efficacy of fall banded urea was not impacted by date of application; this was attributed to better drainage in these areas (Tiessen *et al.,* 2008). In another experiment, it was observed that when urea was applied in fall, barley recorded 792 kg ha^{-1} lower yield and 15 kg ha^{-1} lower N uptake compared to application of urea in spring. However the efficiency of fall applied N was higher when larger urea pellets (2.5 g) were applied at the depth of 15 cm and the relative yield efficiency recorded was 95 per cent compared to spring-applied N. This was mainly attributed to less nitrification when larger urea pellets were applied (Nyborg and Malhi, 1992).

14.20 Methods to Control Nitrification

Nitrification Inhibitors

The oxidation of NH_3 to NO_2 is catalysed by two enzymes AMO and hydroxylamine oxidoreductase (HAO). Most of the nitrification inhibitors act by targeting the AMO enzymatic site. Hydroxylamine is not targeted for developing nitrification inhibitors as it is toxic to *Nitrosomonas* if it accumulates in larger quantities (Subbarao *et al.,* 2006). The AMO has a wider substrate range (Less, 1952). The alternate substrates can affect the activity of AMO by:

(i) Competitive and non-competitive inhibitors: The competitive inhibitors, as the name indicates, occupy the same binding site as NH_3. While the non competitive inhibitors bind to an alternate site. The alternate site is usually a hydrophobic region. The non competitive binding site is still not well studied and well defined which is evident from the presence of the structural diversity in non-competitive inhibitors which affect ammonia oxidation (Keener and Arp, 1993). The metal chelators are also known to possess the ability to inhibit nitrification by inhibiting AMO. Some of the important chelators include thiourea (which is a copper-chelators), potassium cyanide, L-histidine, guanidine etc (Subbarao *et al.,* 2006).

(ii) Suicide or mechanism based inhibitors: In mechanical or suicidal inhibition, the product which is formed by the oxidation of the compound binds to the polypeptides and inhibits the catalysis of AMO. In this mechanism the enzyme is inactivated irreversibly. The examples of mechanism based inhibitors include acetylene, allylsulfide and trichloroethane (Hyman and Wood, 1985).

Apart from targeting ammonia monooxygenase there are a few compounds which inhibit nitrification by targeting hydroxylamine oxidoreductase (HAO), the enzyme which catalyses the formation of nitrite from hydroxylamine. The compounds which target HAO include hydrogen peroxide and phenyl-methyl-, or hydroxyethyl hydrazine. (Terry and Hooper, 1981; Logan *et al.,* 1995; Logan and Hooper, 1995). Less and Simpson (1957) reported that the second step in nitrification, the oxidation of nitrite to nitrate, is inhibited by chlorate by suppressing the *Nitrobacter* population. It must be noted that chlorate inhibits this reaction by suppressing the population of *Nitrobacter* and not by inhibiting their ability to oxidize nitrite. Many pesticides are reported to inhibit the reaction by suppressing the population of *Nitrobacter* namely heptachlor, lindane, chlordane and eptam (Winely and San Clemente, 1970). Chelating compounds are known which indirectly inhibit nitrite oxidation by interfering with flavoprotein-cytochrome and respiratory systems. P-nitrophenol, m-nitrophenol, 2,4-dinitrophenol which have the ability to uncouple the oxidative phosphorylation are also known to inhibit nitrite oxidation by *Nitrobacter* (Subbarao *et al.,* 2006).

Malhi and Nyborg (1988) conducted an experiment to compare the efficacy of thiourea, ATC (4-amino-1,2,4-triazole hydrochloride) and N-Serve 24 E (2-chloror-6-trichloromethyl-pyridine) as inhibitors of nitrification. The experiment was conducted on two soils: Malmo silty clay loam and Breton loam. In the laboratory experiment aqueous ammonia was nitrified in 14 days. The concentration of NO_3^--N was lower in the treatments where inhibitors were added when compared to the control (without inhibitors) which shows that there was a decrease in the rate of nitrification. When placement of fertilizers was compared, it was noticed that nitrification was less in case of ATC and thiourea when the point placement was done when compared to mixing in soil. But this was not so for N-Serve 24 E. ATC and N Serve 24 E were efficient in inhibiting nitrification for a longer duration *i.e.* more than 28 days. In the uncropped field experiment it was noticed that when ATC or thiourea was used along with urea in fall, then the nitrification was depressed by May. It was also noticed that recovery of NH_4-N was higher when spacing was 60 cm or when fertilizer was placed in nests. It was concluded that the nitrification rates for fall applied N was reduced by placing NH_4-based N fertilizers in widely spaced bands or nests along with reduced rates of inhibitors (Malhi and Nyborg, 1988). Nitrification inhibitors also help to control N_2O emissions from soil. Bremner and Blackmer (1978) reported that soil N_2O emission from soils treated with $(NH_4)_2SO_4$/urea were reduced by using nitrapyrin [2-chloro-6(trichloromethyl) pyridine]. The latter reduces the nitrification of NH_4-N and urea N in soils by inhibiting oxidation of NH_4 to NO^- by *N. europaea.*

14.21 Slow and Controlled Release of Fertilizers

The slow and controlled release fertilizers can be applied to increase fertilizer availability to plants, enhance fertilizer effectiveness and curb fertilizer losses (Oertli, 1980; Guertal, 2000). The losses of nutrients by leaching or volatilization are less in case of slow release fertilizer which contain nitrogen, because as the name suggests the fertilizer becomes available slowly. As the fertilizer is released slowly, so the rates of nitrification may reduce because of the less availability of substrate to act upon. The benefits of using slow release fertilizers include economic saving, saving in form of energy and reduction in environmental pollution. The flip side of the use of slow release fertilizers is that due to the continuous release of fertilizer the salts may accumulate in the field when the plants are not present on the field. The slow release fertilizers can be grouped into coated fertilizers, osmocote, sulphur coated fertilizers and sparingly soluble compounds. By using the slow release fertilizers it is possible to synchronise the nutrient supply from fertilizer with the nutrient requirement of the plant.

The rate of release of nutrient from the fertilizer is controlled by coating or encapsulating fertilizer with an osmocote membrane. In sulphur coated fertilizers the rate of nutrient release is governed by thickness of sulphur coating, kind of sealant and kind of microbicide (if microbicide is used). In urea-aldehyde condensates: for ureaformaldehydes the release rate depends upon the degree of polymerization and on urea-formaldehyde ratio; for isobutylidenediurea the same is limited by size of granules and hardness and for crotonylidendiurea it depends upon the size. In addition to these the rate of nutrient release is also affected by the other factors *viz.* soil pH, microbial activity, temperature and moisture (Oertli, 1980).

Brown *et al.* (1988) conducted an agronomic study to evaluate the performance of preplant-sulphur coated urea with preplant-banded and split applications of urea in onion crop. Nitrogen was applied at rate of 56 and 224 kg ha^{-1}. The slow release of N from sulphur coated urea when compared to urea was indicated by the higher concentration of root NO_3-N and soil NO_3-N in the bed centres. The use of sulphur coated urea significantly increased the total and large bulb yield and N uptake in onion bulbs and leaves under low nitrogen conditions as compared to preplant urea treatment; however it was at par with split urea treatments (Brown *et al.,* 1988).

14.22 Factors Effecting Nitrous Oxide Emissions during Nitrification

Carbon Dioxide, N Addition, Temperature and Moisture

Barnard and Leadley (2005) observed that the increase in the CO_2 levels decreased the nitrifying enzyme activity (NEA) in 11 experiments and in other five experiments the net nitrification was increased. They concluded that gross nitrification was not affected by the increased CO_2 levels. Also increase in CO_2 levels did not have any significant effect on N_2O fluxes in field or in laboratory measurements. The net nitrification and gross nitrification were significantly increased when N levels were increased. All the nitrification variables measured increased with the increase in N levels. It was observed that the N additions significantly increased the N_2O efflux under both field and laboratory conditions. Temperature had significant impact on nitrification rate, however the authors did not mention about the effect of temperature on N_2O emissions.

Wrage *et al.* (2004) observed that N_2O production during nitrifier denitrification was low under wet conditions then under dry conditions. Some wet soils recorded negative values for nitrifier denitrification. However, the total N_2O emissions were significantly greater in wetter soils than dry soils. They did not find any significant difference between dry and wet soils for nitrification. Soil texture did not have any significant effect on the sources of N_2O; it was observed that nitrification and nitrifier denitrification were approximately equal in both sand and clay soils. The N_2O production was not affected significantly by the eight fertilizer treatments. It was observed that the total N_2O production was affected significantly with increase in N fertilization. This increase was related higher N_2O production by nitrifier denitrification and other sources. It was found that the use of calcium nitrate instead of ammonium nitrate did not have any significant effects on N_2O production (Wrage *et al.,* 2004). In another study it was observed that nitrifier denitrification did not impact of N_2O generation from wet soils; however, in dry soils about 30 per cent of N_2O was generated from total nitrifying sources (Webster and Hopkins, 1996).

14.23 Summary and Conclusion

Nitrogen is an essential nutrient for plant growth and is found in soil as in anionic form and is also taken up by plants in anionic form. Nitrogen in soil undergoes oxidation and reduction reactions thus forming gases which can result in environmental problems. Nitrogen is the nutrient which has received the most attention in terms of management. Therefore it becomes more important to apply nitrogen in accurate quantities, as both the excess application of N and deficient application of N may not be good from the productivity point of view (Brady and Weil, 2008). It is estimated that the N fertilizer inputs will double in the future to fulfil the food needs of the ever growing population. The process of nitrification which converts ammonia to nitrate also makes the N susceptible to losses by various mechanisms like leaching, denitrification, gaseous losses as N_2O. It has been observed that with the intensification of agricultural activities the rates of nitrification have also increased as a result of which the chances of changes like global warming, eutrophication have also increased. Agronomic practices like time of application have been exploited to control nitrification. Apart from agronomic practices major techniques that are being used for nitrification control is the application of nitrification inhibitors. According to Subbarao *et al.* (2006) ~ 60 per cent of total global N used under agricultural systems is lost every year; equivalent to an economic value loss of US$ 17 billion per annum. So there is a great need to develop methods/techniques to control the losses of N occurring during nitrification. Efforts must also be made to develop more efficient fertilizer, more efficient and economically feasible technology for application of fertilizers and better nitrification inhibitors which function equally well in all type of agro- climatic conditions. Concentrated and focused efforts to manage nitrification in an efficient way can help improve the utilization efficiency of fertilizer, reduce economic losses and safeguard the environment.

References

Anderson IC, Poth M, Homstead J and Burdige D 1993. A comparison of NO and N_2O production by autotrophic nitrifier *Nitrosomonas europaea* and heterotrophic nitrifier *Alcaligenes faecalis*. *Applied and Environmental Microbiology* 59: 3525-3533. 199: 295-296.

Barnard R and Leadley PW 2005. Global change, nitrification and denitrification: A review. *Global Biogeochemical Cycles* 19: 1-13.

Brady NC and Weil RR 2008. Nitrogen and sulphur economy in soils. In: *The Nature and Properties of Soils*, pp. 543-591.

Bremner JM and Blackmer AM 1978. Nitrous oxide: emission from soil during nitrification of fertilizer nitrogen. *Science* 199: 295-296.

Brown BD, Hornbacher AJ and Naylor DV 1988. Sulphur-coated urea as a slow release nitrogen source for onions. *Journal of American Society for Horticultural Science* 113: 864-869.

Davidson EA 1992. Sources of nitric oxide and nitrous oxide following wetting of dry soil. *Soil Science Society of America Journal* 56: 95-102.

Dennis DL, Karlen DL, Jaynes DB, Kaspar TC, Hatfield JL, Colvin TS and Cambardella CA 2002. Nitrogen management strategies to reduce leaching in tile-drained Mid western soils. *Agronomy Journal* 94: 153-171.

Giles J 2005. Nitrogen study fertilizers fears of pollution. *Nature* 433: 791.

Guertal EA 2000. Preplant slow-release nitrogen fertilizers produce similar bell pepper yields as split applications of soluble fertilizer. *Agronomy Journal* 92: 388-393.

Hyman MR, and Wood PM 1985 Suicidal inactivation and labeling of ammonia mono-oxygenase by acetylene. *Biochemistry J.* 227: 719–725.

Keener WK, and Arp DJ 1993. Kinetic studies of ammonia monooxygenase inhibition in *Nitrosomonas europaea* by hydrocarbons and halogenated hydrocarbons in an optimized whole-cell assay. *Appl. Environ. Microbiol.* 59: 2501–2510.

Kowalchuk GA and Stephen JR 2001. Ammonia-oxidizing bacteria: a model for molecular microbial ecology. *Annual Reviews of Microbiology* 55: 485-529.

Kroeze C 1994. Nitrous oxide and global warming. *The Science of the Total Environment* 143: 193-209.

Less H and Simpson JR 1957. The biochemistry of nitrifying organisms. 5. Nitrite oxidation by *Nitrobacter. Biochemistry* 65: 297-305.

Less H 1952. The biochemistry of the nitrifying organisms. 1. The ammonia- Oxidizing systems of *Nitrosomonas. Biochem. J.* 52: 134–139.

Logan MSP and Hooper AB 1995. Suicide inactivation of hydroxylamine oxidoreductase of *Nitrosomonas europaea* by organohydrazines. *Biochemistry* 34: 9257–9264.

Logan MSP, Balny C and Hooper AB 1995. Reaction with cyanide of hydroxylamine oxidoreductase of *Nitrosomonas europaea. Biochemistry* 34: 9028–9037.

Malakoff D 1988. Death by suffocation in the Gulf of Mexico. *Science* 281: 190-192.

Malhi SS and McGill WB 1982. Nitrification in three Alberta soils: effect of temperature, moisture and substrate concentration. *Soil Biology and Biochemistry* 14: 393-399.

Malhi SS and Nyborg M 1988. Control of nitrification of fertilizer nitrogen: effect of inhibitors, banding and nesting. *Plant and Soil* 107: 245-250.

Nyborg M and Malhi SS 1992. Effectiveness of fall versus spring-applied urea on barley. *Fertilizer Research* 31: 235-239.

Oertli JJ (1980) Controlled-release fertilizers. *Fertilizer Research* 1: 103-123.

Poth M and Focht DD 1985. 15N kinetic analysis of N_2O production by *Nitrosomonas europaea*: an examination of nitrifier denitrification. *Applied and Environmental Microbiology* 49: 1134-1141.

Poth M 1986. Dinitrogen production from nitrite by a *Nitrosomonas* isolate. *Applied Environmental Microbiology* 52: 957-959.

Sahrawat KL 1982. Nitrification in some tropical soils. *Plant and Soil* 65: 281-286

Salsac L, Chaillou S, Morot-Gaudry J and Lesaint C 1987. Nitrate and ammonium nutrition in plants. *Plant Physiol. Biochem.* 25: 805–812.

Slangen JHG and Kerkhoff P 1984. Nitrification inhibitors in agriculture and horticulture: *A literature review. Fertilizer Research* 5: 1-76.

Strong DT, Sale PWG and Helyar KR 1999. The influence of soil matrix on Nitrogen mineralization and nitrification. IV. Texture. *Australian Journal of Soil Research* 37: 329–344.

Strong WM and Cooper JE 1992. Application of anhydrous ammonia or urea during the fallow period for winter cereals on the darling downs, Queensland. I. Effect of time of application on soil mineral N at Sowing. *Australian Journal of Soil Research* 30: 695-709.

Subbarao GV, Ito O, Sahrawat KL, Berry WL, Nakahara K, Ishikawa T, Watanabe T, Suenaga K, Rondon M and Rao IM 2006. Scope and strategies for regulation of nitrification in agricultural systems-challenges and opportunities. *Critical Reviews in Plant Sciences* 25: 303-335.

Terry K and Hooper AB 1981. Hydroxylamine oxidoreductase: a 20-heme, 200,000 molecular weight cytochrome c with unusual denaturation properties which forms a 63,000 molecular weight monomer after heme removal. *Biochemistry* 20: 7026

Tiessen KHD, Flaten DN, Bullock PR, Grant CA, Karamanos RE, Burton DL and Entz MH 2008. Interactive effects of landscape position and time of application on the response of spring wheat to fall-banded urea. *Agronomy Journal* 100: 557-563.

Vitousek PM, Aber J, Howarth RW, Likens GE, Matson PA, Schindler DW, Schlesinger WH and Tilman DG 1997. Human alteration of the global nitrogen cycle: Sources and consequences. *Ecol. Appl.* 7: 737–750.

Vitousek PM, Aber J, Howarth RW, Likens GE, Matson PA, Schindler DW, Schlesinger WH and Tilman GD 2000. Human alteration of the global nitrogen cycle: causes and consequences. *Issues in Ecology* pg: 1-16 (http://esa.sdsc.edu/tilman.htm).

Webster EA and Hopkins DW 1996. Contributions from different microbial processes to N_2O emission from soil under different moisture regimes. *Biology and Fertility of Soils* 22: 331-335.

Wild HE, Sawyer CN and McMahon TC 1971. Factors affecting nitrification kinetics. *Water Pollution Control Federation, Annual Conference Issue* 43: 1845-1854.

Winely CL and San Clemente CL 1970. Effects of pesticides on nitrite oxidation by *Nitrobacter agilis. Appl. Microbiol.* 19: 214–219.

Wrage N, Velthof GL, Beusichem ML and Oenema O 2001. Role of nitrifier denitrification in the production of nitrous oxide. *Soil Biology and Biochemistry* 33: 1723-1732.

Wrage N, Velthof GL, Laanbroek HJ and Oenema O 2004. Nitrous oxide production in grassland soils: assessing the contribution of nitrifier denitrification *Soil Biology and Biochemistry* 36: 229-236.

Yuan F, Ran W, Shen Q, and Wang D 2005. Characterization of nitrifying Bacteria communities of soils from different ecological regions of China by molecular and conventional methods. *Biol. Fertil. Soils.* 41: 22–27.

2015, Impact of Global Warming and Climate Change on
    Human and Plant Health   *Pages 163–168*
Editors: Dr. Arun Arya and Prof. V.S. Patel
Published by: DAYA PUBLISHING HOUSE, NEW DELHI

Chapter 15

Role of Horticulture in Mitigating Climate Change

M.K. Yadav[1]*, N.L. Patel[2], Ankita Hazarika[3] and P.R. Singh[2]

[1]*N.M. College of Agriculture,*
[2]*ASPEE College of Horticulture and Forestry,*
Navsari Agricultural University, Dandi Road, Navsari , Gujarat, India
[3]*College of Agriculture, Assam Agricultural University, Jorhat, Assam, India*

ABSTRACT

The prediction of climate change due to human activities began with Svante Arrhenius, a Swedish chemist in 1896. He realized that the industrial revolution that had started and carbon dioxide released was increasing as world's consumption of fossil fuel increased more rapidly. Modern intensive horticulture is also a source it emit greeenhouse gases, though emission of GHG's from horticulture is very low. To keep pace with ever increasing population an increase in food grain production is a must. Significant increase in food grain production may lead to increased GHG's emission. Increased carbon dioxide may enhance crop yields. Certain agricultural and horticultural practices like Micro-irrigation more popularly known s a drip irrigation, use of mulches to protect soil moisture, application of farm yard manure or vermicompost for improving soil heath, water holding

---

* *Corresponding author.* E-mail: manoj_kadali@hotmail.com

capacity of the soils, use of drought tolerant rootstock for horticultural crops and Use of growth regulators may reduce the emission of greenhouse gases and hence retard the climate change process. Various such practices are described in detail.

Keywords: *Horticulture, Mitigation, Greenhouse gases, Climate change.*

I5.1 Introduction

Horticulture contributes to climate change through their influence on the carbon cycle. It is stored large quantity of carbon in vegetation and soil exchange carbon with the atmosphere through photosynthesis and respiration or sources of atmospheric carbon when they are distributed become atmospheric carbon sinks during regrowth after disturbance and can be manage to after true role in the carbon cycle. Mitigating options by the horticulture sector including extending carbon retention in harvested products, substitution and production of biomass for bioenergy, aggressive adaptation of improved cultural practices for becoming a significant net CO_2 reduction from the atmosphere in the future and thus contributing to reduction in climate change. Management stra

There are two main causes for climate change

1. Natural Causes

Natural fluctuation in the sun's intensity.

Volcanic eruption – injecting sulphur dioxide to the upper atmosphere which is highly reflective to sunlight.

Short term cycle like *EL Nino*.

2. Atmospheric Causes

Increase in greenhouse gases through modern intensive horticulture.

Land use changes including deforestation and urbanization.

The climate all over the world has changed the performance of the crop, wild plant livestock and aquatic resources. The impact of global climate change on horticulture can vary over time and across locations dependent on different agro ecologies, farming system, production condition and even a particular plant species.

For the present purpose climate can be considered as a process of global warming, in part attributed to the greenhouse gasses generated by human activity. Greeenhouse gasses are essential to maintain earth temperature at about 33°C without these gases the earth will be as cold as to be uninhabited. Excessive emission of these gases has raised the earth's average temperature in last century by 5°C.Without any curb on the emission of greenhouse gases the average rise in temperature is expected to be between 2-6°C by the end of this century.

15.2 Greenhouse Gases

The main greeenhouse gases (GHG's) in the atmosphere comprise of water vapours, Carbon dioxide, Methane, Ozone and Halocarbons.

Table 15.1: Greenhouse Gases Emission Scenario (Agrawal, 2001)

Greeenhouse Gases	Pre industrial (1750 -1850) Atmosphere Concentration (ppm)	Current Rate of Change in GW (per cent)
Carbon dioxide	280	1.80
Methane	0.8	0.015
CFC-11	0.0	9.5
CFC-12	0.0	17.5
Nitrous oxide	288	0.8

Table 15.2: Major Greeenhouse Gases Causing Global Warming (Bazzaz and Sombroek, 1996)

Gases	Atmospheric Concentration (ppm)	Annual Increase in per cent	Contribution in GW
CO_2	351.3	0.4	55
CFCs	0.000225	5.0	25
CH_4	1.675	1.0	11
Other	–	–	4

GHG's trap sun's heat near the earth's surface and keep the earth warm acting as a natural insulation. If there is no such insulation, sun's heat would escape and normal temperature of earth may go deistical up to -18°C making earth so cold to support any diversity of life (Bharadwaj *et al.,* 2009). The activity of human being's are responsible for changing the concentration of naturally occurring GHG's in the environments.

15.3 Greeenhouse Effect and Global Warming

Sun light pass through the atmosphere warming the earth's surface. In turn the land and ocean release heat of infrared radiation in to the atmosphere balancing the incoming energy. Water vapour, carbon dioxide and such other naturally occurring gases can absorb part of this radiation allowing it to warm. This absorption of heat which keeps the surface of the earth warm enough to sustain our life is called the natural greeenhouse effect. On the other hand, higher concentration of greeenhouse gases in the earth's atmosphere will lead to increased tapping of infra red radiation. The lower atmosphere is likely to warm up, changing the weather and climate. Thus the enhanced greenhouse effect is addition to the natural greeenhouse effect and is due to human activities changing the gas composition of the atmosphere. The enhanced greenhouse effect is often referred to as a global warming.

15.4 Horticultural Management Strategies

Some of the management strategies to overcome fluctuating weather effect under different agroecosystem have been suggested by Bharadwaj *et al.* (2009). Most of the perennial horticultural crops are input intensive and cultivated under irrigated condition. Therefore, irrigation is one of most important input for successful production of quality fruits. A number of technologies have been developed to minimize the impact of water stress up to some extent.

Popularization of micro irrigation system: Micro-irrigation more popularly known s a drip irrigation is one of the very important strategies in reducing the impact of drought. The main feature of micro irrigation is higher level of efficiency in water used coupled with higher yield.

Use of mulches: Soil moisture could be conserved through mulching either with black polythene or locally available mulches; ground covers crops or inter culturing in the orchard to check soil erosion and runoff rain water. Application of farm yard manure or vermicompost for improving soil heath, water holding capacity of the soils is another way to efficiently use and conserved the water is available limited quantity. Area under plant if remain covered with mulches (100 micron) throughout the growth period help in conserving solid moisture, saving water for critical stage during summer and reducing the weed population by 60 per cent with grass mulch and 200 per cent with polythene mulch. This reduces the cost of cultivation. The polythene mulch maintained 29 per cent more soil moisture as compared to un-mulched tree on soil available water content basin.

Use of drought tolerant rootstock for horticultural crops: Root stock provide attractive and environmentally sound protection to the vine and grape under acute shortage of quality water for irrigation and uncertainty of rains.

Use of shade net in nursery: Maintaining the nursery plants under shade net and irrigation using micro jet help in increasing the survival percentage, reducing the irrigation requirement which can later on be used for more frequent irrigation as compared with traditional nursery practices.

Rain water harvesting: Rain water harvesting is another important technique, which compromise of in situ water harvesting by opening small tranches, contour bonding etc. or collecting the rain water in the lower region of the farm of recycling it by providing critical irrigation to crops through micro –irrigation systems.

Use of growth regulators: Certain growth regulating chemicals can also help the plant to grow with limited water spread and increase growth of the roots, which determines the efficiency of absorbing surface after water uptake. In crops like ber it has shown distinct proportional effect of KNO_3 on root growth and development. Use of CCC in grape limits the vigour and indirectly helps in reduced water use.

Selection of variety: In annual crop during the drought period or heavy rains, it is more appropriate to use appropriate variety resistant to water stress or water logging as case may be.

15.5 Mitigation Measures Against Climate Change

15.5.1 Mitigation Measures at Domestic Level

Expending forest protects the existing forest and encourages Afforestation.

Create awareness among people about how the changes in human life pattern and modernization affect the ecology and human life.

Less exploitation, disturb nature only at minimum level.

Renewable energy popularize the non-conventional energy sources by biogas and utilized them to minimum reduce dependence on petrofuels by increasing use of renewable energy.

Approach those projects which are ecologically sound and feasible to reduced GHG emissions.

Analyse regionally the atmospheric changes due to greeenhouse effects and try to solve them accordingly.

Minimize the use of chemicals in horticulture and popularize organic farming.

Crop varieties and date of planting according to environment may be adopted.

15.5.2 Mitigation Measures at International Level

Stabilization of the concentration of greenhouse gasses (GHG's) in the earth's atmosphere is the need of hour. According to IPCC's reports the costs of mitigation to achieve stabilization of the atmosphere at 445ppm of CO_2 equivalent by 2030 would amount to only 3 per cent of global GDP. Companies will have to develop technologies that are low in carbon and GHG's intensity.

The IPCC report calls for development of renewable or carbon free technologies that one is currently available or will be commercializing in coming decade. The reduction of GHGs was discussed in Kyoto at the end of 1997. It was notable that a legal binding protocol the so called Kyoto protocol was adapted in the protocol the target reduction in GHGs in 38 developed countries were explicitly set to reduce emission by 5.2 per cent below 1990 level between 2008- 2012.

The UN report call for developed countries to reduce CO_2 emission by 30 per cent by 2050 and developing countries by 20 per cent from 1990 levels. However, the global warming is not going to be selective. It will affect the countries near the equator far more those in higher latitudes. Emission of CO_2 in the world is going up by over 3 per cent per year. To create climate impact insurance found for low latitude countries. Countries could apply for relief from the fund whenever they suffer a climate impact. A final alternative is to compensate low latitude countries by investing in their economic development. United nation frame work convention (UNFCCC) held on climate change, suggested to provide some compensation to help poor countries mitigate emission. One possibility is that an international fund such as a Global Environmental Facility (GEF) could subsidize adaptation. For example the GEF could provide poor nation with financial and technical support for joining public adaptation such as water project, costal protection of endangered sepses. A well designed economic development programme would bring large benefit directly to people of countries. International development institute such as the World Bank could administer such developmental programme.

The international community must now address both climate change and development as single issue. Technology transfer is required so that energy production is more efficient. At the same time consideration of global warming is a major challenge and its matter of time that attention to climate change should be first priority throughout the world.

15.6 Conclusions

Environmental changes occur mainly due to human activities.

Horticultural scenario is surely going to change due to effect on photosynthesis, input use efficiency, pest population dynamics, land use and land cover changes etc. Change in climate will have several implications in horticulture *i.e.* change in the optimal growing area of particular crops. Shift in the geographical area of forest and natural ecosystems. There is need for intensive research in this regards, so that sustainability of horticulture can be maintained. Nature being our mother, it is duty of everyone to protect Mother Nature.

References

Agrawal, P.K. 2001. Climate changes and Indian Agriculture Report, pp. 26-32.

Bharadwaj, V., Mishra, A. K. and Singh, S.K. 2009. Climate change and management of weather for sustainable agriculture. Agrotech Publishing Academy,Udaipur, India.

Bazzaz. F.and Sombroke, W. 1996. Global Climate Change and Agriculture Production. FAO (pub) pp.141-171.

IPCC (Intergovernmental Panel on Climate Change Working Group I) 2007. Climate change 2007: The Physical Science Basis IPPCC working group I.

2015, **Impact of Global Warming and Climate Change on Human and Plant Health**

Pages 169–178

Editors: **Dr. Arun Arya and Prof. V.S. Patel**
Published by: **DAYA PUBLISHING HOUSE, NEW DELHI**

Chapter 16

Zinc Remediation by Maize and Cowpea Plants using AM Fungi

Darshini Trivedi and Arun Arya*

Botany Department, Faculty of Science,
The Maharaja Sayajirao University of Baroda, Vadodara – 390 002, India

ABSTRACT

Zinc is one of the essential micronutrients required at low concentrations for normal growth and development of plants and is essential for several metabolic processes. In higher concentration it becomes toxic which causes symptoms like stunted growth, intervenal chlorosis in young leaves and the roots become brownish and necrotic. Phytoremediation is a technique to reduce the concentration of toxic metals to sustain better plant life. Mycorrhizal fungi contribute directly to plant establishment in soils contaminated with heavy metals by binding metals to complexes which make the plant better suited to resist the toxic levels.

Zinc is produced as a by product by a unit of Transpek-silox Industry. An effort was made to assess the performance of two crop plants-Cowpea and Maize in soil containing different concentrations of Zinc. Out of the three concentrations *i.e.* 100,500 and 1000 ppm tried, the maximum growth was obtained in AM inoculated plants at 1000 ppm, thus AM can attenuate toxicity of metals by increasing plant growth. The performance of cowpea plants was better than maize plants. Use of cowpea plant can be suggested for remediation of zinc from soil.

Keywords: *Remediation, Zinc, AM fungi, Maize, Cow pea.*

* *Corresponding author.* E-mail: aryaarunarya@rediffmail.com

16.1 Introduction

Mycorrhizae show mutualistic symbiosis between soil fungi and plant root system. AM focus the most common and widely occurring of all the mycorrhizal associations. AM are thought to be ecologically important to most vascular plants are harboring mycorrhizal fungi as an integral and normal component of their root systems and benefited from it. AM are characterized by the formation of unique structures such as vesicles and arbuscules.

AM fungi is known to increase mobilization and transfer of nutrients like phosphorus along with Zinc(Zn), Aluminum(Al), Iron(Fe) etc.AM fungi improves plant growth through increased uptake of P,Zn etc.They are known to increase nodulation and nitrogen fixation in legumes.(Kumar and Jalali). With the development of industries, mining activities, application of wastewater and sewage sludge on land, heavy metal pollution of soils is increasingly becoming a serious environmental problem.

Phytoremediation of soils contaminated by heavy metals has been widely accepted as a cost effective and environmental friendly clean up technology. Complex interactions between roots, microorganisms and fauns in the rhizophere have a fundamental effect on metal uptake and plant growth.(Saleh M. and Saleh AL-Garni).

The present paper is an effort to access the uptake of Zn in different levels of soil by two plants. The paper reports the influence of AM fungi on plant height, root and shoot dry weight, mycorrhizal colonization of cowpea and maize. The association of AM fungi with higher plants has been found to result in higher Zn Absorption than their non-mycorrhizal one.

16.2 Materials and Methods

Soil

Soil taken in pots was a mixture of sand, soil (sandy loamy), cowdung and some amount of leaf litter.

Plants Selected

    ☆ Maize (*Zea mays*)

    ☆ Cowpea (*Vigna unguiculata*)

AM Inoculum (Mixed Consortium)

Brought from K.C.P Sugar Industries, Andhra Pradesh.

Zinc Sample

Taken from Transpek Silox Industries, Atladra, Vadodara.

Growth Conditions

Pot culture experiment was conducted with sandy loamy soil (garden soil), in which few pots were taken as control *i.e.* Without AM and Zinc and in other pots Zinc was incorporated in soil at different concentrations *i.e.* 100, 500 and 1000 ppm.10 replicates were taken for each treatment.

Plate 16.1: Photographs Showing Growth of Maize and Cowpea Plants in Soil Containing Zn.

Plate 16.2: Photographs Showing Growth of Maize and Cowpea Plants in Soil Containing Zn.

AM mixture was added by making holes at 5 different sites in each treatment. 5 seeds per pot were sown *i.e.* both cowpea and maize. Pots were irrigated or watered regularly (until the seeds germinated). Plants in pots were allowed to grow under greeenhouse condition. After that seedlings emerge and attains a height and then after 60 and 90 days. Plant samples were collected washed thorough with water for taking necessary analyses.

Soil Characteristics

Soil pH was recorded using digital pH meter by taking 1:5 (soil: water) and Soil conductivity was measured with help of conductivity meter.

Growth Parameters

Plant height (cm) *i.e.* Shoot and root length (cm), Fresh and dry weight of root and shoot of both the plants were measured.

Mycorrhizal Root Infection

Percentage root colonization was assessed by using root clearing and staining method (Phillips and Hayman, 1970).

Spore Isolation

Spores were isolated by Wet sieving and decanting method as suggested by Gerdemann and Nicolson 1963.

Determination of Zinc

Zinc concentration was analysed using Atomic Absorption Spectrophotometry. During the experiment the soil samples were analysed by Vaibhav Laboratories, Ahmadabad.

16.3 Results

In cowpea fresh weight was highest at Zinc concentration of 100 ppm after 60 days while after 90 days, it was maximum at 1000 ppm of Zinc concentration in AM inoculated plants as compared to non-inoculated plants (Table 16.1). Studies conducted by Al-Garni (2006) on cowpea plants revealed that dry weight of plants increased up to 200 ppm. However, in comparison to non inoculated plants the dry weight was more in inoculated over up to 1000 ppm.

In studies dealing with Maize the dry weight of maize plants increased up to 90 days. Maximum dry weight was obtained at Zinc concentration of 1000 ppm in AM inoculated plants as compared to non-inoculated plants (Table 16.1).

16.4 Effect of AM Fungi on Fresh Weight and Dry Weight

After 60 days at 100 and 1000 ppm conc. of Zn plant growth was more. The results showed more growth in 60 days as compared to 90 days.

16.5 Effect of AM on Plant Height

In Cowpea the maximum height of plants was obtained in plants inoculated with AM Fungi and 100 ppm and 1000 ppm of Zinc compared to non-inoculated plants at same concentration of Zinc (Table 16.3).

Table 16.1: Effect of AM Fungi on Biomass at different Concentration of Zinc in Maize and Cowpea Plant

Plants	Treatment	Fresh Weight (g)				Dry Weight (g)			
		60 Days		90 Days		60 Daya		90 Daya	
		Shoot	Root	Shoot	Root	Shoot	Root	Shoot	Root
	Control	51.7	10.6	14.21	4.56	5.461	1.167	5	1
Maize	M-100	26.3	11.8	17.65	2.101	2.649	0.657	3.95	0.464
	M-500	20.3	5.83	16.5	3.212	2.73	0.81	3	1
	M-1000	66.4	10.21	20.5	2.689	5.55	0.71	12.18	1.701
	Control	24.8	0.858	12.4	0.25	3.863	0.247	1.798	0.063
Cowpea	C-100	39.9	1.085	21.7	0.548	4.953	0.326	4.394	0.085
	C-500	10	0.389	14.9	0.373	1.305	0.098	1.958	0.112
	C-1000	21.4	0.694	27.6	1.222	3.009	0.174	4.669	0.324

Table 16.2: Estimation of Zinc in Soil of the Pots Growing Maize and Cowpea Plants after 30 and 60 Days

Plants	Treatments	Zinc Concentration (ppm)			
		0 days	30 days	60 days	90 days
	Control	140	149	116	100
Maize	M-100	250	238	181	95
Cowpea	C-100	250	170	156	85
Maize	M-500	600	579	394	310
Cowpea	C-500	600	545	431	350
Maize	M-1000	1110	1087	759	700
Cowpea	C-1000	1110	721	563	460

Table 16.3: Effect of AM on Plant Height at different Zinc Concentration in Cowpea

Treatment	Plant Height of Cowpea			
	60 Days		90 Days	
	Shoot	Root	Shoot	Root
Control	106.5	8.0	118.5	9.4
C-100	162.0	9.0	143.2	7.0
C-500	43.25	5.5	88.0	8.0
C-1000	105.5	9.0	164.0	13.8

C-100: 100 ppm; C-500: 500 ppm; C-1000: 1000 ppm.

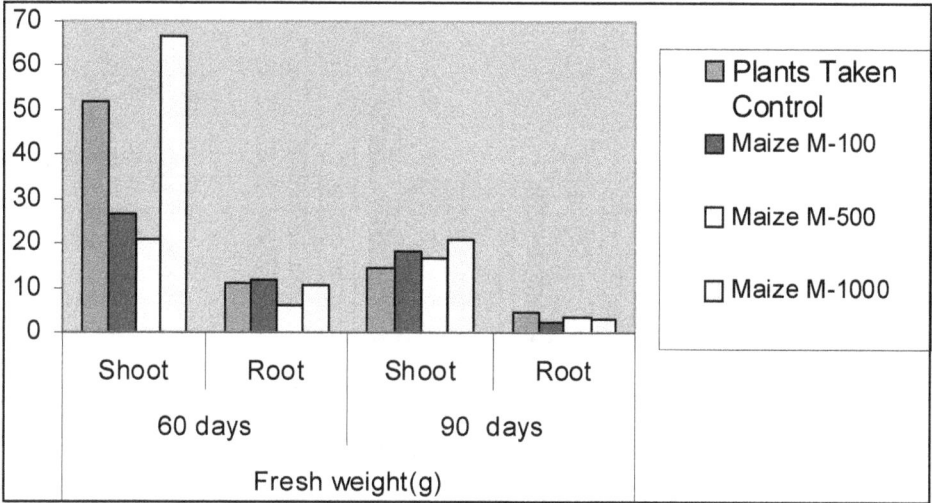

Figure 16.1: Histogram Showing Effect of AM Fungi on Biomass in Maize.

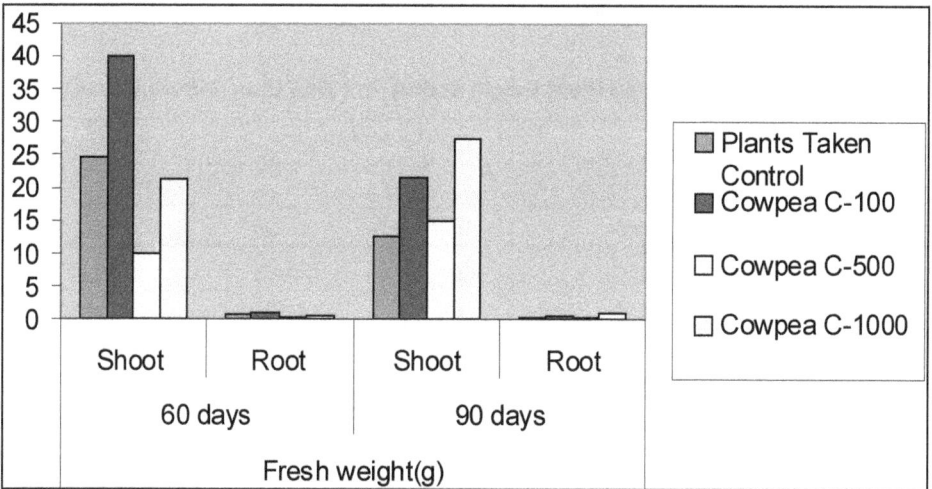

Figure 16.2: Histogram Showing Effect of AM Fungi on Biomass in Cowpea.

In Maize, the maximum height of plants was obtained in plants inoculated with AM Fungi at 1000 ppm Zinc concentration (Table 16.4) Shoot and root length was more than non-inoculated plants in another study at 200 ppm (Al-Garni, 2006).

16.7 Mycorrhizal Association

AM successfully colonized cowpea roots. Percentage root colonization was highest at 1000 ppm of Zn level on 60 day, while it decreased afterwards on 90 day. The AM infection helped in more acquisition of P and better growth of plants.

In maize AM colonized roots, which was highest at 100 ppm concentration of Zinc level which increased slightly on 60 day and then decreased after 90 days.

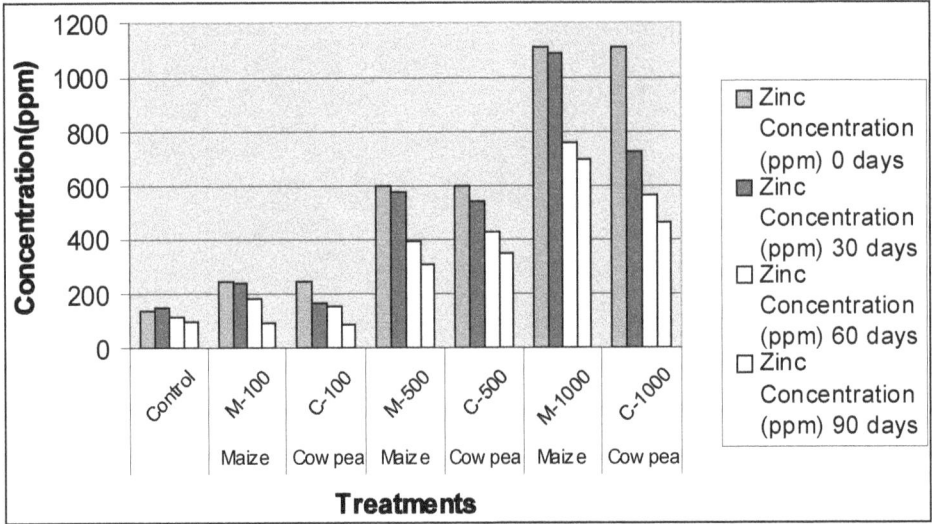

Figure 16.3: Histogram Showing Reduction in Zinc Level in Maize and Cowpea Inoculated with AM Fungi.

Table 16.4: Effect of AM on Plant Height at different Zinc Concentration in Maize

Treatment	Plant Height (cm)			
	60 Days		90 Days	
	Shoot	Root	Shoot	Root
Control	48	14.5	105.2	23.5
M-100	32.25	22.5	72	23
M-500	29.5	9	69	14
M-1000	48.5	12	109.5	24

16.8 Discussion

AM Fungi and *Rhizobium* are common beneficial microbes of leguminous plants. This studies shows that inoculation with AM Fungi can increase dry weight, plant height more than non-inoculated one at all Zinc levels. The results revealed that cowpea and maize plants accumulate higher amount of zinc in their tissues from polluted soil than that in control.

The higher heavy metal concentration in AM infected plants could be explained by the fact that AM infection increased plant uptake of metals by mechanism such as enlargement of Absorbing Area and efficient translocation (Saleh and Al Garni, 2006).

In addition although heavy metals concentration in infected plants were much higher than in noninfected one, the growth parameters of infected cowpea and maize plants were higher than that for non-infected plants grown in heavy metal polluted soil. These results suggest that bacterial –AM-legume tripartite symbiosis can be

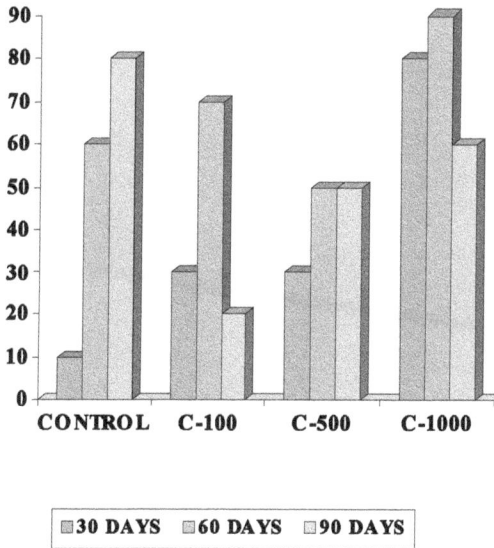

30 DAYS   60 DAYS   90 DAYS

Figure 16.4: Histogram Showing Percentage Root Colonization in Cowpea.

used for remediation of Zn from soil. Most reports note a positive effect of AM inoculation on growth of plants in metal contaminated soil. This protective benefit may be related to adsorptive or binding capability for metals of relatively large fungal biomass associated with host plant.

References

Gerdemann, J.W and Nicolson T.H. 1963. Spores of mycorrhizal Endogone species extracted from soil by wet sieving and decanting. *Transaction of the British Mycological Society* 46: 235-244.

Gautum, A. and Mahmood, I. 2002. A survey of some cultivated legume crops of Aligarh district to determine the occurrence of AM Fungi. *Mycorrhiza News.* 14(1):13-14.

Kumar, R., Gupta, P.P. and Jalali, B.L. 2001. Impact of VA-Mycorrhiza, Azotobacter and Rhizobium on growth and nutrition of cowpea. *J.Mycol.Pl.Pathol.* 31(1):38-41.

Kumar R., Jalali B. L. and Harichand 2002. Influence of AM on growth and nutrient uptake in chickpea. *J.Mycol.Pl.Patho.* 32 (1):11-15.

Phillips, J.M and D.S. Hayman 1970. Improved procedures for clearing and staining parasitic and VAM Fungi for rapid assessment of infection. *Trans.Br.Mycol.Soc.* 55 :158-161.

Saleh M and Saleh Al-Garni 2006. Increased heavy metal tolerance of cowpea plants by dual inoculation of an AM fungi and nitrogen fixer Rhizobium bacterium. *African Jour. of Biotech.* 5 (2) : 133-142.

Çelik S. Dalcý, S. Arcak, 2002. Effects Of Vesicular- Arbuscular Mycorrhizae On The Growth And Uptake Of Some Heavy Metals By Oat (Avena sativa L.).International Conference on Sustainable Land use and Management.2002. page:95 (ÇANAKKALE).

Sharma.A.K, Srivastava P.C. and B.N. Johri. Multiphasic zinc uptake system in mycorrhizal and nonmycorrhizal roots of French bean (*Phaseolus vulgaris* l.) *Curr.Sci.* 76: 228-230.

Susan, I. 2000. *Fungal-Plant Interactions* John Wiley and Sons LTD.

Sharma A.K. and B.N.Johri, B.N. 2002. *Arbuscular Mycorrhizae : Interactions in plants, rhizosphere, and soils* Science Publishers, 309.

2015, Impact of Global Warming and Climate Change on
     Human and Plant Health
Editors: **Dr. Arun Arya and Prof. V.S. Patel**
Published by: **DAYA PUBLISHING HOUSE, NEW DELHI**

*Pages **179–182***

Chapter 17
Environmental Effects on *Rhizobium* Cells: Ultrastructural Studies

Sheuli Dasgupta*

*Department of Microbiology,
Gurudas College, Kolkata – 700054, India*

ABSTRACT

Molecular nitrogen has strong triple covalent bond which makes it unavailable for the use by higher plants. Leguminous plants have the capability to fix the nitrogen with the help of *Rhizobium.* Nitrogen is assimilated in the form of nitrate or ammonia. To observe effect of nitrogenous fertilizers on *Rhizobium* cells the EM studies were conducted. The use of nitrogenous fertilizer caused opaque protoplasmic remnants in the cells. Use of increased fertilizer may reduce the yield of leguminous plant and may lead to global warming.

Keywords*: Leguminous crops, Rhizobium, Nitrogen fixation, Electron microscopy.*

17.1 Introduction

Nitrogen is essential component of many chemicals present in the living cells. Nitrogen can be found in many forms in the biosphere. Molecular nitrogen (N_2) forms 78 per cent of that earth's atmosphere by volume; however, this reservoir of nitrogen is not directly available to plants. Nitrogen has a strong triple covalent bond, and the higher plants do not have the ability to break this bond directly. Plants instead

---

* E-mail: sheulidasgupta@yahoo.co.in

assimilate nitrogen in the form of nitrate or ammonia. Nitrogen is a macronutrient required by plants. Nitrogen deficit in plants will lead to stunted growth. The plants which fix atmospheric nitrogen are called legumes. *Rhizobium,* the nitrogen fixing bacteria fix atmospheric nitrogen in symbiosis with the leguminous plants. To curtail down the dependence on nitrogenous fertilizers, it is advisable to produce more leguminous crops. *Rhizobium* in association with these leguminous crops maintains the nitrogen balance. Beans, peas, lentils and other pulses are good for our consumption because they are rich in protein. If leguminous plants flower and produce seeds then most of the fixed nitrogen is tied up in the plant. Some of it we eat, the rest ends up in compost.

17.2 *Rhizobium*: Entry of Symbiont in Roots

The most important organisms that can fix nitrogen are the prokayotes such as eubacteria and cyanobacteria either in free living form or in symbiotic association with plants. Legumes including peas, lentils and alfalfa can form symbiotic associations for nitrogen fixation with a soil bacterium called *Rhizobium*. For such a symbiotic association to occur, the bacteria and the root hairs of the plant must first recognize each other. This recognition is quite specific and such recognition is mediated by flavonoids which are produced by the roots of host plant. These flavonoids influence a series of genes in *Rhizobium* known as the nod genes, the genes involved in infection and nodule formation. The Rhizobia then binds to the surface of the root hair. The specific binding of the bacteria to the roots trigger a complex chain of events as a result of which the bacteria enters the root cells via an infection thread and stimulate the host cortical cells to divide and form a large root nodule.

A symbiotic phenotype of inhibited or delayed recruitment of *Rhizobium* bacteria to host plant roots, produced fewer root nodules showed lower rates of nitrogenase activity, and a reduction in overall plant yield at time of harvest. The environmental consequences of synthetic chemicals compromising symbiotic nitrogen fixation are increased dependence on synthetic nitrogenous fertilizer, reduced soil fertility, and unsustainable long-term crop yields.

In addition, when the crop is harvested and the plant is cut down to ground level, the root nodules should release all the valuable fixed nitrogen for following crops. The relationship between the legume plant and the nitrogen fixing *Rhizobium* is highly specialized. The *Rhizobium* strain suitable for a certain plant is usually found naturally in the soil, where the plant originally developed. After all they have evolved together.

The establishment of the nitrogen fixing capability of two symbionts Rhizobia and legumes require an exchange of signals between the two partners. In response to flavanoids excreted by host plants, Rhizobia synthesize nod factors (NFs) which elicit at very low concentrations and in specific manners, various symbiotic responses on the roots of the legume hosts. Environmental stimuli play vital role in alteration of some signals, which affect growth, nodule formation and ultimately affect nitrogen fixation by Rhizobia. Among various types of environmental stimuli nitogenous fertilizers form a profound category of environmental stimuli.

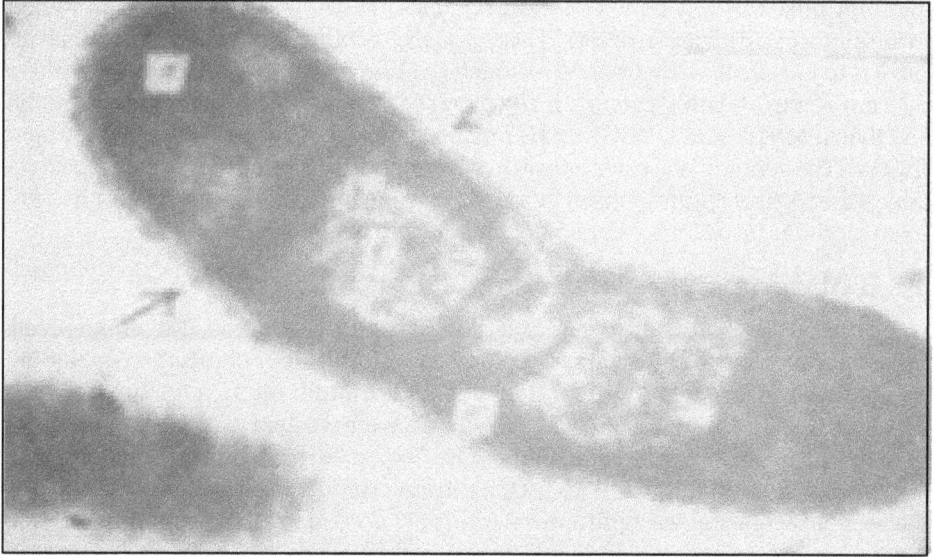

Figure 17.1: Electron Micrograph of Ultrathin Sections of Untreated Control *Rhizobium* Cells.

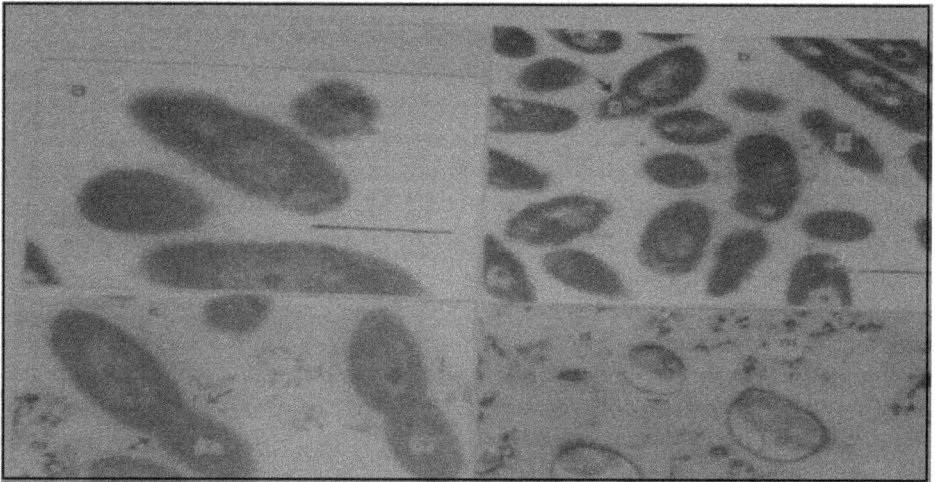

Figure 17.2: Electron Micrograph of *Rhizobium* Cells Treated with Nitrogenous Fertilizers.

Among various components present in the living system nitrogen is an important element and green plants assimilate nitrogen in various forms as nitrate as well as ammonium ions and in other forms too (Davis, 1980). However, as this is not enough for the growth of plants additional amounts of nitrogen is supplied in the form of fertilizers by crop producers. This additional nitrogen sometimes becomes even detrimental to the various soil microbial inhabitants. Plants also get some amounts of nitrogen by the compounds formed during lightning. The enormous energy of lightning breaks nitrogen molecules and enables their atoms to combine with oxygen in the air forming nitrogen oxides. These dissolve in rain, forming nitrates, which are carried to the earth. Atmospheric nitrogen fixation probably contributes some 5– 8 per cent of the total nitrogen fixed. The major conversion of nitrogen into ammonia, and then into proteins, is achieved by microorganisms in the process called nitrogen fixation. The genus *Rhizobium* appears to be numerically most important group which infects the roots of leguminous plants leading to the formation of symbiotic nitrogen-fixing nodules for accomplishment of biologic nitrogen fixation.

17.3 Materials and Methods

To observe whether nitrogenous fertilizers have any effect on *Rhizobium* sp. cells these cells were treated with different nitrogenous fertilizers. For ultrathin sectioning, washed cells were fixed in glutaraldehyde and osmium tetraoxide. Subsequently they were fixed in uranyl acetate. The fixed cells were washed in physiological saline dehydrated in alcohol and embedded in ERL-4602 medium. Ultrathin sections (500-700Å) were cut with glass knives by the ultramicrotome (Reichert Supernova) and the sections were stained with saturated uranyl acetate in 70 per cent alcohol, lead acetate and were observed under electron microscope (Hitachi, 600).

17.4 Results and Discussion

However, on constant application of nitrogenous fertilizers to the soil these natural *Rhizobium* inhabitants lose their structure and undergo conformational changes. The control cells appeared rod shaped with intact cells wall having electron opaque cytoplasm and electron transparent nucleoid. However, cells treated with nitrogenous fertilizers showed electron opaque protoplasmic remnants. The ultrastructural studies and electron microscopic data have revealed how structural modifications take place in these bacterial cells (Dasgupta, 2004, 2005).

Excess of nitrogen fertilizers may cause global warming. Since nitrogen is a major agricultural requirement it must be added to the crops. But the elemental nitrogen volatizes in the form of nitrous oxide in the air which should be curtailed. This nitrous oxide which volatizes in the air causes global warming and it also causes dead zones in the world's oceans. It may, therefore, be concluded that use of nitrogen in the form of fertilizers must be restrained. Though it has several positive effects onto the plants, it also has several negative effects. A balance must be kept between the two.

References

Dasgupta S. 2004. Effect of nitrogenous fertilizers on *Rhizobium* sp. cells. *Annals of Microscopy*. 4: 96-102.

Dasgupta S. 2005. Endosulfan treated *Rhizobium* sp. cells: A study. *Annals of Microscopy*. 5: 85-92.

Davis, B.D., Dulbecco R., Eisen H.N. and Ginsberg H. S. 1980. *Microbiology* 4th *ed.* Wolters and Kluwer, pp. 1216.

2015, Impact of Global Warming and Climate Change on
Human and Plant Health
Editors: **Dr. Arun Arya and Prof. V.S. Patel**
Published by: **DAYA PUBLISHING HOUSE, NEW DELHI**

Pages 183–195

Chapter 18

Impacts of Climate Change on Agriculture in Arid and Semi-Arid Regions: A Review

Alka Awasthi

Programmes Cecodecon F
Sitapur Industrial Area, Jaipur – 302 020

ABSTRACT

Climate change is a defining challenge of the century that is affecting all sectors of the economy as well as the integrity of natural systems on which human development and existence are based. Agriculture is one of the most sensitive sectors in the context of climate change. The fact that 45 per cent of the global population makes its living through agriculture underscores the need for studying impacts of climate change on agricultural productivity and farmer incomes. In arid and semi-arid regions where agriculture is already constrained due to natural conditions, minor changes in seasons and crop productivity can destroy agricultural production systems and the natural resources on which these systems depend.

18.1 Introduction

Arid and semi-arid ecosystems are characterized by limited availability of water and a relatively low primary productivity cover of 30 per cent of the world's area. These regions are home to 20 per cent of the world population including 24 per cent of population of Africa, 23 per cent in Asia, 17 per cent in Americas and Caribbean, 11 per cent in Europe, and 6 per cent population of Australia and Oceania.

Knowledge on climate changes and their impacts are drawn from various sources. Analysis of meteorological or palaeo-meteorological data reveals trends in climate change up to the present. Computer simulation studies model the expected climate changes on the basis of select parameters and assuming different scenarios of GHG emissions and development the models show couple of future scenarios. Modelling studies draw on macro level data over long time scales and are able to generate predictions of average conditions at global or regional scales. Low confidence in regional changes and the large grid scale of most GCM models limit their direct usefulness in local impact assessments (Downing, 1992). Exposure chamber studies examine the effects of one or few parameters in controlled conditions and give only a partial picture of reality as they do not take into consideration the complexity of crop physiology under natural conditions.

Anthropological studies and structured social studies investigate climate change at much shorter time scales, typically according to memory of living populations. In these studies the emphasis is more on micro-locality and household dynamics than on global processes. The contextual interpretation and qualitative approaches of social science research are able to incorporate elements of human ecology, political economy, informal and formal resource rights, cultural aspects, risks and vulnerabilities, and in this sense are able to capture the complexities of real life decisions.

Understanding of local level impacts of climate change is also drawn from traditional knowledge of indigenous peoples and farmers' perceptions. Though indigenous knowledge systems are based on frameworks of analysis and measurement that are very different from mainstream western scientific knowledge some researchers have attempted to bridge the gap between these knowledge systems to add to our understanding of the impacts of climate change at local levels.

Drawing on both scientific and social research along with farmers observations this paper attempts to present the known manifestations of climate change and variability on agriculture based livelihoods in semi-arid and arid regions of the world.

Climate variability has been, and continues to be, the principle source of fluctuations in agricultural and food production especially in the semi-arid tropics. Climate Change is expected to further enhance the vulnerability of arid and semi-arid regions. According to the IPCC 2007/2008 many semi-arid and arid areas are particularly exposed to the impacts of climate change and are projected to suffer a decrease in water availability and quality while extreme weather events such as droughts and floods are projected to increase. For many parts of the arid regions there is an expected precipitation decrease.

Climate in a narrow sense is usually defined as average weather, or more rigorously as the statistical description in terms of mean and standard variability of its properties persisting for 30 years or more. Putting it simply climate is an average of the daily and monthly variations in meteorological phenomena. Climate Change refers to a change in the state of the climate that can be identified (*e.g.*, by using statistical tests) by changes in the mean and/or the variability of its properties, and that persists for an extended period, typically decades or longer. Climate variability refers to the actual conditions that are variations from the calculated mean. Climate variability includes extreme weather events such as droughts and floods. Climate variability is a natural feature that is now seen to be increasing as a result of anthropogenic climate change.

Experiences of climate variability and climate change have been documented from many arid and semi-arid regions of the world. Studying the impact of climate change on populations and livelihoods in the arid and semi-arid regions of Palestine. Rabi *et al.* (2004) have found that within a period of 41 years the mean annual average temperature has risen by 2.3°C. Further, mean annual average rainfall has declined by 22.4 mm and number of rainy days has decreased by 10 days within a span of 65 years (Anonymous, 2009) in the coastal region. In Tanzania (Mongi *et al.*, 2010) both farmers as well as agriculture extension workers noted increased temperatures, declining rainfall and increasing frequency of droughts, and stated that the changes had occurred within the last ten years. They felt that the climate is continuously changing and getting worse over time resulting in food shortages. Peoples' perceptions on climate variability and change in Ethiopia and Mali (Naess *et al.,* 2010) have indicated a trend of declining and more variable rainfall. The regional long term climate trends in the Sahel (Kanji *et al.,* 2006) show that overall rainfall has generally decreased, but while decreasing duration of rainy season is observed in Senegal and Northern Benin, and decline in rainfall in main rainy season but no change in duration of rainy season is seen in Mali and Niger.

According to a study conducted by CUTS International (Chatterjee and Khadka, 2011) on perceptions of farmers about climate change in Afghanistan, India, Bangladesh and Pakistan, people from all countries reported change in duration of seasons. In the case of Afghanistan winter appears to have shrunk considerably. The survey also reveals that the length of summer has increased and the season has extended forward to cover parts of rainy season. Similar weather patterns are observed in India and Pakistan as well. Farmers have reported longer summer season and shorter winter and rainy seasons. In India, summer season has expanded by more than a month. This has shortened other important seasons, especially the rainy season (studies conducted in East Rajasthan, Coastal Maharashtra, and Gangetic plain).

While investigating perceptions of climate change in different regions it is interesting to note that the year of turning point when the climate is perceived as changing varies in different regions. Usually communities in arid regions say that climate change set in earlier during the eighties. This may be because temperature and water are already limiting features for agriculture in this region, and any small change in diurnal temperatures and precipitation patterns are immediately visible and also have a significant impact on plant and animal behavior.

In India almost 53.4 per cent land area comprises arid and semi-arid regions (First NATCOM, GoI, 2004). Semi–arid region in India constitutes parts of Gujarat, northern plains and central highlands (Malwa), and the Deccan plateau. In general, during extreme deficient years of south-west monsoon over the Indian subcontinent, aridity takes over the semi-arid areas and its spatial extent continues deep down south to the peninsula. On an average, about 19 per cent of the country experiences arid conditions every year, of which 15 per cent is in northern India and 4 per cent in the peninsula (First NATCOM, GoI). Climate change is further aggravating the existing problems in both arid and semi arid regions. It is estimated that the arid and semi-arid regions are set to suffer further from water shortages. Communities in the region are particularly vulnerable to climate change not only because of their dependence on climate sensitive sectors such as agriculture, forestry etc. but also due to limited capacities to anticipate and effectively respond to climate change. Climate change projections indicate erratic rainfall and extreme weather, increase in natural hazards, and rising temperature.

Study on climate change and drought vulnerability in Gujarat (Anonymous, 2005) has revealed that the intensity of droughts has increased and return period of droughts has decreased in the last two to three decades. Traditionally, Gujarat has a drought cycle of 5 years, where in 2 years there is moderate rainfall, 2 years less rainfall, and 1 year of good rainfall. Consecutive two years of less rainfall makes it difficult for the communities to sustain their livelihoods. It becomes worse, when there are 3 consecutive of less rainfall. Statistical data show that in last two decades, the intensity of 3 years of consecutive less rainfall is increasing, and thereby creating severe drought situation.

The state of Rajasthan shows warming trends for maximum temperatures and a cooling trend for minimum temperatures. An overall increase in extreme rainfall events and their intensities during the period 1901–2000 have also been observed (Parthasarthy, 2006).

Table 18.1: Rainfall Pattern over Eastern Rajasthan (Semi-arid region)

Season	Years	Trend
Annual	1866 to 1898	Stationery but mostly above normal
	1899 to 1942	Increase
	1943 to 2006	Decrease
Winter	1866 to 1907	Increase
	1908 to 2006	Decrease
Summer	1866 to 1916	Stationery
	1917 to 1947	Decrease
	1908 to 2006	Increase
Summer Monsoon	1866 to 1898	Stationery
	1988 to 1942	Increase
	1943 to 2006	Decrease
Post Monsoon	1866 to 2006	Increase

Source: IITM 2008.

As seen in the abovementioned table that trends for the last half century show decline in annual rainfall and monsoon season rainfall, and increase in rainfall during pre- and post-monsoon seasons in East Rajasthan.

Singh (2011) studying trend of change in annual temperature from 1969 to 2009 states that in arid (west) Rajasthan and adjoining areas, the temperature, in general, is rising. Rate of increase in mean annual temperature in West Rajasthan and adjoining areas is as high as 0.28°C per decade but the change in temperature is not uniform throughout the year.

Analyzing farmers' perception on climate change through structured questionnaires in Bundi district of East Rajasthan, Dhaka *et al.,* found that according to a majority of farmers the region is getting drier with more pronounced changes in timing of rains and frequency of droughts (Dhaka *et al.,* 2010).

Participatory studies conducted by NGOs (Awasthi, 2011) in semi-arid areas of Rajasthan also reveal farmers' perceptions of climate change such as:

☆ Seasonal changes are being observed after the floods of 1982

☆ There is late onset of monsoon, less rainy days, more intense rainfall, sporadic rainfall instead of previously widespread rains, large gaps between rainy days during monsoon

☆ After monsoon temperature rises sharply and remains hot even in September; ponds dry up earlier

☆ Winter starts late, sowing delayed; winter ends earlier and it becomes warm in March

☆ Frequency of drought increased, even consecutive years of drought are now seen.

Crop water requirements were estimated from daily potential evapo-transpiration at ambient and projected air temperature by 2020, 2050, 2080 and 2100 using modified Penman-Monteith equation and then by multiplying with crop coefficients. Crop water requirements in the region varied from 308 to 411 mm for pearl millet, 244 to 332 mm for cluster bean, 217 to 296 mm for green gram, 189 to 260 mm for moth bean, 173 to 288 mm for wheat and 209 to 343 mm for mustard. Further, due to global warming, if the projected temperatures rises by 4°C, by the end of 21st century, water requirement in arid Rajasthan increases from the current level, by 12.9 per cent for pearl millet and cluster bean, 12.8 per cent for green gram, 13.2 per cent for moth bean, 17.1 per cent for wheat and 19.9 per cent for mustard. The increased crop water requirements in the region, resulted in reduction in crop growing period by 5 days for long duration crops, but the crop acreage where rainfall satisfies crop water requirements, reduced by 23.3 per cent in pearl millet, 15.2 per cent in cluster bean, 6.7 per cent in green gram, 13 per cent in moth bean. The study reveals that the impact will be more severe on *rabi* (winter) crops than *kharif* (monsoon) crops, the *rabi* crops being dependent on depleting groundwater resources in the region.

18.2 Climate Change Impacts

It is well acknowledged that current and predicted climate change can have serious implications for food security and livelihoods of people living in the arid and semi-arid regions of the world. Current impacts on agriculture, livestock and forest based livelihoods are documented from many countries. But empirical and scientific studies on impacts of climate change on livelihoods are few and deal with isolated aspects of agriculture, livestock production, groundwater availability etc. The studies do not provide sufficient understanding at farming system level and as such are not guiding the affected communities in undertaking climate change adaptation and mitigation actions for securing their livelihoods and the sustainability of their natural resources.

18.3 Impacts on Agriculture

Scientific studies on the current and projected impact of climate change in dry lands are notoriously few. Although climate change will affect different regions in different ways, for dry lands in general it is projected that climate change will lead to a decrease in water availability and quality while extreme weather events such as droughts and floods are projected to increase (IPCC 2007; Millennium Assessment Report 2005).

Agricultural production, including access to food, in many African countries and regions is projected to be severely compromised by climate variability and change. The area suitable for agriculture, the length of growing seasons and yield potential, particularly along the margins of semi-arid and arid areas, are expected to decrease. This would further adversely affect food security and exacerbate malnutrition in the continent. In some countries, yields from rain-fed agriculture could reduce by up to 50 per cent by 2020 (IPCC 2007). In South Asian region crop yields could decrease by up to 30 per cent by the mid-21st century. As three-fifths of the cropped area is rain fed, the economy of South Asia hinges critically on the annual success of the monsoon. In the event of rainfall failure, the worst affected are the landless and small and marginal farmers whose sole source of income, is from agriculture and allied activities.

In Ethiopia (Mongi *et al.,* 2010) except in drought years, changes in the seasonality, distribution and regularity of rainfall are more of a concern than the overall amount of rainfall. The main rainy season is also seen as becoming progressively shorter – it now starts later and finishes earlier than it used to – and the rains in general are becoming more unpredictable. Ten years ago, rains would normally start in the middle of September and continue until December. Now, it is more common for rains to start in October and continue for a maximum of one month, sometimes lasting for less than 20 days. As a result, farmers are increasingly unsure of receiving enough rain to justify planting staple food crops.

In Palestine (Rabi *et al.,* 2004) – areas previously cultivated with wheat and barley (which also used as fodder) are diminishing leading to further unemployment and increasing costs of raising livestock.

In India, nearly two thirds of the population is rural, whose dependence on climate-sensitive natural resources is very high. Its rural populations depend largely on the agriculture sector, followed by forests and fisheries for their livelihood. Indian agriculture is monsoon dependent, with over 60 per cent of the crop area under rain-fed agriculture that is highly vulnerable to climate variability and change (First NATCOM, GoI, 2004).

Most of the simulation studies have shown a decrease in the duration and yield of crops as temperature increased in different parts of India (Aggarwal *et al.,* 2001). The magnitude of this response varied with crop, region and climate change scenario.

With increase in temperature (by about 2 - 4°C) wheat and rice potential grain yields would reduce in most places. However, if the temperature increases are higher, western India may experience some negative effect on productivity due to reduced crop durations.

Using a combination of historical datasets and regional climate model projections under A2 scenario of Intergovernmental Panel on Climate Change (IPCC) Chaudhari *et al.,* 2014 have studied the impact of climate change on major food crops of India. According to the authors under the A2 scenario during the period 2020-2100, reduction in wheat yields will be 21 per cent in East Rajasthan followed by 18 per cent in West Rajasthan and 14 per cent in East Madhya Pradesh. The climate change scenario may lead up to 39 per cent reduction in rapeseed-mustard and 19 per cent reduction in potato yields. However, the yield change projection uncertainties were large due to the uncertainties associated with the yield model.

The impact of projected climate change by 21st century on water requirements of rain-fed monsoon and irrigated winter crops of arid Rajasthan has been studied by Rao and Poonia (2011).

Singh (2011) studying trend of change in annual temperature from 1969 to 2009 states that in arid western Rajasthan and adjoining areas, the temperature, in general, is rising. The rate of increase in mean annual temperature in western Rajasthan and adjoining areas is as high as 0.28°C per decade. He further states that the change in temperature is not uniform throughout the year and that increasing trend of temperature during February and March may have adverse effect on productivity of winter crops especially in west Rajasthan and adjoining plains of Punjab. In March 2004 the heat wave 40⁺ Celsius temperature reduced the wheat yield by 3.87 per cent in Rajasthan. The dominant *kharif* crop of the Thar Desert pearl millet, which is cultivated during the summer monsoon, is very well adapted to the harsh climate, including drought at various stages of growth. It is a staple food for the vast majority of rural masses, and provides fodder to the large livestock population. Singh has conducted simulation models of impact of temperature rise of pearl millet crop. Pearl millet has a good tillering habit due to which the plant can adjust well to the available soil moisture and nutrients, and can withstand long drought to resume rapid growth upon termination of drought. It can also extract moisture from deeper soil layers and has high moisture utilization efficiency. Its excellent photosynthetic mechanism ensures efficient conversion of solar energy into grain and Stover. Despite these qualities, its productivity is likely to be affected under the climate change scenario as shown by the CERES Millet (crop growth model). Under present conditions crop takes 44-46 days to reach flowering stage. In Western Rajasthan rise in temperature not only influenced crop yield but also its phenology. With each degree C of rise in temperature the days taken to flowering reduced by 1.5 days. Crop matured by 2.5 days earlier with each degree of temperature rise as compared to the present maturity period of 72 to 75 days.

Participatory studies with farmers in semi-arid regions of east Rajasthan have revealed that due to high uncertainty of rainfall and declining soil fertility area under gram, barley and wheat has been greatly reduced. Ground nut crop acreage has been reduced due to very poor rainfall in the region that has created a lot of pests resulting in low yield and very poor returns. Due to late onset of winter the crop sowing period has been delayed in mustard by about 15 days and gram by 10 days.

18.4 Impacts on Livestock

In semi-arid areas livestock rearing is the second most import livelihood, but becomes most important for income and food security in times of drought (which is frequent) and more so under the current situation of climatic variability. Being a ready source of cash and means of meeting daily household nutrition needs livestock are the best insurance against the vagaries of nature due to drought, famine and other natural calamities. Studies from arid and semi-arid regions of India have shown that livestock provide stability to family income (Anonymous, 2007). Livestock assets provide security against external shocks as livestock can be sold easily in times of stress for immediate income. For marginal and sub-marginal farmers who are net food buyers the ownership of a single animal is crucial for resisting hunger and poverty (Singh *et al.*, 2002). In general crop-livestock mixed systems make households less vulnerable to climate risks. Though livestock is relatively robust to multiple stresses, climate variability and emerging climate change in semi-arid regions pose considerable risks to livestock systems.

Throughout the GHA (Greater Horn of Africa), livestock production is a vital component of the economy. In Eritrea and Sudan, for example, 57 to 62 per cent of the agricultural GDP respectively is coming from livestock. In Somalia this share goes up to 88 per cent. For poor farmers, livestock are recognized as essential assets for their livelihoods. Temperatures are rising and rainfall patterns are expected to become even more erratic. An increase in extreme climate events, such as droughts and floods, is anticipated (Christensen *et al.*, 2007, KNMI, 2006). An increasing uncertainty of the onset and duration of seasons leave livestock farmers in the GHA susceptible to extreme climate events. They are often unable to appropriately plan their livestock management (Notenbaert *et al.*, 2010). Expected impacts of the observed climatic trends include reduced agricultural productivity (of food, feed and livestock products), higher disease prevalence (within crops, livestock and humans alike) and reduced freshwater availability (Thornton *et al.*, 2009a, b).

In the Palestine study (Rabi *et al.*, 2004) the impacts on livestock which constitute the main source of income, include increase in livestock mortality by 10 per cent, 48 per cent decline in milk productivity, breeding season delayed by at least one month, increase in water demand, and people resorting to livestock selling to buy water tankers for the remaining livestock. Socio-economic impacts include increased water costs, increased migration and social instability, reduction in population relying on livestock, shift out of farming, less expenditure on basic needs affecting nutrition levels.

Participatory studies on perceptions of pastoralists in Ethiopia have shown that conflicts over natural resources are perceived as the greatest stressor (Naess *et al.*, 2010). Not surprisingly, many of the stressors for pastoralists relate to their mobility and the availability of pasture. Struggles over access to water and pasture are central, as illustrated in the following quote:

"Previously, conflict was mainly about heroism and raiding of cattle.

Now the conflict between Barona and neighbouring clans is about securing

pasture and water and also for territorial integrity and power."

(Obbo Arero Jateni, 80 years old, Dire Worde, agro-pastoral area)

In India climate change poses formidable challenge to the development of livestock sector. The anticipated rise in temperature between 2.3 and 4.8°C over the entire country together with increased precipitation resulting from climate change is likely to aggravate the heat stress in dairy animals, adversely affecting their productive and reproductive performance, and hence reducing the total area where high yielding dairy cattle can be economically reared. The predicted negative impact of climate change on Indian agriculture would also adversely affect livestock production by aggravating the feed and fodder shortages (Sirohi and Michaelowa, 2007). Unpublished farmers testimonies on climate change mention declining milk productivity of cattle and declining fertility as major consequences of frequent fodder and water shortages in Rajasthan.

18.5 Impact on Livelihoods

For farmers practicing mixed farming, adjusting to climate variability involves shifting the emphasis between cropping and animal husbandry – in years of good rainfall farmers prioritize crop production, in low rainfall years live stock assume more importance together with other trades. In Ethiopia (Mongi *et al.,* 2010) farmers are increasingly unsure of receiving enough rain to justify planting staple food crops. In arid regions changes in climate patterns have clear implications for cropping and livestock based livelihoods through longer hunger gaps, increasing uncertainty in livelihood activities.

Limited options of alternative livelihoods and widespread poverty continue to threaten livelihood security of millions of small and marginal farmers in the arid and semi arid regions (State of the Environment Report 2009). Due to 2-3.5°C of temperature rise accompanied by 7-25 per cent of precipitation change, farmers may be losing net revenue between 9-25 per cent that may adversely affect GDP by 1.8-3.4 per cent (Kumar and Parikh, 1998, Sinha *et al.,* 1998). There will be serious consequences for food security in the South and India stands to lose a massive 125 Mt equivalent to some 18 per cent of its rain fed cereal production potential (Fisher *et al.,* 2001).

18.6 Relevance of Climate Information from different Disciplines

Global climate policies give more weight to scientific studies including macro level models to drive policy making. Social and anthropological studies, as well as studies by development practitioners focus on micro level impacts of climate change on societies. These studies are better able to incorporate the complexities of real life and are therefore of direct relevance to decision making at the level of communities, yet this evidence is largely ignored in climate policy making at international and national level.

Reliable climate information is crucial for farmers' responses to the challenge of climate change. Major issue with climate change information is that climate change predictions give only long term changes in mean values whereas farmers are concerned with current variability arising as a result of climate change.

Experience in the field has showed that available meteorological data and farmers' perceptions cannot always be correlated. Rainfall data is recorded at only one or two places in an administrative block, whereas actual amount of rainfall varies widely even between different villages. Therefore, it is not possible to possible to validate farmers estimates of rainfall received with available meteorological data. Further farmers may be more concerned with available soil moisture than total rainfall in a month. In Rajasthan, farmers measure rainfall by how many *aangal* rain has fallen, meaning to which depth has water in filtered into the soil measured by '*aangal*' or finger width. Soil infiltration can be very different if the same amount of rainfall is received in low intensity form for few hours, or if it is received at high intensity within one hour.

Longitudinal research studies have been compiled for rain-fed farming areas of Tamil Nadu, India. As part of the study, researchers compiled and analyzed 40 years of secondary data and cross checked it with the communities' own experiences. They conducted a statistical analysis of rainfall, determining how much it varied from year to year, how likely it was to fall in a given month, what the trend was, and ultimately a new length for the growing period. Using this data, researchers came up with alternative cropping suggestions, which they then presented for farmers in a workshop. Instead of accepting the researchers' suggestions for alternative cropping patterns, farmers expressed concerns about factors besides rainfall: what about the need for dry spells in the crop period for optimum growth? What about the ability of the crop to withstand excess moisture? What about the timing of pests etc? Therefore, it is less useful to do rainfall analysis as an isolated academic exercise. It can become meaningful only if the analysis is grounded in the vast reserve of real life experiences of farmers. Climate change can influence many factors in the micro environment. It is not enough to provide only temperature and rainfall data to farmers, they also consider duration of dry spells, sunshine availability, stored soil moisture and dew for crop growth.

It is recognized that climate science does not always address the questions people are asking (Kempton *et al.,* 1995, Glantz, 1996). In West Africa seasonal forecasts refer to total rainfall quantity, while timing of onset and end of the rainy season and distribution of rainfall during the season are more relevant parameters for African farmers (Ingram *et al.,* 2001). East African pastoralists are more concerned with water availability than with precipitation *per se* (Mahmood and Little, 2000).

Another important issue with climate change studies is the skepticism regarding farmers' perceptions. Development practitioners and farmers use terms that are not recognized and easily defined by trained scientists leading to a gap in communication. There are many works that have used statistical methods to bridge the gap between 'perceptions' and meteorological terms. Simelton *et al.,* 2011 have used statistical and structured social surveys to understand the usage of common terms regarding climate change. A common observation of farmers is that rainfall used to start earlier than now, but no evidence of this is found in meteorological records. The study explained that according to meteorological terminology onset of rainfall would mean the day rainfall starts after a long dry period, but according to farmers onset of rainfall is actually understood as the day when the soil horizon is moist to a depth of an underarm's length because this is the time when they start planting. With climate change there are gaps in first few showers so that accumulated soil moisture reaches depth of underarm's length after many showers, therefore the farmers perceive rainfall as starting later due to climate change.

Climate Change is a complex phenomenon that is still unfolding and is therefore only partially understood. The challenge of climate change can be best understood, and addressed, by multi-disciplinary approaches incorporating indigenous knowledge systems.

References

Aggarwal, P.K., Nagarajan, S. and Udai Kumar. 2001. Climate Change and Indian Agriculture: Current State of Understanding and Future Perspectives. Report submitted to the Indian Council of Agricultural Research, New Delhi.

Awasthi, A. 2011. Climate Change Negotiations: Perspectives of Smallholder Farmers, UNFCCC COP 16. Presentation, at http://regserver.unfccc.int/seors/attachments/get_attachment?code=FA8TVIPA42CK9A71MYUIG27CDRSBFQ4G

Chatterjee, B., Khadka, M. (eds.) 2011. *Climate Change and Food Security in South Asia*, CUTS International.

Christensen JH, Hewitson B, Busuioc A, Chen A, Gao X, Held I, Jones R, Kolli RK, Kwon WT, Lap rise R, Magana Rueda V, Mearns L, Menendez CG, Räisänen J, Rinke A, Sarr A and Whet ton P 2007. Regional Climate Projections. In: Climate Change 2007: The Physical Science Basis. Contribution of Working Group I to the Fourth Assessment Report of the Inter governmental Panel on Climate Change [Solomon, S, D. Qin, M. Manning, Z. Chen, M. Marquis, K.B. Averyt, M. Tignor and H.L. Miller (eds.)]. Cambridge University Press, Cambridge, United Kingdom and New York, NY, USA.

Community level climate change adaptation and policy issues, prepared by Kyoto University Graduate School of Global Environmental Studies, March 2005.

Dhaka, B.L., Chayal, K., Poonia M.K. 2010. Analysis of Farmers' Perception and Adaptation Strategies to Climate Change, Libyan Agriculture Research Center Journal Internation 1 (6): 388-390

Downing, T.E. 1992. *Climate change and vulnerable places: global food security and country studies in Zimbabwe, Kenya, Senegal and Chile.* Environmental Change Unit, University of Oxford

Fisher G, M. Shah, H van Velthuizen, and F. Nachtergaele 2001. Global Agro-ecological Assessment for Agriculture in the 21st Century, International Institute for Applied Systems Analysis, Austria, pp. 27-31, cited in Down to Earth, August 15, 2001, pp. 42-43.

India first National Communication to UNFCCC 2004. MoEF, Government of India

Ingram, K., Ron coli C, Kirshen, P., Flit croft, I. 2001. Opportunities and constraints to using seasonal precipitation forecasting to improve agricultural production systems and livelihood security in the Sahel-Sudan region: a case study of Burkina Faso. In: Proceedings of the International Forum on Climate Prediction, Agriculture, and Development. New York, April 26-28, 2000. International Research Institute for Climate Prediction, Palisades, NY, p 265 -277

K.N. Chaudhari, M.P. Oza and S.S. Ray (n.d.) Impact of Climate Change on Yields of Major Food Crops in India. ISPRS Archives XXXVIII-8/W3. Workshop Proceedings: Impact of Climate Change on Agriculture

Kanji, S.T., Verchot, L. and Mackensen, J. 2006. *Climate Change and Variability in the Sahel Region: Impacts and Adaptation Strategies in the Agricultural Sector*. UNEP and ICRAF.

Kempton,W., Boster J, Hartley J 1995. Environmental values in American culture. MIT Press, Cambridge.

KNMI 2006. Climate change in Africa. Changes in extreme weather under global warming, Royal Netherlands Institute of Meteorology, http://www.knmi.nl/africa_scenarios/.

Kumar, K. S. and Parikh J. 1998, Climate Change Impacts on Indian Agriculture: The Ricardian Approach, In: Measuring the Impact of Climate Change on Indian Agriculture, edited by A Dinar, *et al.,* Washington DC: World Bank. [Technical Paper No. 402]

Lantz, M.H. 1996 Currents of change: *El Niño's* impacts on climate and society. Cambridge University Press, Cambridge

Mahmud, H., Little, P. 2000 Climatic shocks and pastoral risk management in northern Kenya. *Pract Anthropol* 22:11-14

Mongi, H., Majule, A.E., and J.G. Lyimo. 2010. Vulnerability and adaptation of rain-fed agriculture to climate change and variability in semi-arid Tanzania. *African Journal of Environmental Science and Technology*.4(6): 371-381.

Naess, L.O., Sullivan, M., Khinmaung, J., Crahay, P., and Otzelberger, A. 2010. *Changing Climate Changing Lives*. ACF International, IDS, Tear fund, IER, A-Z CONSULT, ODES.

Notenbaert, A, Mude, A, van de Steeg J, and Kinyangi. 2010. Journal of Geography and Regional Planning 3(9): 234-239. Options for adapting to climate change in livestock dominated farming systems in the Greater Horn of Africa. *Journal of Geography and Regional Planning* Vol. 3(9), pp. 234-239.

Parthasarthy, B. 2006. Inter-annual and long-term variability of Indian summer monsoon rainfall. The Proceedings of the Indian Academy of Sciences. *Earth and Planetary Sciences* 93: 371-385.

Rabi, A., Carmi, N., Abu Jamus, N.O. and Abu Saadah, M. 2004. *Expected Impact of Climate Change on Population and livelihood in Arid and Semi-Arid Areas*: Case Studies from Palestine.

Rao, A.S. and Poonia. S. 2011. Climate Change Impact on crop water requirements in Arid Rajasthan. *Journal of Agro-meteorology* 13(1): 17-24.

Report of the Working Group on Animal Husbandry and Dairying, 11th Five Year Plan 2007-12, Planning Commission, GoI.

Simelton, E., Claire, H. Q, Philip A, Nnyaladzi, Andrew J. D, Jen Dyer, Evan D. G. F, David M, Staffan R, Susannah S, Lindsay C. S. 2011. African Farmers Perception of Erratic Rainfall. Centre for Climate Change Economics and Policy Working Paper No. 73, Sustainability Research Institute Paper No. 27.

Singh, D.V. 2011. Effect of climate change on pearl millet productivity in Thar Desert, *in CAZRI DEN News* Vol. 10(1): 2-3.

Singh, R.B., Kumar, P. and Woodhead, T. 2002. Smallholder Farmers in India: Food Security and Agricultural Policy, FAO.

Sinha, S.K., M. Rai, and G.B. Singh 1998, Decline in Productivity in Punjab and Haryana: A Myth or Reality? Indian Council of Agricultural Research (ICAR) Publication, New Delhi, India, 89.

Sirohi, S. and A. Michaelowa, 2007. Sufferer and cause: Indian livestock and climate change. Climate Change, 85: 285-298.

The Arab Water Council 2009. *Vulnerability of arid and semi-arid regions to climate change – Impacts and adaptive strategies.*

Thornton PK, Jones PG, Alagarswamy G, Andresen J 2009a. Spatial variation of crop yield response to climate change in East Africa. *Global Environ. Change,* 19(1): 54-65.

Thornton PK, van de Steeg, J.A., Notenbaert, A., Herrero, M 2009b. The impacts of climate change on livestock and livestock systems in developing countries: a review of what we know and what we need to know. *Agric. Sys.,* 101: 113–127.

2015, Impact of Global Warming and Climate Change on
     Human and Plant Health
Editors: **Dr. Arun Arya and Prof. V.S. Patel**
Published by: **DAYA PUBLISHING HOUSE, NEW DELHI**

Pages 196–200

Chapter 19

Occurrence of Seed Mycoflora, and Effect of Heavy Metal Pre-treatment in Germination of Rajmah

Nimisha Yadav and Arun Arya*

Department of Botany, Faculty of Science,
The M.S. University of Baroda, Vadodara – 390 002, Gujarat, India

ABSTRACT

Many fungi are serious parasites of seed primordia, maturing and stored seeds and grains and their invasion can result in various abnormalities including, reduced yield of seed in both quantitatively and qualititatively, discolorations, decrease germinability, mycotoxin production and total decay. Contaminated seeds can often result in poor seedling vigour, resulting in an un-healthy crop. Healthy seed is the foundation of healthy plant; a necessary condition for good yields. The fungal pathogens play a major role in the development of diseases on many important field and horticulture crops; resulting in severe plant yield losses. Seed diseases may cause seed abortion, reduction in size and seed rot.

Extensive use of fungicides has resulted in accumulation of toxic compounds potentially hazardous to humans and environment and also in the increase of resistance in the pathogens. The increasing awareness of fungicide-related hazards has emphasized the need of adopting biological methods. As an alternative disease control method can the heavy metals be used as disease controlling chemicals.

---

* *Corresponding author.* E-mail: aryaarunarya@rediffmail.com

Use of heavy metals was tried in germination of Rajmah seeds. essential contents as well as reduces seed germination. Among the sample studied the fungi, *Aspergillus niger, A. fumigatus, A. flavus, Chaetomium globosum* and *Penicillium citrinum* were present.

Keywords: *Occurrence, Seed mycoflora, Heavy metals, Pretreatment Germination Rajmah.*

19.1 Introduction

The deterioration of seeds by storage fungi has been studied in different ways. As regards the characteristics of the seedlings raised from the deteriorated seeds, slow growth, meagre amount of chlorophylls, total free sugar and am into acid have been reported in them besides stimulated activities of respiratory enzymes and those of amino acid degradation (Pyare, 1991. Rao *et al.,* 1989).

French bean botanically described as *Phaseolus vulgaris* L. is a protein rich crop. It is also known as *Rajmash* or *Rajma* (Hindi) or haricot bean or kidney bean or common bean or snap bean, navy bean. The French bean, *P. vulgaris,* is an herbaceous annual plant domesticated independently in ancient Mesoamerica and the Andes, and now grown worldwide for its edible bean, popular both dry and as a green bean.18.3 million tonnes of dry French beans and 6.6 million tonnes of green beans were grown worldwide in 2007. Brazil and India are the largest producers of dry beans while China produces, by far, the largest amount of green beans, almost as much as the rest of the top ten growers altogether. Similar to other beans, the French bean is high in starch, protein and dietary fibre and is an excellent source of iron, potassium, selenium, molybdenum, thiamine, vitamin B6, and folic acid. It is valued for its protein rich (23 per cent) seeds. Seeds are also rich in calcium phosphorus and iron.

After harvesting seed are stored at different storage conditions and if these storage conditions are not proper, various microorganisms like bacteria, viruses, nematodes and fungi interact with these seeds. Among these microorganisms fungi play dominant role in decreasing quality and longevity of the seeds. Several seed-borne fungi, including species of *Alternaria, Aspergillus, Fusarium* and *Penicillium,* have been detected as seedling pathogens of cereals (Fahim *et al.,* 1983; Gulya, *et al.,* 1979). Dayal *et al.,* 2001 assayed the nitrate reductase and urease besides the estimation of total free amino acid in the seedlings of rajma raised from the seeds deteriorating due to storage fungi at varying RH levels.

The seed treatment of rajmah was done with Thiram or Captan at the rate of 4.5 g per Kg of seed and requirement of *Rhizobium* as bio-fertilizer is 200 g for every 28 Kg of seed mixed with 1250 ml of cooled off boiled rice starch or water. The seed to be dried in shade before sowing for 30 minutes. Irrigation: Critical stage for irrigation is 20-25 days after the seed is sown. Under sufficient rainfall conditions, the number of irrigations can be reduced. Plant protection: Anthracnose or any leaf diseases can be controlled by normal spraying of Diathene M-45 or Captan in every 10 to 15 days interval at the rate of one g/litre of water in case of Captan and 2 g/litre of water in case of Diathene M-45.

Fungal occurrence in seed can be seen by blotter method or agar plate method. In the present study blotter method was preferred as sugar fungi belonging to Mucorales group may show over growth wchich masks the growth of other fungi. Total 88 per cent seeds were attacked by fungi. The percentage germination of Rajmah with heavy metal treatment was recorded.

19.2 Study of Seed Mycoflora

Fungal occurrence in Rajmah was observed with the help of blotter method. Out of 100 seeds taken, 88 seeds were attacked by fungus (cottony growth), whereas bacteria were present in 30 seeds. The untreated seeds of Rajmah showed presence of 8 different fungi and one mycelium without any conidia. The percentage occurrence of the fungi is recorded in Table 19.1. As *Aspergillus* has many species, some of its species were seen on the same seed *i.e.* different fungal colonies on single seed which leads to poor or no germination of seed Resulting occurrence of mycoflora was responsible for decline in per cent germination.

Table 19.1: Percentage Occurrence of different Fungi on Treated and Untreated Seeds of Rajmah

Fungi	Percentage Occurrence	
	Untreated	Treated
Aspergillus flavus	26	15
A. fumigatus	28	20
A. niger	30	14
A. tereus	4	4
Fusarium oxysporum	6	2
Monilia sitophila	4	4
Penicillium citrinum	16	10
Rhizopus stolonifer	6	–
Unidentified Fungal mycelium (white)	4	2

According to Arya and Chauhan (1995) the seeds of *Cicer arietinum* L. (chickpea) harboured 14 species of fungi and in case of pigeon pea, 10 species of fungi were isolated from seeds of one year old pigeon pea (Arya and Mathew, 1991). In case of Rajmah the presence of eight fungi was recorded along with one unidentified white coloured fungal colony. In case of Rajmah seeds occurrence of four different species of *Aspergillus* was observed while in case of chick pea the number of *Aspergilli* was 3 in freshly harvested seeds and 4 species after 2 years of storage. It has been observed that different species of *Aspergilli* are common post harvest pathogens. Their number usually increases with time.

Both the varieties and their mutants showed the incidence of seed-born fungi. Total twenty two fungi were recorded from both the varieties (Swapnil *et al.,* 2011). Out of which the *Macrophomina phaseolina* showed its quantitative dominance which were followed by *Aspergillus niger* and *Fusarium oxysporum*. Ten mutants and one control sample of variety Waghya showed the occurrence of different fungi while eight mutants and one control sample of variety Varun was studied for fungal susceptibility. Broad pod mutant of variety Waghya showed maximum association of fungi while Long pod mutant showed minimum detection of fungi. Broad pod mutant of variety Waghya showed the occurrence of seven different fungi which includes *Macrophomina phaseolina, Aspergillus niger, Fusarium equsiti, Curvularia lunata, Alternaria dianthi, Rhizopus stolonifer.* Small leaf mutant of variety Varun was associated with maximum number of fungi whileDwarf and Branched mutant showed minimumdetection of fungi.

19.3 Study of Pretreatment with Heavy Metals

Table 19.2 represents effect of pretreatment with heavy metals on seed germination. Comparisons were made with reference to control which are free from heavy metals *i.e.* soaked in distilled water whereas rest seeds are soaked for 8 hours in different concentration of different heavy metals. Results of this experiment shows that complete seed germination take nearly 9-10 days in control seeds.

As copper is an essential as micronutrient for life process but exposure to longer period or higher concentration declines the seed germination capability *i.e.* the germination gradually decline from 500 ppm to 1000 ppm and excess concentration (2000ppm) of $CuSO_4$ hampers the germination of Rajmah. The results showed 0 per cent germination in $Mn\ Cl_2$ and $CuSO_4.$

Table 19.2: Effect of Heavy Metal Pretreatment on Percentage Seed Germination

Heavy Metal Compounds		Day 6	Day 9
Control		52	64
$CuSO_4$	(500 ppm)	40	40
	(1000 ppm)	12	12
	(2000 ppm)	0	0
$MnCl_2$	(500 ppm)	12	36
	(1000 ppm)	52	64
	(2000 ppm)	0	0
$MnSO_4$	(500 ppm)	12	16
	(1000 ppm)	28	28
	(2000 ppm)	20	40
$ZnSO_4$	(500 ppm)	40	44
	(1000 ppm)	48	52
	(2000 ppm)	40	64

19.4 Conclusion

Seed is a potential carrier of unwanted pests. Every nation has a sovereign to protect its environment and food security and take appropriate measures according to the international Plant protection convention. Seed mycoflora study resulted into occurrence of seed borne fungal flora including *Aspergillus* spp. and *Penicillium citrinum*. Field fungi like *Fusarium, Chaetomium* and *Alternaria alternata* were recorded

Results indicated that less concentration (500 ppm) of manganes have less germination rate compared to 1000 ppm concentration having more germination rate. But higher concentration (2000 ppm) further resulted in decline in germination rate. Ultimately it was concluded that less or higher concentration of Manganese chloride hampered the seed germinability. Chemicals containing Manganes chloride and Copper sulphate can not be used as seed treatment chemicals.

References

Arya A. And Mathew D.S. 1991 Seed Mycoflora of pigeon pea. *Acta Bot. Indica* 19: 102-103

Arya A. And Chuhan R. 1998 Seed mycoflora of chick pea. *Indian Bot Soc.* 123-125.

Dayal, S. Sanja V. K. Kumar M. Singh S.P. Kumar V. And Prasad B.K. 2001 Effect of storage fungi on nitrate reductase and urease activities in rajma seedlings. *Indian Phytopath.* 54 (3): 370-372

Fahim M.F., Barakat F.M., Aly H.Y. 1983. Influence of storage conditions on sorghum grains and the associated fungi. Egyptian Society of applied microbiology, *Proc V Conf Microbiol*, Cairo, Vol. III, *Plant Pathology.* 78.

Gulya T.J., Martinson C.A., Tiffang LH. 1979. Ear rotting fungi associated with opaque-2 maize. *Plant Dis Rep.* 63: 370–373.

Pyare, K. (1991). *Studies on the influence 0/storagemoulds on the growth and biochemical constituents 0/sesame seedlings.* Doctoral thesis, Magadh University, Bodh Gaya. pp. 62-81.[Vol. 54(3) 2001].

Rao, R.N., Singh, R.N., Narayan, N., Kumar, S. And Prasad, B.K. (1989). *Indian Phytopath.* 42: 538-543.

Swapnil E. Mahamune and Rajendra B. Kakde 2011. Incidence of seed-borne mycoflora on French bean mutants and its antagonistic activity against *Trichoderm harzianum. Recent Research in Science and Technology,* 3(5): 62-67.

2015, Impact of Global Warming and Climate Change on
    Human and Plant Health
Editors: Dr. Arun Arya and Prof. V.S. Patel
Published by: DAYA PUBLISHING HOUSE, NEW DELHI

Pages 201–211

Chapter 20

Climate Change and Plant Diseases

R.K. Srivastava and R.K. Singh*

Department of Mycology and Plant Pathology, N.D. University of Agriculture and Technology, Crop Research Station, Bahraich – 271 801, U.P.

ABSTRACT

Control of plant diseases is essential to improve food security in the context of climate change, plant protection professionals must work with societal change, defining its key processes and influencers to effect change. More specifically, there is a key role to play in improving food security. Plant pests and diseases could potentially deprive humanity of up to 82 per cent of the attainable yield in the case of cotton and over 50 per cent for other major crops and, combined with postharvest spoilage and deterioration in quality, these losses become critical, especially for resource-poor regions. Actual average losses for rice in the period 2001–2003 totalled 37·4 per cent, comprising 15·1 per cent to pests, 10·8 per cent to pathogens and 1·4 per cent to viruses, with the remaining 10·2 per cent accounted for by weeds. Forests protect cities and villages against avalanches, landslide etc. They influence climate and air quality. With increasing temperature tree to tree infection has increased in European forests. Each year an estimated 10–16 per cent of global harvest is lost to plant diseases. In financial terms, disease losses cost US$220 billion. There are additional postharvest losses of 6–12 per cent; these are particularly high in developing tropical countries lacking infrastructure and consequently are difficult to estimate. As attested by the infamous 19th century Irish potato famine or the Bengal famine, devastations from plant diseases can be far reaching and alter the course of society and political history.

Keywords: *Climate change, Plant diseases, Health effects.*

* *Corresponding author.* E-mail: rksingh05@gmail.com

20.1 Introduction

Growing concern for food security, impact of climate change and variation in agriculture, has caused significant impact on society. The effect of impending global warming on cropping patterns caused changes in agricultural output. Air pollution affects the agriculture need for developing meteropathological models.

Despite the technological advances that have taken recently in the field of rainfed agriculture, the year to year variation in food grain production has remained the consonance with the weather variability in terms of onset of the monsoon and its behavior during the kharif crop growing season. The recent all India draught condition during past few years has demonstrated the need for the strengthening the application of meteorological knowledge towards tactical decision making by the farmers to minimize the crop loss. An integrated and holistic approach to agro-meteorology has been chosen as the theme for the crop protection to develop the principles of management through interaction and interdisciplinary mode which has been the weakest link in the past. Pest plant and people have lived in a close association for centuries based on both synergistic and antagonistic relationship. However, the alliance assumed for more significance after the beginning of systematic agriculture. Guesstimates reveal that the share of different pests in annual total loss of agriculture production stands at 45, 30, 20, and 5 per cent due to weeds insects, diseases and other pest respectively. The losses caused by the pests in India are enormous. The available estimated loss figure varies from one to the other source. However, it has been estimated that approximately 10 per cent of the total loss is caused by weeds (33 per cent) which was followed diseases (26 per cent) (Mehta,1976).

Whenever, plants are distributed either by pathogen or some environmental conditions one or more of the aforesaid functions are dislocated the plant become diseased. Thus a disease in plant can be defined as "any disturbance brought about by a pathogen or an environmental factor which interferes with manufacture translocation or utilization of food mineral nutrients and water in away that affected plant changes in appearance and/or yield less than a normal and healthy plant of the same variety." (Agrios, 1978).

The estimates have evinced that there are 30,000 of recorded plant diseases and approximately 5000 of them have been noticed in India. The large number of diseases would naturally cause local losses. In India, the crop losses due to various pests including diseases range from 10 to 30 percent under normal condition (Pal, 2002). Initiation and the optimal development of a phyto-disease are conditioned by the presence of susceptible plant infective pathogen and the environment. Weather and climate are the integral parts of agricultural production system, which comes under environmental factor. Weather is an important component not only in crop production but also for crop protection and other agricultural activities.

Meteorological parameters have a pronounced impact on growth and intensity of attacks by pests, insects and crop diseases. Adverse weather warning s for the same would be useful to farmers for estimating the nature of impending attack based on the nature of these impending attacks based on the current conditions and past experiences, so that they can take suitable remedial action and reduce monetary losses.

The damaging effects caused by pests and diseases to crops are well known. Reduction in losses caused by pests and diseases by timely effective control measures considerably add to agricultural production of India. Incidence and spread of many pests and diseases are influenced by weather. Several studies on actual field observations have been carried out in India at various SAUs, ICAR institutes and co-ordinated projects on different crops relating weather parameters such as maximum and minimum temperature, relative humidity, vapour pressure rainfall, number of rainy days, sunshine hours and wind speed and experimental microclimate parameters for the incidence and population density and pests and diseases of different crops. A number of multiple regression equation models are available.

Region wise compilation of such information in respects of major crop classified according to different agro-climatic zones is required. A clear understanding of dynamics of plant disease incidence as influenced by different biotic and climatic factors of environment is important. Relationship of climate factors to plant diseases is used in formating meteorological models. For this knowledge of climatic conditions conducive for induction of sporulation, dispersal and penetration is required. Some of the earlier work on forecasting of several epidemic of brown rust of Northern India has shown weather chart would be a very useful tool in disease prediction. For instance, synoptic situation favourable for southerly flow of air as indicated by 700 hpa weather charts would improve the chance of prediction of the spread of the disease to central India.

In fact, there is no reliable estimate of crop losses in India yet, it may run in to crores of rupees annually. By thumb rule, each crop of economic significance is attacked at least once by one disease. A crop like sorghum is infested by as many as twenty diseases. However, the estimates reveal that wheat rusts destroy the grain of worth Rs. 6 crores (Rangaswami and Madhhavan, 1999). In Rice, blast and bacterial blight, bunchy top in banana, tristeza in citrus, black arm in cotton, late blight and ring rot in potato, viral diseases in vegetables and they devastate these crops and cause monetary losses of crores of rupees every year in India.

According to world meteorological organization (WMO), climate change can adversely impact global environment, agricultural productivity and quality of human life. More importantly in developing countries, it will be difficult for farmers to carry on farming in the increased temperatures. Recognizing this it is necessary that India should address the issue of climate change and focus on providing better environment to improve quality of human life. Climate change affects everyone, but the worst sufferers would be the hundreds of millions of small and marginal farmers. The rise of global temperature on account of climate change would affect agriculture, while in temperate latitudes a rise in temperature would help countries increased food productivity, and it will have adverse effect in India and countries in tropics. The monsoon accounting for 75 per cent of India's rainfall significantly impacts countries' agriculture and livelihood of tens of millions of small farmers. Climate change is likely to intensify the variability of monsoon dynamics leading to a rise in extreme seasonal aberration's such as increased precipitation and devastating floods in the some part of country as well as reduced rainfall and prolonged droughts in other areas. It will also have an impact on the incidence of pests and diseases. Frequent change in weather parameters more importantly temperature and precipitation (rainfall) would not only threats food productions but also access, stability and utilization of food resources.

"The Indian Council of Agricultural Research (ICAR) has estimated that annual wheat output may decline by four to five million tons with every degree Celsius rise in temperature." The frequent change in particularly low rainfall and warming has posed serious threat to sustainable agriculture. Low rainfall has resulted in groundwater depletion because of draining of water for irrigation and warming has affected the yield of crops and increased the disease incidence.

20.2 Plant Diseases and Weather Parameters

Singh *et al.* (2009) found that the intensity of *Colletotrichum* leaf spot disease of turmeric was maximum at 25-28°C and more than 80 per cent RH ant it was also noted that variation in temperature showed positive co-relation with disease intensity. Similarly Narayana *et al.* (2006) developed disease prediction model in groundnut rust epidemics also found temperature variation were significantly and positively correlated with rust severity. There was also accordance with findings of Singh *et al.* (1994). Ann *et al.* (1994); Singh *et al.* (1994); Degaonkar and Kiratwar, (1997); Gadre *et al.* (2002); Gadre and Joshi, (2003); Dubey, (2003); Moity, (2003); Singh *et al.* (2009), and several other workers correlated diseases severity with environmental factors and found positive correlation results.

Fungal pathogens can be important selective forces on plants in habitats favorable to fungal development, but estimating the magnitude of selection by fungal pathogens in nature is difficult. The development of fungal disease in plants has at least three important control points: the prevalence of fungal inoculum, the environment during infection, and disease development as modified by plant defenses (Jones, 1998; Van der, 1975). Most field studies on wild plants have not measured pathogen abundance but have focused on disease expression (Froelich and Snow, 1986; Burdon, 1987; Dix and Webster, 1995; Garcia *et al.*, 1996). Atmospheric moisture is generally the single most important environmental factor influencing the incidence and severity of fungal diseases on plants (Burdon, 1987; Burdon, 1991; Nan *et al.*, 1991: Schmiedeknecht, 1995; Venier *et al.*, 1998). High relative humidity and several hours of free surface water are critical for both spore germination and successful infection (Huber and Gillespie, 1993; Cooke and Whipps, 1993; Harrison *et al.*, 1994). In addition, infection (*i.e.*, invasion of plant tissue by the fungus) and disease (*i.e.*, the expression of symptoms such as lesions or necrosis) (Agrios, 1978) on plants due to air-borne fungi are favored by temperatures of 15–40°C (Cooke and Whipps, 1993; Griffin, 1994). Field studies on plant pathogens have demonstrated that the growth of fungi is favored by high moisture and moderate temperatures (Froelich and Snow, 1986; Griffin, 1994; Colhoun 1973; Taylor, 1979; Rowan *et al.*, 1999) and that low relative humidity and extreme temperatures inhibit growth and spore germination (Harrison *et al.*, 1994; Juniper, 1991). Other studies on soil fungi also show that prevalence differs among habitats and seasons and correlates positively with moisture and negatively with temperature (Harrison *et al.*, 1994; Johansen and Rushforth, 1985). This apparent positive relationship between moisture and fungal growth and abundance may result from the high surface-to-volume ratio of fungi, making them vulnerable to water loss. The positive association between habitat moisture and the incidence and severity of disease suggests that the selective pressures imposed on plants by pathogenic fungi may vary among habitats. However, the relationship of habitat characteristics to inoculum abundance, infection and disease is poorly understood. This study focuses on abiotic correlates of inoculum abundance. It is not an attempt to describe the distribution of particular species of fungi, but rather to obtain an independent predictor of the prevalence of fungal inoculi in different habitats. To do this, we examined fungi in the air and on leaf surfaces rather than trying to separate opportunistic and obligate pathogens from non-pathogens. First, it would be impractical to identify all fungi sampled, and second there is no reason to expect that pathogens and non-pathogens differ in their abiotic requirements for growth. The growth and sporulation of many pathogens is favored by high humidity, and temperature requirements are similar to other mesophiles, with ranges between 5 and 50°C and optima between 20 and 30°C (Jones, 1998; Cooke and Whipps, 1993; Harrison *et al.*, 1994). Thus, characterizing the weather conditions favorable for fungi in general may be useful in predicting the extent of selection imposed by pathogenic fungi on plants in different habitats.

Climate change is affecting plants in natural and agricultural ecosystems throughout the world (Stern 2007). However, little work has been done to model the effects of predicted twenty-first-century climate change on plant disease epidemics (Garrett *et al.*, 2006). Changing weather (*e.g.* temperature, rainfall) can induce severe plant disease epidemics (Coakley *et al.*, 1999; Chakraborty 2005), which threaten food security if they affect staple crops (Luo *et al.*, 1995; Chakraborty *et al.*, 2000; Anderson *et al.*, 2004) and can damage landscapes if they affect amenity species (Brasier and Scott 1994; Bergot *et al.*, 2004). Severity of human, animal and plant disease epidemics is greatly affected by climatic factors, especially temperature and rainfall (Wint *et al.*, 2002; Fitt *et al.*, 2006a; Thomson *et al.*, 2006; Bosch *et al.*, 2007). Therefore, weather-based forecasts have been developed to guide control strategies for many important diseases worldwide (Wint *et al.*, 2002; Garrett *et al.*, 2006; Thomson *et al.*, 2006). There is now an opportunity to link weather-based plant disease forecasts with recent climate change models, to predict the effects of climate change scenarios on the distribution and severity of plant disease epidemics. Much discussion on the impact of climate change on plant disease epidemics has used qualitative, rule-based reasoning, which cannot easily accommodate the complex host–pathogen–environment interactions involved (Coakley *et al.*, 1999; Anderson *et al.*, 2004). Modeling approaches have included those matching existing climates in one region with climates predicted for another (Brasier and Scott 1994) or combinations of simulation models for crop growth and disease development (Luoet *et al.*, 1995). Before 1999, no work had used predicted climate variables generated by new more sophisticated general circulation models (GCMs); most studies had relied on predictions of fixed changes in temperature and rainfall (Coakley *et al.*, 1999). Recently, GCMs have been used to predict the increase in range of *Phytophthora cinnamomi* (Bergot *et al.*, 2004). Whereas empirical modeling has been used to produce disease epidemic models for combining with climate change predictions (Chakraborty *et al.*, 1998; Coakley *et al.*, 1999), few models have been based on datasets including both regional and seasonal variations that are sufficiently extensive to allow both model development and validation. Phoma stem canker (blackleg, *Leptosphaeria maculans*) is an internationally important disease of wild and cultivated brassicas; during each growing season, it causes yield losses of millions of tonnes of *Brassica* oilseed and vegetable crops in Europe, North America, Australia and Africa (Fitt *et al.,* 2006b). It has spread across North America and Eastern Europe in the last 20 years and now threatens 10 M ha of highly susceptible oilseed and vegetable Brassicas in China, mostly grown by subsistence farmers. Temperature and rainfall affect not only the development of the pathogen (Huang *et al.,* 2005) but also the resistance response of the host (Huang *et al.,* 2006). Globally, the most severe epidemics occur in oilseed rape (*Brassica napus*) growing areas of Australia, with their Mediterranean climate, where susceptible crops can be destroyed by the disease (Howlett *et al.,* 2001; Sprague *et al.,* 2006). However, much of the world's oilseed rape crop is grown in cooler climates. In the UK, the most severe phoma stem canker epidemics occur in southern England; in Scotland, where the climate is colder, phoma leaf spotting does occur but damaging phoma stem cankers do not subsequently develop. To illustrate the effects of predicted climate change on the range and severity of plant disease epidemics, weather-based models predicting the development of disease epidemics were combined with climate change models to generate scenarios for the future severity of epidemics in India.

It was predicted that global warming can increase the range and severity of plant diseases of worldwide importance within the next 20 years. The effects of climate change may be on the pathogen, the host or the host–pathogen interaction (Coakley *et al.,* 1999; Huang *et al.,* 2005, 2006; Garrett *et al.,* 2006). The long-term effects of human-made environmental change on plant diseases may be masked by short-term seasonal fluctuations (Bearchell *et al.,* 2005; Fitt *et al.,* 2006c). To ignore such effects may result in devastating epidemics on staple food crops, with far-reaching socioeconomic consequences, or on important plants in natural ecosystems, threatening wildlife (Luo *et al.,* 1995; Chakraborty *et al.,* 2000; Anderson *et al.,* 2004). The evidence that climate change will increase the range and severity of diseases like phoma is supported by the observations that disease epidemics are currently most severe in crops growing regions with Mediterranean climates (*e.g.* in Australia or France; Howlett *et al.,* 2001; Fitt *et al.,* 2006a). There is a need for new cultivars destined for the future market to be tested under such climates. Whereas some predicted effects of climate change may be anticipated by qualitative reasoning (Coakley *et al.,* 1999), others, such as the small effect on the date of phoma leaf spotting in autumn, may not. It is important to recognize that climate change effects on complex host–pathogen–environment interaction may also decrease severity of epidemics (Chakraborty *et al.,* 1998). These predictions about the effects of climate change on range and severity of diseases were possible only because of the availability of both knowledge about the epidemiology of this monocyclic disease and datasets on epidemic development and weather for a range of sites and growing seasons. To improve accuracy of predictions, there is a need to improve the model by further validation against data obtained in a wide range of climates, such as those predicted, and by incorporating into the model a weather based crop growth model to describe the effects of climate change on crop growth that influence disease development (Steed *et al.,* 2007). A high priority over the next decade should be the collation of accurate disease and weather data and development of models to forecast the effects of climate change on other plant diseases to provide the necessary foresight for strategic adaptation to climate change. These models can guide policy and practice to counter such emerging threats to delicately balance natural and agricultural ecosystems.

Probably there is no other human activity in the world, which is more influenced and affected by weather than agriculture. There is no doubt that weather and climate is the single largest determining factor of agriculture and allied activities. Weather manifests its influence on agricultural operations and far production through its effects on soil as well as on every phase of crop growth. Weather and climate is important starting from crop planning and variety selection to the final stage of processing and preservation of the produce. Even if all other factors are normal, only a season of favourable weather can increase the production to the extent of 40 per cent while on the other hand spells of unfavourable or aberrant weather can reduce the yield from 10 to 100 per cent. Aberrant or hazardous weather can reduce the yield directly by creating uncongenial condition for the growth and development as well as causing mechanical or physical damage of the crop or plant. At the same time, the spell of weather, which is unfavourable for crop, may be favourable for a rapid growth of pests and diseases and there can be a total crop failure. This has happened many times in the past over different parts of the country. In India, each year there is extensive crop damage due to floods in one part of the country and a severe drought ruining crops in another part. In one season, crops are damaged by a severe attack of pests and diseases; in the other season, the damage can even be more due to some other factors. The total annual pre-harvest losses for various crops in the country as a whole are estimated to range between 10 and 100 per cent average is considered an appropriate figure for all the crops.

To minimize loss from weather hazards with a view to sustain the growth of agriculture the following steps and to be followed for a comprehensive planning in the field.

A. Identity the weather hazards.

B. Identity the period and frequency of their occurrence along with the standing crops frequently affected by such hazards during each period.

C. Locate the areas of occurrence of the hazards on maps,

D. Analyse the nature of such hazards, intensity and nature of effect on crops,

E. Suggest remedial measures, agro-technique or agronomic manipulation to over come the problem to the optimum possible extent.

F. Cultivation of environmental friendly crops: Jute, Mesta and Sisal

Because of environmental problems (artificial fibre produce long time pollution) and increased paper consumption, these applications of above mentioned fibre crops has drawn tremendous attention in the world. Moreover, leaf and crop trash remains in the field to be recycled as organic materials, thereby reducing demand for supplementary chemical fertilizers for subsequent crops. The fibres are biodegradable, environmentally benign and renewable, and provide reliable employment in many rural areas. Jute is mainly cultivated in West Bengal, Bihar, U. P., Assam and Orissa whereas mesta is mainly cultivated in Bihar, U. P., Bengal and A. P. Sisal is mainly cultivated in Orissa. Jute and mesta are drought tolerant as well as cultivated in flood prone areas.

Mesta (*Hibiscus cannabinus*) core is strong and absorbent and it can be used to clean up oil spills as well as chemicals. For its low density, once oil is absorbed, the product floats on the surface, which makes collection easier. *H. cannabinus* (Kenaf) core is also non-toxic, nonabrasive and is more effective than classical remediants, like clay and silica (Sameshima, 2000). Kenaf can absorb CO_2 and NO_2 3-5 times faster than forests, and its deep roots can improve the soil. It can clean the environment efficiently (Lam, 2000). In some Japanese cities, kenaf is planted on road sides by government to improve the air quality.

References

Agrios, G.N., 1978. *Plant Pathology* New York: Academic Press.

Anderson, P. K., Cunningham, A. A., Patel, N. G., Morales, F. J., Epstein, P. R. and Daszak, P. 2004 Emerging infectious diseases of plants: pathogen, pollution, climate change and agrotechnology drivers. *Trends Ecol. Evol.* 19, 535–544. (doi: 10.1016/j.tree.2004.07.021)

Ann, P.J., Huang. H.C., and Chen. M. F 1994. Effects of environmental factors on disease incidence of mango anthracnose. *Pl. Path. Bull.* 3: 34-44.

Anon. 2006 GENSTAT release 9 reference manual. In Release 9 reference manual. Oxford, UK: VSN International. Bearchell, S. J., Fraaije, B. A., Shaw, M. W. and Fitt, B. D. L. 2005 Wheat archive links long-term fungal pathogen population dynamics to air pollution. *Proc. Natl Acad. Sci.* USA 102, 5438–5442. (doi: 10.1073/pnas.0501596102)

Bearchell, S. J., Fraaije, B. A., Shaw, M. W. and Fitt, B. D. L. 2005. Wheat archive links long-term fungal pathogen population dynamics to air pollution. *Proc. Natl Acad. Sci.* USA 102, 5438–5442. (doi: 10.1073/pnas.0501596102)

Bergot, M., Cloppet, E., Pe´rarnaud, V., De´que´, M., Marc,ais, B. and Desprez-Loustau, M.-L. 2004 Simulation of potential range expansion of oak disease caused by *Phytophthora cinnamomi* under climate change. *Global Change Biol.* 10, 1539–1552. (doi: 10.1111/j.1365-2486.2004.00824.x)

Bosch, J., Carrascal, L. M., Duran, L., Walker, S. and Fisher, M. C. 2007. Climate change and outbreaks of amphibian chytridiomycosis in a montane area of Central Spain; is there a link? *Proc. R. Soc.* B 274, 253–260. (doi: 10.1098/rspb.2006.3713)

Brasier, C. M. and Scott, J. K. 1994. European oak declines and global warming: a theoretical assessment with special reference to the activity of *Phytophthora cinnamomi. EPPO Bull.* 24: 221–232.

Burdon, J.J., 1987. *Diseases and plant population biology. Cambridge: Cambridge University Press.*

Burdon, J.J.,1991. Fungal pathogens as selective forces in plant populations and communities. *Aust J Ecol*, 16: 423-432.

Chakraborty, S. 2005. Potential impact of climate change on plant–pathogen interactions. *Aust. Plant Pathol.* 34: 443–448. (doi: 10.1071/AP05084).

Chakraborty, S. *et al.,* 1998. Potential impact of climate change on plant diseases of economic significance to Australia. *Aust. Plant Pathol.* 27: 15–35. (doi: 10.1071/AP98001).

Chakraborty, S., Tiedemann, A. V. and Teng, P. S. 2000. Climate change: potential impact on plant diseases. *Environ. Pollut.* 108: 317–326. (doi: 10.1016/S0269-7491(99)00210-9).

Coakley, S. M., Scherm, H. and Chakraborty, S. 1999. Climate change and plant disease management. *Annu. Rev. Phytopathol.* 37: 399–426. (doi: 10.1146/annurev.phyto.37.1.399).

Colhoun, J., 1973. Effects of environmental factors on plant disease. *Annu Rev Phytopath,* 11: 343-363.

Cooke, R.C., and Whipps, J.M., 1993. *Ecophysiology of Fungi. Oxford: Blackwell Scientific Publications.*

Dix, N.J., and Webster, J., 1995. *Fungal Ecology. New York: Chapman and Hall.*

Fitt, B. D. L., Brun, H., Barbetti, M. J. and Rimmer, S. R. 2006a. World-wide importance of phoma stem canker (*Leptosphaeria maculans* and *L. biglobosa*) on oilseed rape (*Brassica napus*). *Eur. J. Plant Pathol.* 114, 3–15. (doi: 10.1007/s10658-005-2233-5)

Fitt, B. D. L., Evans, N., Howlett, B. J. and Cooke, M. 2006 b. Sustainable strategies for managing *Brassica napus* (oilseed rape) resistance to *Leptosphaeria maculans* (phoma stem canker). Dordrecht, *The Netherlands: Springer.*

Fitt, B. D. L., Huang, Y.J., van den Bosch, F. and West, J. S. 2006c Coexistence of related pathogen species on arable crops in space and time. *Annu. Rev.Phytopathol.* 44: 163–182.(doi: 10.1146/annurev.phyto.44.070505.143417)

Froelich, R.C., and Snow, G.A., 1986. Predicting site hazard to fusiform rust. *For Sci,* 32: 21-35.

Garcia-Guzman, G., Burdon, J.J., and Ash J.E., Cummingham, R.B., 1996. Regional and local patterns in the spatial distribution of the flower-infecting smut fungus *Sporisorium amphilophis* in natural populations of its host *Bothriochloa macra. New Phytol,* 132: 459-468.

Garrett, K. A., Dendy, S. P., Frank, E. E., Rouse, M. N. and Travers, S. E. 2006 Climate change effects on plant disease: genomes to ecosystems. *Annu. Rev. Phytopathol.* 44, 489–509. (doi: 10.1146/annurev.phyto.44.070505.143420).

Grade, U.A., Joshi, M. S., and Mandokhot, A.M. (2002). Effect of weather factors on incidence of Alternaria leaf blight, wheat rust and powdery mildew of mustard. *Ann. Pl. Protec. Sci.* 10: 337-339

Griffin DH, 1994 ed: *Fungal Physiology New York: Wiley-Liss.*

Harrison, J.G., Lowe, R., and Williams, N.A., 1994. Humidity and fungal diseases of plants – problems. *In: Ecology of plant pathogens (Edited by: JP Blakeman, B Williamson) Wallingford: CAB International,* 79-97.

Howlett, B. J., Idnurm, A. and Pedras, M. S. C. 2001. *Leptosphaeria maculans,* the causal agent of blackleg disease of *Brassicas. Fungal Genet. Biol.* 33, 1–14. (doi: 10.1006/fgbi.2001.1274)

Huang, Y.-J., Fitt, B. D. L., Jedryczka, M., Dakowska, S., West, J. S., Gladders, P., Steed, J. M. and Li, Z.-Q. 2005 Patterns of ascospore release in relation to phoma stem canker epidemiology in England (*Leptosphaeria maculans*) and Poland (*Leptosphaeria biglobosa*). *Eur. J. Plant Pathol*. 111, 263–277. (doi: 10.1007/s10658-004-4421-0)

Huang, Y.-J., Evans, N., Li, Z.-Q., Eckert, M., Chevre, A.M., Renard, M. and Fitt, B. D. L. 2006. Temperature and leaf wetness duration affect phenotypic expression of Rlm6- mediated resistance to *Leptosphaeria maculans* in *Brassica napus*. *New Phytol*. 170, 129–141. (doi: 10.1111/j.1469- 8137.2006.01651.x)

Huber L, and Gillespie, T.J., 1993: Modeling leaf wetness in relation to plant disease epidemiology. *Ann Rev Phytopath*, 30: 553-577.

Jones, D., 1998. *The Epidemilogy of Plant Diseases. Boston: Kluwer Academic Publishers*.

Johansen, J. R., and Rushforth S. R., 1985. Cryptogamic soil crusts: seasonal variation in algal populations in the Tintic Mountains, Juab county, Utah. *Great Basin Nat*, **45:** 14-21.

Juniper, B.E., 1991. The leaf from the inside and the outside: a microbe's perspective. *In: Microbial Ecology of Leaves (Edited by: JH Andrews, SS Hirano) New York: Springer-Verlag*, 21-42.

Lam T. B. T., 2000. Structural details of kenaf cell walls and fixation of carbon dioxide. *Proceedings of the 2000 international kenaf symposium*, Hiroshima, Japan, OCT. 13-14, pp. 81-90.

Luo, Y., Tebeest, D. O., Teng, P. S. and Fabellar, N. G. 1995 Simulation studies on risk analysis of rice leaf blast epidemics associated with global climate change in several Asian countries. *J. Biogeogr*. 22: 673–678. (doi: 10.2307/2845969).

Mehta, P.R. (1976). *Pesticides Annual* 14: 17.

Moity, S.S. (2003). Effect of meterological factors on production of conodia of *Helminthosporium* spp. and *Alternaria* spp.: the incidence of foliar blight of wheat. *Ann Pl. Protec. Sci*. 11: 315-317.

Narayana. L. Raut, S.P. and Gadre, U.A. (2006). Linear disease prediction model in groundnut rust epidemics. *Ann Pl. Protec. Sci*. 14 (1): 173-176.

Nan, Z. B., Skipp, R. A., and Long, P. G. (1991). Fungal invasion of red clover roots in a soil naturally infested with a complex of pathogens: Effects of soil temperature and moisture content. *Soil Biol Biochem*, **23:** 415-422.

Pal, M. (2002). Basics of Agriculture, First edition, Jain Brothers, New Delhi, 677-1033.

Rangaswami, G. and Madhhavan, A. (1999). Diseases of crop plants in India. Fourth edition. Prentice Hall of India, New Delhi (India).

Rowan, N.J., Johnstone, C.M., McLean, R.C., Anderson, J.G., and Clarke, J.A., 1999: Prediction of toxigenic fungal growth in buildings by using a novel modelling system. *Appl Environ Microbiol*, 65: 4814-4821.

Sameshima, K., 2000. Improvement of kenaf core oil absorption property by heattreatment at 200-500°C. *Proceedings of 3rd annual America kenaf society conference, Corpus* Christi, TX, February, pp. 64-72.

Singh, S.K., Naraian, U. and Singh, M. (1994). Effect of rainfall atmospheric temperature and relative humidity on Alternaria leaf spot of castor

Singh, A., Verma, K.S, and Mohan, C., 2009. Effect of weather parameter on *Colletotrichum gloesporiodes* causing anthracnose of guava. *Pl. Dis. Res.* 24 (1): 38-40.

Singh, A.K. (2009). Effect of weather parameters on management of Colletotrichum leaf spot of turmeric with fungicides and varietal resistance. *J. Mycol. and Pl. Path.*

Schmiedeknecht, M., 1995. Environmental tolerance range of *Meliolales* as mirrored in their horizontal and vertical distribution patterns. *Microbiol Res*, 150: 271-280.

Sprague, S. J., Balesdent, M.-H., Brun, H., Hayden, H. L., Marcroft, S. J., Pinochet, X., Rouxel, T. and Howlett, B. J. 2006. Major gene resistance in *Brassica napus* (oilseed rape) is overcome by changes in virulence of populations *of Leptosphaeria maculans* in France and Australia. *Eur. J. Plant Pathol.* 114, 33–40. (doi: 10.1007/s10658-005-3683-5)

Steed, J. M., Baierl, A. and Fitt, B. D. L. 2007 Relating plant and pathogen development to optimise fungicide control of phoma stem canker (*Leptosphaeria maculans*) on winter oilseed rape (*Brassica napus*). *Eur. J. Plant Pathol.* 118, 359–373. (doi: 10.1007/s10658-007-9137-5)

Stern, N. 2007. The economics of climate change: the Stern review. Cambridge, UK: Cambridge University Press. Sun, P., Fitt, B. D. L., Gladders, P. and Welham, S. J. 2000 Relationships between phoma leaf spot and development of stemcanker (*Leptosphaeriamaculans*) onwinter oilseed rape (*Brassica napus*) in southern England. *Ann. Appl. Biol.* 137, 113–125. (doi: 10.1111/j.1744-7348.2000.tb00043.x)

Taylor, E.C., 1979: Seasonal distribution and abundance of fungi in two desert grassland communities. *J Arid Environ*, 2: 295- 312.

Thomson, M. C., Doblas-Reyes, F. J., Mason, S. J., Hagedorn, R., Connor, S. J., Phindela, T., Morse, A. P. and Palmer, T. N. 2006 Malaria early warnings based on seasonal climate forecasts from multi-model ensembles. Nature 439, 576–579. (doi: 10.1038/nature04503).

Venier, L.A., Hopkin, A.A., McKenney, D.W., and Wang, Y., 1998. A spatial, climate-deterimined risk rating for Scleroderris disease of pines of Ontario. *Can J For Res*, 28: 1398-1404.

Van der Plank, J.E., 1975: *Principles of Plant Infection. New York: Academic Press.*

Wint, G. R. W., Robinson, T. P., Bourn, D. M., Durr, P. A., Hay, S. I., Randolph, S. E. and Rogers, D. J. 2002 Mapping bovine tuberculosis in Great Britain using environmental data. *Trends Microbiol.* 10, 441–444. (doi: 10.1016/S0966-842X(02)02444-7).

2015, Impact of Global Warming and Climate Change on
    Human and Plant Health
Editors: **Dr. Arun Arya and Prof. V.S. Patel**
Published by: **DAYA PUBLISHING HOUSE, NEW DELHI**

Pages 212–218

Chapter 21

Role of Microorganisms on Seedling Growth and Survival of *Terminalia bellerica* (Gaertn.) Roxb.

Hiral Buch, Vijay Mane and Arun Arya*

*Department of Botany, Faculty of Science,
The Maharaja Sayajirao University of Baroda,
Vadodara – 390 002, Gujarat, India*

ABSTRACT

Association of Mycorrhizal fungi is known to help the plants in alleviating the stress from deficiency of nutrients, and availability of water. A well known medicinal plant *Terminalia bellerica* locally known as 'Baheda' belongs to the family Combretaceae. The study was conducted over a period of one year to explore the effects of different seed treatments on germination and growth of the seedlings. Seeds were subjected to 10 different treatments by using three different gums obtained from *Anogeissus*, *Prosopis* and Gum Arabic. Percentage Seed germination was observed after 15 days of sowing. The highest percentage of germination and biomass was observed in the seeds treated with AM fungi and bacteria as compared to the seeds treated with only bacteria or AM fungi. The percentage of increase in plant biomass was more in seeds treated with microorganism.

* *Corresponding author.* E-mail: aryaarunarya@rediffmail.com

An effort was made to find out effect of water stress and the survival rate of *T. bellerica* seedlings by reducing water supply. When the potted plants were not watered for 10 days, the plants which were treated with AMF survived, while all others died.

Keywords: *Seedling germination, Growth, Terminalia bellerica, AM fungi.*

21.1 Introduction

Mycorrhizae form symbiotic association between plant roots and certain fungi. These associations have evolved with plants since the colonization of dry land by plants as an act of survival mechanism and higher plants. Both endure the existing environments of low soil fertility, periodic drought, diseases, extreme temperature and other natural stresses. AM fungi cannot complete their life cycle without host plants. Benefits derived by plants from this relationship includes increase nutrient uptake, more water absorption, increased nutrient mobilization, production of feeder roots, longevity as well as stress tolerance etc. (Manoharachary *et al.,* 2008). AM fungi are naturally occurring fungal component of soil microbiota in most of terrestrial ecosystems as well as in some aquatic plants (Stenland and Charvat 1994, Manoharachary *et al.,* 2008). However, it is important to distinguish between specificity (the ability to colonize), effectiveness (plant response to colonization), and ineffectiveness (the amount of colonization) because AM are widely different in these abilities depending on the environment. They do have wide host ranges, however and are capable of long term relationship with many different plants. In order for this partnership to work at at least four elements must be in place: a) Appropriate root morphology, b) Fungal structures able to penetrate the plant cell, c) Extra radical mycelia which are root like vegetative fungal structure growing in the soil and d) Soil condition.

AM fungi are the dominant component of the rhizosphere soil and transfer many assimilates to the roots. This alters the root exudation pattern and hence changes the microbial population dynamics of the rhizosphere and rhizoplane regions (Brundrett, 2004). The fungi responsible are classified in the phylum Glomeromycota, of order Glomales. They are assumed to be unculturable and, except for germination, wholly dependent on photosynthetic plants. AM fungi were used to be classified as Vesicular Arbuscular Mycorrhizae (VAM) but researches have uncovered that a major suborder did not form thin walled, lipid filled vesicles, so they are now referred to as AM associations. There is no evidence for specificity between plants and AM fungi (Smith and Read 1997). Mycorrizal fungi act as providers and protectors for plants. For example N,P,K are deficient in certain soils and can be increased in plant intake by mycorrhizae (Norland, 1993). In addition to phosphorus, AM mycelium also enhances the uptake of nitrogen in the form of NO_3 (Frey and Schuepp, 1993; Morte *et al.,* 2000) and also increases the potassium content in the plants (Azcon and Barea, 1992; Maksoud *et al.,* 1994). Other essential nutrients such as Ca, Mg, S, Fe, Zn, Al and Na have been shown to be increased in plants with AM fungi (Daft and Hacskaylo, 1976). Two mycologist Gupta and Germida (1988) found ample evidences for the involvement of fungi in the soil aggregation and stability of aggregates. Soil water repellency by AM fungi was due to presence of hydrophobins- a group of proteins usually present in filamentous fungi but not reported to occur in AM fungi (Rellig, 2005; Rellig *et al.,* 2010).

21.2 *Terminalia bellerica* (Gaertn.) Roxb.

T. bellerica belongs to Family Combretaceae commonly known as 'Baheda'. It is a large deciduous tree common on plains and lower hills and also grown as an avenue tree. In traditional Indian Ayurvedic medicine, its fruits are used in the popular Indian herbal rasayana called triphala. Its leaves are considered as a good fodder for cattle.

21.3 Materials and Methods

a) Isolation and Identification

The AM spores were isolated from rhizospheric soil of certain grasses of Baria division by Wet Sieving and decanting technique (Gerdemann and Nicolson, 1963). The bacteria *P. fluorescence* was isolated from the soil by serial dilution method (Aneja, 2009).

b) Mass Multiplication of AM Fungi

Funnel tube method was followed to germinate maize seeds in sterilized soil. The AM spores were taken in 3 different treatments 1, 5 and 25 spores per funnel. In each funnel 3 seeds were sown. The plants were later on transferred to pots/field. Mixture of sand + soil (50: 50) was used after sterilization for 1 h. After cooling down the pots (30 cm dia) 3/4th were filled with this mixture. AM spores were sterilized by keeping them in 200 ppm Streptomycin sulphate solution for 5 min. After disinfection, the spores were thoroughly washed in sterilized distilled water and were used as inoculant for initiating mycorrhization in maize plant. Add AM spores (20 spores in each pot) in center and one layer of soil was spread over to it. Five maize seeds were sown in center and again covered with 1 layer of soil. The pots were watered and kept in greeenhouse under constant observation. After 90 days, soil containing AM spores was used to inoculate nursery seedlings. Further after 90 days, the maize roots were cut in to pieces of 1cm and were mixed with the soil for preparing the AM inoculum and this incoulum was further filled in 150 pots containing garden soil and fresh soil without any fertilizer for mass multiplication of AM spores.

c) Seed Pelleting

Seed inoculation with *Pseudomonas fluorescence* was done by mixing the culture in 10 per cent jaggery and 40 per cent gum Arabic to form slurry to which seeds were added. With the result a uniform coat of the *P. florescence* was formed around the seeds. The inoculated seeds were dried in shade and were sown to observe their performance.

21.4 Results and Discussion

The saplings of *T. bellerica* were raised after treatment with bacterium (*P. fluorescence*) and AM fungus (*Glomus fasciculatum*) and a combination of Bacteria

Table 21.1: Showing Biomass Study of *Terminalia bellerica* Saplings Produced after Sowing Pelleted Seeds in 3 Gums

Type of Gum Used	Treatment	Shoot Length (cm)	Root Length (cm)	No. of Leaves	Shoot F.W. (g)	Shoot D.W. (g)	Root F.W. (g)	Root D.W. (g)
				15 Days				
	Control	20	8	9	1.87	0.26	0.28	0.03
G1	Bacteria	18	9	8	2.10	0.33	0.32	0.05
	AM	18	5	7	1.76	0.23	0.27	0.03
	AM+ Bacteria	19	8	8	1.95	0.31	0.34	0.05
G2	Bacteria	19	8	8	2.19	0.34	0.36	0.05
	AM	20	7	9	2.17	0.38	0.29	0.05
	AM + Bacteria	20	7	9	2.02	0.33	0.38	0.07
G3	Bacteria	22	10	8	2.34	0.40	0.39	0.06
	AM	22	8	9	2.44	0.32	0.33	0.05
	AM + Bacteria	23	10	9	2.35	0.42	0.40	0.06
				30 Days				
	Control	20	10	11	3.43	0.62	0.46	0.07
G1	Bacteria	25	11	11	3.89	0.79	0.72	0.13
	AM	22	10	10	3.16	0.52	0.46	0.06
	AM+ Bacteria	26	10	12	4.16	0.90	0.68	0.13
G2	Bacteria	26	8	12	3.54	0.67	0.39	0.07
	AM	27	10	12	4.11	0.80	0.58	0.11
	AM+ Bacteria	27	11	12	4.15	0.88	0.71	0.15
G3	Bacteria	30	12	14	3.85	0.90	0.54	0.10
	AM	31	16	12	3.43	0.71	0.46	0.11
	AM+ Bacteria	28	13	12	3.20	0.60	0.50	0.10
				45 Days				
	Control	30	14	12	3.4	0.91	0.63	0.2
G1	Bacteria	28	18	13	4.0	1.37	0.87	0.33
	AM	28	21	14	3.29	1.01	0.75	0.20
	AM+ Bacteria	27	20	13	3.3	1.12	1,03	0.34
G2	Bacteria	28	19	15	4.01	1.26	1.14	0.37
	AM	32	25	18	4.0	1.48	1.16	0.48
G3	Bacteria	31	24	15	5.01	1.65	1.33	0.53
	AM	30	28	18	4.40	1.67	1.27	0.48
	AM + Bacteria	33	27	18	4.0	1.07	0.80	0.26

Contd...

Table 21.1–*Contd...*

Type of Gum Used	Treatment	Shoot Length (cm)	Root Length (cm)	No. of Leaves	Shoot		Root	
					F.W. (g)	D.W. (g)	F.W. (g)	D.W. (g)
				60 Days				
	Control	30.3	22	16	3.89	1.34	1.27	0.43
G1	Bacteria	31.3	24	17	4.18	1.50	1.58	0.65
	AM	31.6	25.6	18	3.38	0.97	1.48	0.42
	AM+ Bacteria	29.6	21.6	16	3.32	1.35	1.09	0.27
G2	Bacteria	29.6	26.6	17	4.35	1.62	1.50	0.56
	AM	33.6	28	18	4.27	1.16	1.21	0.42
	AM + Bacteria	32	28	18	4.16	1.39	1.38	0.64
G3	Bacteria	34.6	30	19	5.20	1.86	2.87	1.10
	AM	36	31	18	4.38	1.30	1.75	0.74
	AM + Bacteria	35	35.6	19	4.84	1.56	2.61	0.90
				75 Days				
	Control	33.3	30.3	17	3.9	1.24	1.73	0.671
G1	Bacteria	33.3	29.6	16	3.1	1.60	2.03	0.94
	VAM	28	33	17	3.1	1.16	1.71	0.85
	AM+ Bacteria	28.3	30.3	16	2.95	1.6	1.56	0.30
G2	Bacteria	30.3	31	17	4.58	1.41	1.29	0.48
	AM	33.0	28.6	18	4.1	1.55	1.56	0.66
	AM + Bacteria	32	34.6	18	4.88	2.28	2.81	0.989
G3	Bacteria	35	32	17	3.84	1.65	1.52	0.77
	AM	36	33	17	3.56	1.37	1.55	0.788
	AM + Bacteria	38	40	18	5.23	2.19	3.30	1.396
				90 Days				
	Control	38.0	33.3	17	5.30	2.35	2.25	0.834
G1	Bacteria	37.3	27.3	16	4.67	2.17	2.20	0.859
	AM	33.8	31.6	17	4.89	2.34	2.68	0.879
	AM+ Bacteria	33.6	30.3	16	4.21	2.11	2.35	0.867
G2	Bacteria	35.2	31.3	17	5.12	2.35	3.15	1.125
	AM	34.2	33.9	17	5.32	2.39	3.20	1.128
	AM + Bacteria	37.1	32.6	18	5.46	2.47	3.49	1.136
G3	Bacteria	35.1	33.1	17	4.12	1.89	2.95	1.122
	AM	34.4	35.2	17	4.97	2.39	3.15	1.135
	AM + Bacteria	39.5	40.7	18	5.25	2.45	3.19	1.142

Contd...

Table 21.1–*Contd...*

Type of Gum Used	Treatment	Shoot Length (cm)	Root Length (cm)	No. of Leaves	Shoot		Root	
					F.W. (g)	D.W. (g)	F.W. (g)	D.W. (g)
				105 Days				
	Control	34.0	33.9	17	5.93	2.24	2.39	0.839
G1	Bacteria	35.2	29.6	17	4. 76	2. 71	2.63	0.853
	AM	33.6	30.3	17	4.21	2.11	2.35	0.867
	AM+ Bacteria	33.8	31.6	17	4.89	2.34	2.68	0.879
G2	Bacteria	35.2	32.3	18	5.29	2.38	3.51	1.145
	AM	34.2	33.9	17	5.40	2.51	3.40	1.153
	AM + Bacteria	37.1	32.6	18	5.64	2.51	3.55	1.178
G3	Bacteria	35.1	35.1	18	4.67	1.94	3.21	1.189
	AM	34.4	37.2	18	5.12	2.93	3.46	1.167
	AM + Bacteria	39.5	40.7	18	5.55	2.78	3.34	1.219
				120 Days				
	Control	30.0	37.0	19	3.17	1.18	3.02	2.26
G1	Bacteria	–	–	–	–	–	–	–
	AM	42	42	20	2.92	1.142	3.31	1.65
	AM+ Bacteria	48	51	19	3.80	1.97	4.32	2.02
G2	Bacteria	–	–	–	–	–	–	–
	AM	48	59	20	3.98	2.72	5.81	2.13
	AM + Bacteria	50	60	21	3.21	1.751	3.35	1.70
G3	Bacteria	–	–	–	–	–	–	–
	AM	38	49	20	3.79	2.59	5.38	3.39
	AM + Bacteria	44	57	21	3.17	1.63	3.22	1.630

G1: Gum of *Anogeissus latifolia* G2: Gum of *Prosopis juliflora*, G3: Gum of *Acacia arabica*, F.W.: Fresh Weight; D.W.: Dry Weight.

and AM fungi. The treatment was done in seeds pelleted with three different types of gums. The use of gum helped in binding the bacteria *Pseudomonas* and small pieces of roots containing AM fungi. Shoot length and number of leaves in *T. bellerica* seedlings increased up to 120 days. In comparison to control more shoot and root length was observed in three treatments. More shoot length was observed in treatment with gum Arabic. Treatment with AM + bacteria showed better results. AM fungi are known to help in water balance in plants in an experiment when the watering was stopped for 10 days, all the plants treated with bacteria failed to survive and physiological wilting was caused in control plants. However, the saplings inoculated with AM fungi and AM fungi + bacteria survived in the water deficit conditions.

References

Aneja, K.R. 2009. *Experiments in microbiology plant pathology and biotechnology,* 4th ed. New Age International Publishers, pp: 604.

Brundrett M. 2004 Diversity and classification of mycorrhizal associations. *Biol Rev* 79: 473-495.

Daft M. and E. Hacskaylo. 1976. Arbuscular mycorrhizas in the anthracite and bituminous coal wastes of Pennsylvania. *Journal of Applied Ecology.* 13: 523 – 531.

Gerdemann J.W, Nicolson T.H 1963. Spores of mycorrhizal endogone species extracted from soil by wet sieving and decanting. *Trans. Brit. Mycol. Soc.* 46:235 – 244.

Manoharachary C., Swarupa Rani S. and Kunwar I.K. 2008. Arbuscular Mycorrhizal Fungi Associated With Some Plants of Apocynaceae. *J Mycol Pl Pathol* 38(1): 91-92.

Manoharachary C., Ameetha P. Kunwar I.K. and Reddy V.S. 2008 Arbuscular Mycorrhizal Fungi in Relation to Plant Growth in *Dalbergia sissoo. J Mycol Pl Pathol* 38(1): 97-101.

Norland M.R. 1993. Soil factors affecting mycorrhizal use in surface mine reclamation. US Department of Interior, Bureau of Mines, Information Circular/9345 pp.21.

Smith S.E and Read D.J. 1997. *Mycorrhizal Symbiosis.* 2nd edn. Academic Press, London. pp 1-589.

Part IV

Changing Environment: Ecofriendly Options

वृक्ष जीवन का सौन्दर्य

छायी हैं खुशियां जीवन में
फैला है सौन्दर्य जहां में।
मिलने को आतुर हैं बूंदें
स्वर्ण रश्मियों से पल भर में।
वृक्ष उगाती है यह धरती
सबका पालन यह करती है।
ज्ञान दिया जब उन ऋषियों ने
सबका पालन यह करती है।
नई सुबह कल फिर आयेगा
सबका जीवन सुधर सकेगा।
मानव फिर अपना हित तजकर
सबका पालन हार बनेगा।
भरी दोपहरी में रह कर के भी
जो शीतलता को बांट रहे हैं।
जीवन संघर्ष का सही तरीका
वह हमसे कहीं बेहतर जान रहे हैं।

2015, Impact of Global Warming and Climate Change on
   Human and Plant Health

Pages 221–232

Editors: **Dr. Arun Arya and Prof. V.S. Patel**
Published by: **DAYA PUBLISHING HOUSE, NEW DELHI**

Chapter 22

Tree Cover and Urban Environment: Trees in Urban Areas in Gujarat State

H.S. Singh

IFS, Addl PCCF,
Social Forestry, Gujarat State
Aranya Bhavan, Sector-10, Gandhinagar – 382 001, Gujarat, India

ABSTRACT

Science related to tree and forest cover in growing urban areas with respect to human environment is drawing interest of the scientists as well as urban development planners. Minimum level of tree cover in a city has to be standardized in the developing countries, although some of the developed countries have developed norms. Gujarat State in India has taken initiatives in this direction. Total tree enumeration in all urban areas (eight municipal corporations and 159 municipalities) in Gujarat State, a first initiative of its kind in the country, was done in 2011 with an objective to plan urban forestry for improving environment. Tree cover and density, tree cover per inhabitant, and Carbon store value and scope of Carbon sequestration near the source of its emission have been studied. Of the eight major cities in the state, the tree density and canopy cover in Gandhinagar is highest in India and it may be listed amongst the greenest cities in the world. Out of 167 urban areas, 31 cities/towns meet the minimum norm of tree cover whereas the rest have inadequate or poor tree cover. Tree cover in the growing urban areas have been critically examined and concluded that it should be treated as environment infrastructure in the urban areas to address growing environmental issues. It has been found that 6.54 m trees in the urban areas in Gujarat store about 1.36 m tons of Carbon which may increase substantially after improving tree cover under a planned urban forestry. The future planning of urban forestry and the scope of improving greenery to achieve a minimum standard of tree cover in the urban areas has been also highlighted.

Keywords: *Green cover, Tree cover, Urban forestry, Urban trees.*

22.1 Introduction

A century ago, just about one tenth of the global population was living in the cities, which now exceeds 50 percent (Singh *et al.,* 2011). At present, the urban areas in the world account for over three fourth of Carbon emission, sixty per cent residential water and about three fourth of industrial wood consumption. The most explosive urban growth is expected in Asia which is likely to have the largest urban population in the world. Urban areas in the developing countries have multi-faceted environmental problems. In the context of rapid urbanization and growing environment problems, social awareness campaign for the governance of urban systems that counter the menace of defacement of patches of tree cover and sensitize the issues of environment, hygiene, sanitation and solid waste management. To establish such governance, there is a need of developing urban environment science which may provide scientific information for judicious decision on matter pertaining urban environment.

Science of tree cover, dotted in the urban landscape, is interesting to understand sustainable urban growth. Trees in the urban landscape play a role in improving air and water quality; bring cooling effect; reduce pollution and suspended particle matter; provide benefits relating to biodiversity; reduces stress; and is seen to benefit the health of urban people and provide so many other environmental and social benefits (Rowantree and Nowak, 1991). In other words, trees are major environment capital assets and infrastructure in cities that require care and maintenance. Recognizing the roles of urban trees, the **Green India Mission (GIM),** one of the missions under India's National Action Plan on Climate Change (NAPCC) has aimed to enhance tree cover in urban and peri-urban areas in over two hundred thousand hectares (Anon. 2011)**.**

Immense ecological and economic benefits from urban forest have been documented in some countries. Urban land in the United States currently occupies about 28 m hectare which store approximately 704 m tonnes of Carbon in trees with an estimated annual net Carbon sequestration of around 22.8 m tonnes. Besides directly storing carbon, urban trees also reduce carbon dioxide emissions by cooling ambient air and allowing residents to minimize annual heating and cooling (Nowak *et al.,* 2006). There are 583,000 people and 1.47 m trees in Portland's city, Oregon, USA and the tree canopy cover is about 26 per cent of the total city area with tree density of about 43 trees/ha. The annual environmental benefits provided by the entire urban forest canopy cover exceed $ 38 m (Anon., 2007). About 1.92 m trees in Washington, D.C. cover 28.6 percent of the city area and store about 526,000 tons of Carbon value, removing about 16,200 tons of Carbon and about 540 tonnes of air pollution annually.

The value of tree cover in the city has been estimated at $3.6 billion in the city (Nowak *et al.*, 2006).

It has been found that 2.4 m trees in the central part of Beijing removed 1261.4 tonnes of pollutants and 772 tons particulate matter with an aerodynamic diameter smaller than 10 μm (PM_{10}) from the air in 2002 (Jun Yang *et al.*, 2004). The tree planting in the urban areas of China has been proposed by the municipal governments, considering tree cover as infrastructure to sequester Carbon, alleviate air pollution and maintain environment of the cities (Ming and Profous, 1993 and Jun *et al.*, 2004). It has been estimated that a full grown Pipal tree (*Ficus religiosa*), for instance, is estimated to give 600 kg of oxygen in 24 hours (Goel and Singh, 2006). Also, the large healthy trees remove more air pollution and Carbon annually than small healthy trees. These studies reveals relevance and necessity of adequate tree cover to maintain the human environment in the urban areas.

To design urban forestry programme, it is necessary to know the status of tree cover. As a result, the tree counting, a first such exercise was planned in all urban areas in Gujarat State, India. Tree counting in all Municipal Corporations areas and Municipalities areas was done in 2011 by the Social Forestry Wing of the Gujarat Forest Department with objectives to estimate the number of trees and status of tree cover in the urban areas; to create a baseline information to monitor trend of tree cover; to find out the preferences of tree species in urban areas for using such information for planning urban forestry; to assess changes in tree cover in future; to assess Carbon store and also scope of Carbon sequestration; and to develop plan for improving tree cover in the cities and towns.

22.2 Study Area

The tree enumeration and survey have been done in the urban areas in Gujarat State in the western region of India. The urban areas in Gujarat cover about 1.77 per cent of the state's geographical area against about 3 per cent earth's urbanized terrestrial surface in the world (Singh *et al.*, 2010)[1]. There are 8 Municipal Corporations, each of them surrounded by the buffer zone of villages under the Urban Development Authority, and 159 Municipalities in Gujarat State. Total tree counting was done in all cities and towns-eight municipal corporations and 159 municipalities which cover a total area of 345,762 ha. Villages under Urban Area Development Authorities around the eight municipal corporations were also covered to estimate tree cover.

Compared to other parts of India, the urbanization in Tamil Nadu, Maharashtra and Gujarat is relatively fast. The urban population in Gujarat has increased to 42.6 per cent (25.7 m) of the state's total in 2011, from 28.1 percent in 1971. With the present trend, forecasts suggest that, by 2021, about 35 m people constituting nearly half of the state population would be residing in the urban Gujarat (Urban Department, Gujarat State, 2011).

22.3 Methodology Adopted

The Government has established Social Forestry Division in each district under Social Forestry Wing of the Forest Department to implement programmes of tree plantation outside forest areas. Also, every taluka (administrative unit under district) in Gujarat has social forestry organization. The Social Forestry Wing decided to conduct total tree counting in all municipal corporation and municipalities in Gujarat State for proper planning of the urban forestry. The areas under urban development authorities of the Municipal Corporations were also included in this exercise. Within the boundaries of municipal corporations (*Mahanagar Palikas*) and municipalities (*Nagarpalikas*), all trees standing in the parks and garden; compounds of schools, colleges and institutions; along streets and roads; cantonment area; forest lands; individual premises and any other such areas were counted in different diameter classes. In the area of Urban Development Authority, tree counting was done in randomly selected villages (about 11 per cent of total villages) and the data was projected to estimate the tree population. The enumerations of trees were done from September to December, 2011, and verification of counting was done in the December 2011 and also in January 2012. The regional heads of the social forestry circles had coordinated tree counting in the circle (zone). The counting in cities at district level was organized and conducted by the Dy. Conservator of Forests whereas Assistant Conservator of Forest and Range Forest Officers coordinated and conducted tree counting in municipalities. The teams for each urban area consisting of foresters, forest guards, members of Non-government Organization, students were constituted, and they were guided and trained about the methodology to count and measure the girth of the trees above 10 cm at breast height (GBH). In certain cities, municipalities and educational institutions also joined the department in this exercise. About 2160 forest staffs, employees of municipal corporations, students and member of NGOs worked to count trees. Labourers were also employed to help the staff. In this tree census, trees of each species were counted in the eleven girth classes (10 to 30 cm, 30⁺to 45 cm, 45⁺ to 60 cm, 60⁺ to 75, 75⁺to 90 cm, 90⁺ to 120 cm, 120⁺ to 150 cm, 150⁺to 200cm, 200⁺ to 250 cm. 250⁺to 300 cm and above 300 cm.).

    Tree cover in Gandhinagar was also assessed using the recent remote sensing data. The results of tree cover and tree population in Gandhinagar helped to develop an equation between tree cover and number of trees, which was applied to estimate tree cover. A study based on remote sensing data has estimated 3,075 ha of tree cover over in Gandhinagar which is equivalent to 8.67 lakh trees (both estimates are of the year 2011) and 53.9 per cent of geographical area. Thus, about 282 trees, distributed in the different girth classes, may be equivalent to one hectare of tree cover (about 40 per cent canopy cover) in Gujarat. The calculations regarding tree cover in the cities presented in this report are based on this equation (Anon, 2011).

22.4 Results and Discussion

Initiatives in Gujarat State

    Tree plantation in cities and towns has been done throughout the state. First planned urban a forestation in Gujarat was done in Gandhinagar, after a decision of the Government to develop area as the State Capital. The Forest Department was first to step in to start tree plantation in a planned manner in mid 1960s. Ravines, areas around Sabarmati River, road sites dividing the sectors were also planted. Subsequently, open areas of the sectors were also afforested. From 1971 to 2011, a total of about 3.50 m seedlings were planted (Anon, 2011).

Van Mahotsav was celebrated in all towns and cities. The park and garden wings of the municipalities carried out tree plantation. In 2005, Gujarat Urban Development Mission (GUDM) of the Urban Department and Forest Department initiated *Nagar Nandan Van Yojana* in 129 municipal corporations and municipal areas in 2005-06 to plant 834,000 tall seedlings and also to distribute about 633,000 seedlings to the people. This programme helped to improve tree cover in several towns and has been one of the successful a forestation programme to improve urban ecology and environment. Subsequently, this plan was extended to more towns. The Urban Department also implemented plantation schemes separately to improve greenery in the cities. Tree plantation in campaign mod was also carried out in some cities such as Ahmedabad, Surat and Vadodara. Through notification of the Gujarat Urban Development Department in November 2009, the Government emphasized Urban Green Plan to improve tree cover. The provision for creation of *Panchavati, Smriti Van, Sanskritic Van,* Oxygen park, green guard etc have been made to involve people, NGOs and civic society to plant trees in the open space. Urban Authorities are advised to earmark certain percent of budget for raising trees.

Gujarat State initiated a programme to establish cultural forest (*Sanskritic Van*) near urban and sub-urban centre in 2004 during celebration of State Level *Van Mahotsav* at Gandhinagar in 2004, *Mangalya Van* at Ambaji in 2005, *Tirthankar Van* at Taranga in 2006, *Harihar Van* at Somnath in 2007, *Bhakti Van* at Chotila in 2008, *Shyaml Van* at Shamlaji in 2009, *Pawak Van* at Palitana in 2010, *Virasat Van* at Pawagadh in 2011 and Govind Guru Smriti Van at Mangadh in 2012. To connect people with trees, *Nakshatra Van, Rashi Van, Panchavati, Ashok Van, Dhanvantri Van, Smruti Van, Ajivika Van* etc. have been created in these cultural forests. The *Sanskritic Van*s have become important recreational and educational parks for people and added to environmental values of the urban area. Urban Authorities also established such forest at several sites. The Government also initiated tree plantation in campaign mod during the last three years in several cities such as Ahmedabad, Surat, Vadodara.

22.4.1 Tree Cover in the Municipal Corporation Areas Tree

Tree counting revealed that Gandhinagar was the greenest city in India in the terms of tree density as well as the area under tree cover. Over all, Gandhinagar, Bhavnagar and Vadodara may be called as green cities having tree densities higher than the average density of the eight municipal corporations. Other municipal corporations-Surat, Ahmedabad, Rajkot, and Jamnagar have tree densities below the average (Table 22.1). Junagadh has less tree density within the municipal corporation boundaries but dense forest in a large area adjoining the city improves tree cover status.

Average tree cover in the eight municipal corporations is 8.78 per cent of geographical area, and the average tree density and tree cover per inhabitant are 24.8 trees/ha and 9.5 m^2, respectively, which is below the average in the developed countries.

Villages around the major cities-Municipal Corporations have been included in Urban Development Zone under the Urban Development Authority. Urbanization in these villages is fast, and some of them have turned sub-urban areas and they are

Table 22.1: Tree Population in Municipal Areas-Eight Major Cities in Gujarat

Sl.No.	Municipal Corporation	Human Population	Geographical Area in ha	Number of Trees (above 10 cm GBH)	Tree Cover (in ha)	Tree Cover/ Inhabitant (m²)	Tree Density per Hectare	Tree Cover per cent of Geographical Area
1.	Ahmedabad	5,570,590	46,985	617,090	2188	3.9	13.1	4.66
2.	Surat	4,462,000	39,549	333,970	1184	2.7	8.4	3.00
3.	Vadodara	1,666,700	16,264	747,190	2650	15.9	45.9	16.29
4.	Gandhinagar	208,300	5,700	866,670	3075	147.6	152.0	53.90
5.	Rajkot	1,287,000	10,400	137,520	488	3.8	13.2	4.69
6.	Bhavnagar	593,770	5,320	475,950	2106	35.5	89.46	21.35
7.	Junagadh	320,250	5,670	76,690	272	8.5	13.5	4.80
8.	Jamnagar	529,310	3,434	45,880	1877	3.1	13.4	4.74
	Total	14,637,920	133,322	3,300,960	13,840	9.5	24.8	8.78

Note: (i) Victoria Park, a forest area, is within Bhavnagar Municipal Areas and a total of 212,500 trees of Victoria Park are included in the above figures. Similarly, trees in Inroad Park (forest area) are included in the tree population in Gandhinagar.

integrating fast with the municipal corporations. The tree populations in 433 villages of the eight Urban Development Authorities, covering an area of 310,261 ha, was about 6.36 m with average density of 20.5 trees/ha.

In Gujarat, it is difficult to increase tree cover beyond a limit in the municipal corporation areas due to non-availability of the space. The gap of tree cover can be compensated by increasing tree cover in sub-urban areas under the respective Urban Development Authorities. There are several lakes/ponds in the villages around the municipal corporation areas. These wetlands should be preserved and their status should be improved by planting suitable trees like *Ficus* sp., *Azadirachta indica*, *Syzizium cumini*, Rayon, Mango-*Manigera indica* and Deshi babool-*Acacia nilotica*. Some of these lakes and ponds can be developed as recreational and tourist sites by enriching areas raising suitable species of trees. Average tree density in Anand district in Gujarat, one of the greenest districts is over 68 trees/ha. Rural areas in Nadiad, Mehsana and Gandhinagar have an average tree density above 50 trees/ha (Anon, 2009)[10]. It is possible to achieve this level of tree cover under an intensive agro-forestry and horticulture programmes in the villages around the Municipal Corporations.

22.4.2 Tree Cover in Municipalities (Nagar Palikas)

Geographical area of 159 municipalities in Gujarat State is about 212,440 ha and a total of 32,44,160 trees having Girth at Breast Height (GBH) above 10 cm were counted in these municipalities with average tree density of about 15.27 trees per hectare and tree cover of about 5.4 per cent of geographical area. Eight urban areas had tree density above 50 trees/ha, 23 between 30 to 50 trees/ha, 58 between 10 to 30 tree/ha and the rest of the urban areas had very poor tree cover below 10 trees/ha.

Table 22.2: Summary of Tree Cover in the Urban Areas (Mahanagar Palikas+Nagar Palikas)

Descriptions	Total/Average of Eight Municipal Corporations	Total of 159 Municipalities	Total/Average of Urban Area
Area (ha)	133,322	212,440	345,762
Human population	14,637,920	11,062,100	25700,000
Tree population	3,300,980	3,244,160	6,545,140
Average tree density (trees/ha)	24.76	15.27	18.93
Tree cover (ha)	13,840	11,504	25,344
Average tree cover of total area	10.38 per cent	5.42 per cent	7.33 per cent
Tree cover per inhabitant (m²/inhabitant)	8.8	10.6	9.5

Eight municipal corporations support about 3.30 m trees, whereas 159 municipalities have 3.24 m trees. Thus, about 6.54 m trees grow in cities and towns in Gujarat state.

22.4.3 Tree Species: Commonly Present in Cities

Over 235 tree species, including exotic trees were encountered during the enumeration in urban areas of Gujarat. The main tree species in order of their decreasing number are Neem (*Azadirachta indica*), *Prosopis chilensis,* Deshi Babool (*Acacia nilotica*), Amaltas or Garmalo (*Casia fistula*), Asopalav and Pendula (*Polyalthia longifolia* and *Polyalthia pendula*), Peltrofurum (*Peltroforum ferruginieum*), Kasid (*Casia siamia*), Gulmohar (*Delonix regia*), Sirus (*Albizia lebbek*), Kanji (*Holopetelia integrifolia*), Saptparni (*Alstonia scholaris*), *Eucalyptus* sp. Sharu (*Casuarina equisetifolia*), Pipado (*Ficus tsila*), Ardusa (*Ailanthus excela*), and Mango (*Mangifera indica*). Gorad (*Acacia senegal*) is dominant in the forest of Victoria Park and Indroda Park. The number of *Ficus* species-Banyan (*Ficus benghalensis*), Pipad (*Ficus tsila*) and Peepal (*Ficus religiosa*) is also good but their presence in all cities or dominance in some cities are felt due to their dimension. Neem is the most dominant tree in towns and cities in North Gujarat, Saurashtra and Kachchh. Number of *Asopalav*, a species planted in gardens and compounds, is also high. Population of grazing resistant trees like *Prosopis,* Kasid, Amaltas, Saptaparni and Kanji is increasing due their success in plantations.

Size of trees in Ahmedabad city was relatively large. In eight municipal corporations, about 61.8 per cent trees belonged to 10-60 cm girth (GBH) class whereas 30.3 per cent trees had girth between 60 cm and 120 cm, 6.6 per cent between 120 cm and 200 cm and about 1.3 per cent trees had girth above 200 cm.

22.4.4 Comparison of Tree Cover in the Urban Areas in Gujarat with other States and Countries

In India, Gandhinagar, Bangalore, Chandigadh, New Delhi, Guwahati, Dehradun, Bhubaneswar, Shillong, and Nagpur area few cities which have relatively good tree cover and natural environment (Chaudhary and Tewari, 2011). FSI (2011) has estimated 11.9 per cent of the geographical area under tree and forest cover in Delhi, and 14.9 per cent in Chandigadh. A good tree crown cover has also been estimated in Bangalore. The tree cover in Gandhinagar (53.9 per cent) is highest amongst all cities in India. Other two cities –Bhavnagar and Vadodara in Gujarat have good tree cover, comparable with the green cities in the world. The rest of urban areas have poor tree cover.

The average tree cover in the 20 main metropolitan areas in the USA has been estimated about 27.1 per cent of geographical area (Nowak *et al.*, 2006). The average of tree crown coverage was 26.74 per cent in the urban and sub-urban areas in cities of Japan (Anon 2003). Average woodland cover has been estimated about 18.5 per cent of the geographical area within municipal limits of 26 large European cities (average tree cover-104 m²/inhabitant) (Chaudhary and Tewari, 2011). In 439 cities in China, the overall green space was 380,000 ha or 20.1 per cent of the urban area and 40 per cent of them had more than 30 per cent green cover in 1991. Subsequently, the tree cover in these cities increased to 23.0 per cent, (6.52 m²/inhabitant) by 2000 and then to 32.5 per cent of the urban area by the end of 2006 and the country is at target to achieve the green cover. This reveals that urban forestry in China has been successful (Singh *et al.,* 2010; Wang, 2009). There is no authentic information that ranks the green cities of the world, but the tree cover in Gandhinagar-capital city of Gujarat is at the level of tree cover in Atlanta-city which has highest tree density in USA (Nowak *et al.,* 2006).

The experts in Germany, Japan and other countries proposed a standard of 40 square meters (m²) urban green space in high quality per capita for reaching a balance between carbon dioxide and oxygen, to meet the ecological balance of human well-being (Singh *et al.*, 2010). Currently, developed countries have tended to adopt a general standard of green space of 20 m² green space per capita. World Health Organization suggests ensuring at least a minimum availability of 9 m² green open space per city dweller. Due to growing environment problems, several cities in Asia have planned to improve tree and green cover and the success is impressive in several cities in China. Manila aims to achieve a target of four trees per person. In India, success in Delhi is appreciable, as tree cover and gardens have expanded and improved during the last decade but the three other mega-city-Kolkata, Mumbai and Chennai have poor tree cover, far below the standard. Among medium and major cities in India, except about a dozen cities, majority of them have poor tree cover.

Majority of the cities, including the green cities, in India have tree cover below the minimum optimum norm. In Gujarat, three municipal corporations and about a dozen municipalities have tree cover above 15 per cent of the geographical area. Due to the lack of space and demand of the land for other purpose, it may be difficult to achieve the tree cover of 15 per cent of city area or 20 m² per inhabitants in the megacities like Ahmedabad and Surat. To develop Carbon sink near source of emission, improving tree cover in the surrounding villages-areas under Urban Development Authorities in such cities is necessary.

22.5 Carbon Storage and Sequestration

In the Indian forest, as a thumb rule, one cubic meter of growing stock in the natural forest is equivalent to 1.1 tons of Carbon (including Carbon in soil) (David *et al.*, 2001; Kishwan *et al.*, 2009). In the Trees outside Forest (TOF), one cubic meter of growing stock may be equivalent to about half tone of Carbon. On an average, one tree store about 181 kg of Carbon in the urban areas in USA (David *et al.*, 2001), although it depend of the climatic and edaphic condition of the area and composition of tree species. In Washington DC, on an average, one tree store about 274 kg of Carbon (Nowak *et al.*, 2006)). The rate of annual Carbon sequestration has great variation with increase in the girth of tree. A study reveals that a good size tree of *Prosopis chilensis* store Carbon at rate of about 7.5 kg every year. The size of trees in urban areas of Gujarat is relatively higher than tree in the rural area. Average of eight municipal corporation indicates that 1.31 per cent of total trees have GBH over 200 cm. In Ahmedabad, about 3.70 per cent trees have GBH above 200 cm. This reveals that some cities in Gujarat have relatively more old trees than in the other areas.

In India, the growing stock of 5,068 m stems of the TOF is about 1,548 m cubic meters. In Gujarat, about 268 m trees have growing stock of 118 m cubic meter. It has been estimated that, on an average one tree in Gujarat may store about 207 kg of Carbon against India's average of 208 kg Carbon/tree (FSI,2011, Anon.,2009 and Kishwan *et al.*, 2009). The average size of trees in urban areas in Gujarat, particularly in Ahmedabad is high, so the Carbon store value.

Also, as per the present growing stock, the annual potential production of timber and fuel wood in Gujarat are 3.1 m cubic meter for timber and 2.12 m tones for firewood (FSI 20011). The Carbon sequestration rate may be estimated about 7.86 kg Carbon/tree/year for the average of India. If the growing stock of the TOF in Gujarat is co-related to the number of trees, the Carbon sequestration rate has been estimated higher than the average of the country. If it is assumed the average Carbon store of about 208 kg Carbon/tree in Ahmedabad and 207 kg Carbon/tree for the rest of cities, the Carbon store in different municipal corporation may be estimated about 179,000 Carbon tone for Gandhinagar, 128,000 Carbon tone for Ahmedabad, 155,000 Carbon tones for Vadodara, and 99,000 Carbon tones for Bhavnagar and 122,000 tonnes for the rest of other four municipal corporations. Thus, total Carbon store in the trees in eight municipal corporations may be about 683,000 tones (equivalent to 2.51 m tones of CO_2).

It has also been estimated that the trees in eight municipal corporations and 159 municipalities store about 1.36 m tone. This figure may be about 2.67 ms tones, if peripheral villages of the municipal corporations are included in the estimate. Annual Carbon di-oxide absorption rate of these trees in municipal corporations and municipalities has been estimated about 197,000 tones of Carbon dioxide. Doubling the tree cover in urban areas in Gujarat may result into two fold increase in rate of Carbon sequestration and reductions in energy consumption.

22.6 Conclusions

Urban environment science found that urban areas, dotted with lakes, tree groves and surrounded with villages had less threat of abrupt rise of temperature in summer, but the situation is changing fast in all fronts-rise in concentration of green-house gases, global rise of temperature, local increase in temperature due to expansion of concrete jungles. In several big cities like Ahmedabad, the surrounding satellite areas are urbanizing at unprecedented rate and they are gradually merging with the big cities. The expanding concrete jungle is changing landscape. There is possibility of development of hot island temperature in the concrete jungle within the large urban landscape in which possibility of abrupt rise in temperature in extreme summer, beyond the tolerance of human beings, in the absence of adequate tree cover is not ruled out. Thus, maintaining adequate tree and green cover has become unavoidable in the urban areas for better life of its citizen.

Many forests at the urban fringe such as Victoria Park in Bhavnagar, Indroda Park in Gandhinagar and parks and garden in Vadodara are threatened by urbanization and they need protection. This is possible when true value of such forests is accounted scientifically to counter the forces of urbanization. The American Forests Urban Ecosystem Analysis (Green City) offers a means of quantification of the environmental benefits of urban trees. American Forests has transferred Urban Ecological Analysis methods to local communities so that they can recognise, measure, and advocate better natural resource policies in their communities. Benefits include energy conservation, urban heat island reduction, storm water runoff, air pollution reduction, particulate pollution reduction, noise and glare control, carbon sequestering and urban recreation/enjoyment. The analysis has been developed to help local people demonstrate the value of trees in their community to its leaders, putting tools into the hands of local people, so that they can make community-based decisions that affect long-term planning, management, and funding. Such method should also be tested in the countries in the Asian region, including urban areas in Gujarat. The tree census in urban areas of Gujarat, provide accurate data which may be used to develop science of urban forestry. If such exercise are done in other cities in the country and the enumeration data are linked with environmental parameter of the area, it may be possible to develop environment science of the urban forestry.

References

Anon. 2003. A Study on Counting Method for Urban Tree Cover Area Using from Natural Vegetation Data. *Journal of the Japanese Institute of Landscape Architecture.* TESHIROGI JUN (Organization for Landscape and Urban Greenery Technol. Dev., JPN). Journal Code: F0408A. 66(5): pp 859-862.

Anon. 2007. *Portlands Urban Canopy.* Portlands Parks and Recreation City Nature Urban Forestry: pp. 79.

Anon. 2009. *Tree wealth of the non-forest areas of Gujarat*-Tree census in non-forest area-2009. Gujarat Forest Department, pp. 56-67.

Anon. 2011. *Green India Mission*, Ministry of Environment and Forests, Government of India, New Delhi.

Anon. 2011. Tree cover in Gandhinagar Capital City Area using remote sensing technique. Gujarat State, Gandhinagar (report).

Nowak, D.J., Crane, Daniel E. 2001. Carbon storage and sequestration by urban trees in the USA. USDA Forest Service, Northeastern Research Station, 5 Moon Library, SUNY ESF, Syracuse, NY 13210, USA. Environmental Pollution 116 2002 381-389.

FSI 2011. India State of Forest Report. The Forest Survey of India, Dehradun. MoEF, New Delhi.

Goel, Abhineety and R. B. Singh 2006. Sustainable forestry in mega-cities of India for mitigating Carbon sequestration: A case study of Delhi. *Advance in Earth Science.* Vol 21 (2):144-150.

Jun Yang, Joe McBride, Jinxing Zhou, and Zhen yuan Sun 2004. The urban forest in Beijing and its role in air pollution reduction. *Urban Forestry and Urban Greening*, Elsevier. 3, pp. 65-78.

Kishwan, Jagdish, Rajiv Pandey and V. K. Dadhwal, 2009. India's Forest and Tree Cover: Contribution as a Carbon Sink (Technical Report). Indian Council of Forestry Research and Education, Dehradun, Uttarakhand.

Ming, S. and Profuse, G. 1993. Urban forestry in Beijing. *Unasylva* 44(173): 13-18.

Nowak, D.J., Hoehn, R.E. III, Crane, Daniel E., Stevens, Jack C., Walton, Jeffrey T. (2006). Assessing urban forest effects and values, Washington, D.C.'s urban forest. Resour. Bull. NRS-1. Newtown Square, PA: U.S. Department of Agriculture, Forest Service, Northern Research Station. 24 p.

Nowak, David J.; Hoehn, Robert E. III; Crane, Daniel E.; Stevens, Jack C.; Walton, Jeffrey T.,2006. *Assessing urban forest effects and values*, Washington, D.C.'s urban forest. Resour. Bull. NRS-1. Newtown Square, PA: U.S. Department of Agriculture, Forest Service, Northern Research Station. 24 p.

Chaudhary, P. and Tewari, V.P. 2011. *Urban forestry in India: development and research scenario.* Interdisciplinary Environmental Review, 12 (1).

Rowantree, R.A., Nowak, D.J. 1991. Quantifying the role of urban forests in removing atmospheric carbon dioxide. *Journal of Arboriculture.* 17(10): 269-275.

Singh, Shanker, V., Pandey, D.N., and Chaudhary, P. 2010. *Urban Forest and Open Green Spaces-Lesson for Jaipur*, Occasional paper no, 1/2010. RSPCB, Rajasthan, India.

Wang, X.J. 2009. Analysis of problems in urban green space system planning in China. *Journal of Forestry Research.* 20 (1):79-82.

2015, Impact of Global Warming and Climate Change on   *Pages 233–241*
    Human and Plant Health
Editors: Dr. Arun Arya and Prof. V.S. Patel
Published by: DAYA PUBLISHING HOUSE, NEW DELHI

Chapter 23

Energy Conservation and Global Warming

H.V. Bhavnani[1] and Abhipsa R. Makwana[1]

*[1]Parul Institute of Engineering and Technology
At and PO Limda, Waghodia Vadodara – 391 760, India
[2]Department of Civil Engineering, Faculty of Technology and Engineering
The Maharaja Sayajirao University of Baroda, Vadodara – 390 001, India*

ABSTRACT

Current environmental issue is to control greeenhouse gases control to minimize global warming. And control of the same is not in our hand and thus it requires mass awareness and action. If everyone takes one necessary step we can have big impact on reduction of greeenhouse gases. One step needed is replacement of conventional incandescent bulbs with compact fluorescent lamps. By replacement of all household bulbs by compact fluorescent bulbs, we can reduce greeenhouse gases emission by almost about 28 per cent, which really contributes a lot. Along with this compact fluorescent bulb has several other advantages like long life, less energy consumption, companionable efficiency etc. thus it saves money as well. This paper covers various facts related to compact fluorescent bulbs.

Keywords: *Conventional incandescent bulb, Compact fluorescent bulbs.*

23.1 Global Warming

Global warming is the increase in the average surface air temperature of the planet that is a result of the buildup of heat-trapping or "greenhouse" gases in the atmosphere. Scientists contend that anthropogenic additions of greenhouse gases, mainly CO_2, greatly enhance the natural warming of the earth. Use of fossil fuels (*e.g.* driving a car, drawing electricity from a coal-fired power plant, heating a home with oil or natural gas) is the main human source of CO_2 and other greenhouse gases released into the atmosphere. The second most important source of heat-trapping gases is land-use changes, such as deforestation. The concentration of CO_2 since pre-industrial times has increased by 31 per cent. Also, agricultural activities such as growing rice and raising cattle has had a large influence in the 151 per cent rise of atmospheric methane. With these considerable increases in greenhouse gases, more heat from the sun and earth's surface is trapped in the atmosphere, causing the phenomenon known as global warming.

23.2 Status of Global Warming

The Intergovernmental Panel on Climate Change (IPCC) predicts a warming in the range of 2.5 to 10.4 degrees Fahrenheit by the year 2100. During the 20th century the global average surface temperature has increased about 1 degree Fahrenheit. This is likely to have been the largest increase of any century during the past 1,000 years. As a result we are already observing various effects of this warming. Snow and ice cover are decreasing, including a widespread retreat of mountain glaciers in non-polar regions during the 20th century. Sea level is rising and the heat content of the oceans has increased. Precipitation and extreme weather events also appear to be increasing.

The United States is the world leader, producing almost 25 per cent of the total CO_2 emissions worldwide. China shows the most rapid increase in CO_2 emissions, and Canada is the world leader in per capita CO_2 emissions (truly a dubious distinction). Graph shown below, compares the CO_2 emissions of the top 10 nations.

23.3 What Can Individuals Do?

Every individual can take action and do their part to curb global warming. Probably the most significant environmental decision a consumer can make is which car to drive. For every single gallon of gasoline burnt, 20 pounds of CO_2 enters the atmosphere. A person can help by choosing a highly fuel-efficient car or an electric or hybrid car and by carpooling, walking, biking, and using public transit. Also, a person contributes to global warming whenever they use electricity. Buying energy-efficient appliances and reducing daily energy use can make a difference as well. Compact fluorescent light bulbs use one-quarter the energy and last 5-10 times longer than standard incandescent light bulbs. Halogen lights are very inefficient. Avoid purchasing halogen lighting and choose compact fluorescent lights instead. And join organizations like the Coalition for Clean Air that are fighting for clean renewable energy resources, and clean, fuel-efficient cars and trucks. For more what you can do as an individual, check out our Energy Tips.

23.4 Introduction to Energy Saving Bulbs

The incandescent light bulb was invented in 1809. Unfortunately, very little has changed about the light bulb since the turn of the 20th century. The device still wastes

95 percent of the electricity it consumes. Thus the bulbs sold to consumers today are designed to self-destruct.

Incandescent lights are a safety hazard and an environmental hazard, since they produce massive CO_2 emissions from the coal power plants used to power these bulbs. They're incredibly cheap to purchase up front, but astonishingly expensive to use over time. It produces ten times as much carbon dioxide that contributes to global warming. Want to warm the climate? Turn on the lights! So why, then, are so many people still using incandescent light bulbs? Primarily because they have no idea what it costs to actually operate them. The fact that these light bulbs are secretly slipping money out of your pocket every time they're used seems to go unnoticed by most consumers. All they see is the price tag at the store. And there, incandescent lights look really cheap.

When trying to limit the size of your household environmental footprint the first place to start is the light bulb. The average house has about 30; there are a couple in every room and their energy consumption can add up to a staggering total in your neighborhood, city and across the country. Utilities are promoting the use of energy efficient light bulbs (compact fluorescent light bulbs, CFL's) and some jurisdictions around the globe have gone the extra step and are banning the sale of the traditional incandescent bulb in an effort to reduce greeenhouse gas emissions.

23.5 Which Bulbs Do I Replace?

Compact Fluorescent Light Bulbs come in many shapes and sizes, and can be installed nearly anywhere that incandescent lights are used. Making the right choice on which bulbs you should replace will maximize your savings. How much you will save will depend on the bulb wattage and the number of hours it is turned on. Start replacing bulbs that are left on for extended periods of time, either with purpose or by accident such as a bulb in the kitchen or garage.

As CFLs are not available in one size fits all, consult with your retailer for specific recommendations. You can also use the chart below as a guide to where they go and what to replace.

23.6 How to Start Saving without Spending Anything?

Turn off lights, whenever you leave a room or don't need them, even for just a few minutes. Contrary to popular belief, less energy is consumed when lights are turned on and off as you come and go than if a light is left on all the time.

You can lower overall energy demand by concentrating bright light where you need it rather than evenly lighting the entire room – this is called 'task lighting'. Opening your blinds during the day is a free way to brighten up a room.

For any light that must be on all night (*e.g.* stairways), replace the bulbs with the lowest wattage bulbs that you're comfortable with or consider a compact fluorescent or a nightlight.

Decorate your home with illumination in mind - lighter colours reflect light, so use them in areas you want to be bright.

Keep light fixtures clean – a cleaner bulb is a brighter bulb.

23.7 An Early Compact Fluorescent Lamp

The parent to the modern fluorescent lamp was invented in the late 1890s by Peter Cooper Hewitt. The Cooper Hewitt lamps were used for photographic studios and industries. George Inman later teamed with General Electric to create a practical fluorescent lamp, sold in 1938 and patented in 1941. The modern CFL was invented by, an engineer with General Electric, in response to the 1973 oil crisis. Development of fluorescent lamps that could fit in the same volume as comparable incandescent lamps required the development of new, high-efficacy phosphors that could withstand more power per unit area than the phosphors used with older, larger lamps.

TYPES OF RECENT COMPACT FLUOROSCENT BULBS

23.8 Integrated CFLs

Integrated lamps combine a tube, an electronic ballast and either an Edison screw bayonet fitting in a single CFL unit. These lamps allow consumers to replace incandescent lamps easily with CFLs. Integrated CFLs work well in many standard incandescent light fixtures, which lowers the cost of CFL conversion. Special 3-way models and dimmable models with standard bases are available for use when those features are needed.

There are two types of bulbs: bi-pin tubes designed for conventional ballasts and quad-pin tubes designed for electronic ballasts and conventional ballasts with an external starter. There are different standard shapes of tubes: single-turn, double-turn, triple-turn, quad-turn, circular, and butterfly.

23.9 CFL Power Sources

CFLs are produced for both alternating current (AC) and direct current (DC) input. DC CFLs are popular for use in recreational vehicles off-the-grid housing. CFLs can also be operated with solar powered street lights, using solar panels located on the top or sides of a pole and luminaries that are specially wired to use the lamps.

COMPARISON WITH INCANDESCENT LAMPS

23.10 Lifespan of CFL

The average rated life of a CFL is between 8 and 15 times that of incandescents.CFLs typically have a rated lifespan of between 6,000 and 15,000 hours, whereas incandescent lamps are usually manufactured to have a lifespan of 750 hours or 1,000 hours. Some incandescent bulbs with long lifetime ratings have been able to "Voltage.2C_light_output.2C_and_lifetime" "Incandescent light bulb" trade voltage for lifespan, slightly reducing light output to significantly improve the rated number of hours. The lifetime of any lamp depends on many factors including operating voltage, manufacturing defects, exposure to voltage spikes mechanical shoc, frequency of cycling on and off, lamp orientation and ambient operating temperature, among other factors. The life of a CFL is significantly shorter if it is only turned on for a few minutes at a time: In the case of a 5-minute on/off cycle the lifespan of a CFL can be up to 85 per cent shorter, reducing its lifespan to "close to that of incandescent light bulbs".

23.11 Energy Efficiency

The chart shows the energy usage for different types of light bulbs operating at different light outputs. For a given light output, CFLs use 20 to 33 percent of the power of equivalent incandescent lamps.

Table 23.1: Electrical Power Equivalency

Electrical Power Equivalent for Different Lamps	
CFL's (W)	Incandescent (W)
9-13	40
13-15	60
18-25	75
23-30	100
30-52	150

23.12 Efficacy and Efficiency

If a building's indoor incandescent lamps are replaced by CFLs, the heat produced due to lighting will be reduced. At times when the building requires both heating and lighting, the heating system will make up the heat. If the building requires both illumination and cooling, then CFLs also reduce the load on the cooling system compared to incandescent lamps, resulting in two concurrent savings in electrical power

A typical CFL is in the range of 17 to 21 per cent efficient at converting electric power to radiant power based on 60 to 72 lumens per watt source efficacy, and 347 lumens per radiant watt luminous efficacy of radiation for a tri-phosphor spectrum. The luminous efficacy of CFL sources is typically 60 to 72 lumens per watt, versus 8 to 17 lm/W for incandescent lamps.

While CFLs require more energy in manufacturing than incandescent lamps, this embodied energy is more than offset by the fact that they last longer and use less energy than equivalent incandescent lamps during their lifespan.

23.13 Cost Savings

While the purchase price of an integrated CFL is typically 3 to 10 times greater than that of an equivalent incandescent lamp, the extended lifetime and lower energy use will more than compensate for the higher initial cost. CFLs are extremely cost-effective in commercial buildings when used to replace incandescent lamps.

Table 23.2: Average Cost of Energy Saved Per Annum

Incandescent (W)	CFL's (W)	Average Cost of Energy Saved Per Annum
25	5	1440
50	11	2790
60	15	3240
75	20	3960
100	25	5535
150	32	7425
200	40	10080

Starting time: Incandescent reach full brightness a fraction of a second after being switched on.

23.14 Health Issues

According to the European Commission Scientific Committee on Emerging and Newly Identified Health Risks (SCENIHR) in 2008, the only property of compact fluorescent lamps that could pose an added health risk is the ultraviolet and blue light emitted by such devices. The worst that can happen is that this radiation could aggravate symptoms in people who already suffer rare skin conditions that make them exceptionally sensitive to light. They also stated that more research is needed to establish whether compact fluorescent lamps constitute any higher risk than incandescent lamps.

If individuals are exposed to the light produced by some single-envelope compact fluorescent lamps for long periods of time at distances of less than 20 cm, it could lead to ultraviolet exposures approaching the current workplace limit set to protect workers from skin and retinal damage. The UV received from CFLs is too small to contribute to skin cancer and the use of double-envelope CFL lamps "largely or entirely" mitigates any other risks.

ENVIRONMENTAL ISSUES

23.15 Mercury Emissions

To prevent the release of mercury into the environment, it is advisable to take CFLs to a local recycling facility instead of throwing them away. CFLs, like all fluorescent lamps, contain small amounts of mercury as vapor inside the glass tubing. Most CFLs contain 3 – 5 mg per bulb, with some brands containing as little as 1 mg. Because mercury is poisonous, even these small amounts are a concern for landfills waste incinerators where the mercury from lamps may be released and contribute to air and water pollution.

In areas with coal-fired power stations, the use of CFLs saves on mercury emissions when compared to the use of incandescent bulbs. This is due to the reduced electrical power demand, reducing in turn the amount of mercury released by coal as it is burnt.

Net mercury emissions for CFL and incandescent lamps, based on EPA FAQ sheet, assuming average emission of 0.012 mg of mercury per kilowatthour and 14 per cent of CFL mercury contents escapes to environment after land fill disposal.

23.16 Broken and Discarded Lamps

Due to health and environmental concerns about mercury, it is unlawful to dispose of fluorescent bulbs as universal waste. Spent lamps should be properly disposed of, or recycled, to contain the small amount of mercury in each lamp, in preference to disposal in landfills. The retail price includes an amount to pay for recycling, and manufacturers and importers have an obligation to collect and recycle CFLs. Safe disposal requires storing the bulbs unbroken until they can be processed. The U.S. Environmental Protection Agency (EPA) recommends that, in the absence of local guidelines, fluorescent bulbs be double-bagged in plastic before disposal or sealed glass jar as the best repository for a broken bulb.

The first step of processing CFLs involves crushing the bulbs in a machine that uses negative pressure ventilation and a mercury-absorbing filter or cold trap to contain mercury vapor. The crushed glass and metal is stored in drums, ready for shipping to recycling factories.

The percentage of fluorescent lamps' total mercury released when they are disposed of in the following ways: municipal waste landfill 3.2 per cent, recycling 3 per cent, municipal waste incineration 17.55 per cent and hazardous waste disposal 0.2 per cent.

Follow these guidelines to ensure proper handling, use and disposal of CFLs:

As with any light bulb, be careful when removing it from packaging, during installation, or when replacing it. Always screw and unscrew the light bulb by its base (not the glass) and never forcefully twist the CFL into a light socket.

Always recycle burned-out and broken CFLs. Follow these guidelines for cleaning up a broken CFL:

Open a window and leave the room for 15 min. to let the powder settle and vapors dissipate.

Using rubber gloves carefully scoop up the bulb fragments and powder with stiff paper or cardboard and place in a sealable plastic bag, such as a freezer bag.

Pick up small glass shards with tape (such as duct tape or packaging tape). Wipe hard surfaces down with a damp paper towel or a disposable wet wipe and place the used towel or wipe in the plastic bag. Do not use a vacuum cleaner to clean up broken CFLs.

Seal all cleanup materials, including gloves and paper, in the plastic bag. Double-seal the plastic bag inside a second plastic bag and store outside or in the garage.

Take the sealed plastic bags and any other burned-out CFLs to your local solid waste management district or community household hazardous waste collection program for recycling.

23.17 Efforts to Encourage Adoption

Due to the potential to reduce electric consumption and pollution, various organizations have encouraged the adoption of CFLs and other efficient lighting. Efforts range from publicity to encourage awareness, to direct handouts of CFLs to the public. Some electric utilities and local governments have subsidized CFLs or provided them free to customers as a means of reducing electric demand (and so delaying additional investments in generation).

More controversially, some governments are considering stronger measures to entirely displace incandescents. These measures include taxation, or bans on production of incandescent light bulbs that do not meet energy efficiency requirements

23.18 Negative Aspects about CFL'S

Under the heading "we just can't do anything right when it comes to the environment!" there are some negatives out there associated with CFL'S.

Some environmentalists say that there is no true energy savings to these light bulbs as a certain amount of energy is consumed in their complicated manufacturing process. The energy expended in the manufacturing stage is called 'embodied energy'. Because many more components and energy are used in the production of each bulb in comparison to the basic incandescent bulb, some argue there is no overall savings. In the end, energy production costs offset the energy saving abilities of the CFB unit, rendering them irrelevant.

Under some conditions CFL'S have a longer life than incandescent; they take a moment to brighten up, and can take much longer in very cold temperatures. Consumers have found that CFL'S wear out faster when turned on and off for short amounts of time. They make the most sense in areas where the light is left on for longer periods, like the porch light, kitchen or hall.

Care must be taken with the disposal of CFL'S. Each bulb contains a small amount of mercury (5 mg compared to 25 mg for a watch battery and 500 for dental amalgam). Since most bulbs break when thrown out with the trash this is a potential health concern.

Certain countries advise consumers to check with their municipal waste management programs for proper disposal and recycling. Broken incandescent bulbs inside a house or an office do not pose any environmental hazard beyond that of broken glass. However, like other fluorescent lamps, broken CFL'S release mercury vapors, and require special handling to clean up. In the United States the EPA warns against vacuum cleaning, suggesting instead that you vacate the room and open windows for fifteen minutes to allow any mercury vapor to air out, then clean up the breakage while wearing protective gloves, and use double plastic bags for all broken pieces. They suggest using duct tape to pick up small pieces.

There are now several locations where you can properly dispose of compact fluorescent bulbs that have burned out. Like all other hazardous household items, extra care must be taken in their disposal in order to preserve the environment. In addition the bulbs may also trigger migraines for some of its members. It may be due to either the flickering, or the low intensity of the light, causing eye strain.

However most of us are already adjusting to the new lights. We're saving money on hydro bills and learning to take old bulbs in with other hazardous waste, batteries and unused paint. In the end the savings are substantial.

References

"http://www.energystar.gov"

www.energystar.gov

"http://www.1000bulbs.com/3-Way-CFL/"

3-Way Compact Fluorescents - 1000Bulbs.com

"http://www.nef.org.uk/energysaving/lowenergylighting.htm"

The National Energy Foundation - Low Energy Lighting - How to Save with CFLs

"http://www.rightlight6.org/english/proceedings/Session_8/Performance_ Standard_and_Inspection_Methods_of_CFL/f013guan.doc"

Performance Standard and Inspection Methods of CFL

"http://www.eia.doe.gov/emeu/reps/enduse/er01_us.html"

"US Household Electricity Report". US Energy Information Administration

www.eia.doe.gov

http://www.1000bulbs.com/3-Way-CFL/

http://en.wikipedia.org/wiki/U.S._Environmental_Protection_Agency

2015, **Impact of Global Warming and Climate Change on Human and Plant Health**

Pages **242–257**

Editors: **Dr. Arun Arya and Prof. V.S. Patel**
Published by: **DAYA PUBLISHING HOUSE, NEW DELHI**

Chapter 24

Rainwater Harvesting towards Conserving the Elixir of Life

Bina Rani[1]*, Upma Singh[2]** and Raaz Maheshwari[3]***

[1]Department of Engineering Chemistry and Environmental Engineering, PCE, Sitapura, Jaipur, Rajasthan, India
[2]School of Applied Sciences, Gautam Buddha University, Greater Noida, U.P., India
[3]Department of Chemistry, SBRMGC, Nagaur, Rajasthan, India

ABSTRACT

Water is one of the five basic elements from which creation emanates. The evolution of human culture and civilization has evolved around river systems. Water has been described as elixir of life and cleaner of sins. In other words, mankind can't do without water. Only 2.5 per cent of all water in the globe is freshwater, the rest being sea/salt water of which, as per WHO estimates – only 0.007 per cent is readily available for human consumption. The remaining is frozen in glaciers or polar ice caps or is deep within the earth beyond our reach. This indicates that freshwater on the Earth is finite and also unevenly distributed. This is when 70 per cent of our body is comprised of water and water is essential to most our activities. Unfortunately, with a galloping population, urbanization and ever-increasing demand on it, water sources world over are fast depleting. The need for conserving water has therefore become imperative. Global water consumption has increased almost 10 fold since 1900, and many parts of the world are now reaching the limits of their supply. World population is expected

E-mail: *binaraj_2005@rediffmail.com; **drupmasingh@gmail.com;
**drraazecoethics151260@gmaiil.com

to increase by 45 per cent in the next 30 years, whilst freshwater runoff is expected to increase by 10 per cent. UNESCO has predicted that by 2020 water shortage will be serious worldwide problem. 1/3rd of the world's population is already facing crisis due to water shortage and poor potable water quality. Many of our rivers, wet lands and bays are degraded partly because of water extracted, polluted surface runoff and storm water flushed into them. "Be under no illusions: the impact of general water shortage is going to hit our cities. In the 21st century, wars will be fought over water." – says former UN Secretary General Kofi Anan. According to the International Water Management Institute (IWMI) about 250km3 of water are extracted for irrigation each year in India, of which the rain put back around 100km3 only, resulting in gradual depletion of the aquifers. To cope up with global water scarcity, UN's General Assembly in 1992, declared March 22nd as World Water Day to create awareness amongst individuals and communities. Access to safe water should be a basic human right and it can be ensured by adopting various measures like watershed management through rehabilitation of existing systems like tanks, construction of check dams and Rainwater Harvesting (RWH). Among the different water management techniques, RWH is gaining favour but it's yet to become an essential part of day to day life. It's true that if each one of us uses the water judiciously and augment the water resources by becoming the custodian rather than the absolute owner, the water resources can well be protected for the future mankind. The traditional as well as new strategies of water management and conservation must be implemented wholeheartedly to save our BLUE PLANET – THE MOTHER EARTH.

***Keywords**: RWH, Artificial recharge, Erosion, Urbanization, Sustainable development, Check dams, Percolation tanks, Nadi-Talab-Johad-Baori-Kund.*

24.1 Introduction

The importance of water as a vital resource to the life system and an essential component of societal development cannot be overemphasized. Recognising the importance of water resource development many ancient civilizations emphasized on various mechanisms of water appropriation, collection and distribution. In earlier times the state took care of the water supply by developing and maintaining several ingenious and indigenous ways of string rain and floodwaters. The maintenance of water quality and the means of regenerating the natural resource were crucial factors for sustainability, especially in the dry areas. The ancient man relied on water structures like ponds, lakes, tanks, wells, *baodies*, small *kutcha bunds, tankas, kund, khadins, gulls*, and *ahars*. Rainwater was meticulously conserved and stored at various places for irrigation purposes in these water bodies. As early as 300 BC Megasthenes, the Greek ambassador in court of Chandragupta Maurya, mentioned in his memoirs – the whole country is under irrigation and very prosperous because of the double harvests which they are able to reap each year because of irrigation. Simple techniques can be used to reduce the demand for water. The underlying principle is that only part of the rainfall or irrigation water is taken up by plants, the rest percolates into the deep groundwater, or is lost by evaporation from the surface. Therefore, by improving the efficiency of water use, and by reducing its loss due to evaporation, we can reduce water demand (http://edugreen.teri.res.in/explore/water/conser.htm)

The remnants of many of these age-old structures are architectural and engineering marvels and show that these water bodies were carefully developed, maintained and sustained over the ages. These traditional water-harvesting systems remained environmentally viable and sustainable until they were subjected to large-scale abuses as in the recent times. With recurring droughts year after year especially in Rajasthan, AP, Gujarat, MP and Orissa, environmentalists are emphasising the need to revive and revert back to the water harvesting systems which existed in earlier times. Ironically, the States currently experiencing drought have been endowed with several water harvesting systems which have now fallen into disuse due to neglect over the decades. In Rajasthan, tankas, talabs and baodies traditionally performed the jobs of collecting and storing runoff water. The tankas (underground tanks for drinking water) were one of the most reliable methods of water harvesting in desert towns. Its water was used judiciously to avoid shortage in summer. In the event of scarce rainfall, water from nearby talabs, nadis or village ponds was used to fill up the tankas. Rooftop harvesting was a common feature in villages and towns across the Thar Desert. The technique of rooftop harvesting involves collecting rainwater that falls on the sloping hose roofs through a pipe into an underground tanker built in courtyard. The locals created numerous other water bodies, these included johad, bandha, sagar, samand and sarovar. Wells including kua, kohar (owned by a community) and stepwells (baodis or jhalaras) were also important water sources. Stepwell is a unique form of underground well architecture very common in Rajasthan and Gujarat. Recent reports have indicated that most of the houses in Dwarka, Gujarat, are practising rooftop rainwater harvesting and thus the city is in far better condition than the rest of the drought-prone areas in the State,

In the history of irrigation of South India, tanks played a prominent role. At the end of the first 5-year plan AP had 58,527 tanks with an irrigated area of over 26 lakh acres. Many of these tanks were part of a system that helped store and conserve the rain runoff in such a manner that the water overflow from the tanks in the head was collected in lower ends so that no wastage of water took place. The storage tanks enriched the water table through percolation. Tank irrigation suffered a steady decline in the decades following the 1950s. This was mainly due to the British policy of considering small tanks as being un-remunerative and an unnecessary burden on the State. As the first FiveYear Plan emphasised on 'Grow-more-Food', tank beds were given to individual to cultivate more food thereby causing irreparable damage to these water bodies. Traditional water harvesting systems practised in other Stats include the Phad system in MS, Haveli system in MP, Khadin in Rajasthan, and Ahar-pyne in Bihar.

Over the years rising populations, growing industrialisation, and expanding agriculture have pushed up the demand for water. Efforts have been made to collect water by building dams and reservoirs and digging wells. Water conservation has become the need of the day. The idea of groundwater recharging by harvesting rainwater is gaining importance in many cities.

In the forests, water seeps gently into the ground as vegetation breaks the fall. This groundwater in turn feeds wells, lakes, and rivers. Protecting forests means protecting water 'catchments'. In ancient India, people believed that forests were the 'mothers' of rivers and worshiped the sources of these water bodies. The Indus Valley Civilization, that flourished along the banks of the river Indus and other parts of western and northern India about 5,000 years ago, had one of the sophisticated urban water supply and sewage systems in the world. The fact that the people were well acquainted with hygiene can be seen from the covered drains running beneath the streets of the ruins at both Mohenjodaro and Harappa. Another very good example is the well-planned city of Dholavira, on Khadir Bet, a low plateau in the Rann in Gujarat. One of the oldest water harvesting systems is found about 130km from Pune along Naneghat in the Western Ghats. A large number of tanks were cut in the rocks to provide drinking water to tradesmen who used to travel along this ancient trade route. Each fort in the area had its own water harvesting and storage system in the form of rock-cut cisterns, ponds, tanks and wells that are still in use today. A large number of forts like Raigad had tanks that supplied water.

Rainwater harvesting essentially means collecting rainwater on the roofs of the building and storing it underground for later use. Not only does this recharging arrest groundwater depletion, it also raises the declining water table and can help augment water supply. Rainwater harvesting and artificial recharging are becoming very important issues. Realizing the importance of recharging groundwater, the CGWB (Central Groundwater Board) is taking steps to encourage it through rainwater harvesting in the capital and elsewhere. A number of government buildings have been asked to go in for water harvesting in Delhi and other cities of India. All we need for a water harvesting system is rain, and a place to collect it! Typically, rain is collected on rooftops and other surfaces, and the water is carried out to where it can be used immediately or stored. We can direct water run-off from this surface to plants, trees or lawns or even to the aquifer.

Some of the benefits of rainwater harvesting are as follows:

☆ Increases water availability

☆ Checks the declining water table

☆ Is environmentally friendly

☆ Improves the quality of groundwater through the dilution of fluoride, nitrate, and salinity

☆ Prevents soil erosion and flooding especially in urban areas

The most important step in the direction of finding solutions to issues of water and environmental conservation is to change people's attitudes and habits – this includes each one of us. Conserve water because it is the right thing to do. We can follow some of the simple things that have been listed below and contribute to water conservation.

☆ Try to do thing each day that will result in saving water. Don't worry if the savings are minimal – every drop counts! We can make a difference.

☆ Remember to use only the amount of water we actually need.

☆ Encourage the family to keep looking for new ways to conserve water, in and around home.

☆ Make sure that our home is leak-free. Many homes have leaking pipes that go unnoticed.

☆ Do not leave the tap running while we are brushing our teeth or soaping our face.

☆ See that there are no leaks in the toilet tank. We can check this by adding colour to the tank. If there is a leak, colour will appear in the toilet bowl within 30 minutes (Flush as soon as the test is done, since colouring may stain the tank).

☆ Avoid flushing the toilet unnecessary.

☆ For a group of water-conscious people and encourage friends and neighbours to be a part of this group. Promote water conservation in community newsletters and on bulletin boards. Encourage friends, neighbours and co-workers to contribute.

24.2 Extracting the Elixir – Groundwater Development in India

During the past five decades, there has been phenomenal growth of groundwater abstraction structures in India. Their number has increased from 4 million in 1951 to about 18 million in 1998-99 while in the same period irrigation potential created from groundwater has increased from 6 to 30 million hectares. Groundwater extraction however, varies over space. It is intensive in the alluvial Indo-Gangetic plains of Punjab, Haryana, UP, UK and in parts of hard rock terrain of southern states. Although over-exploitation of this resource in pockets of the country has created serious problems, yet a large portion of the available groundwater still remains untapped, particularly in the north-eastern areas, where precipitation is high and the demand for irrigation is low. This is also true of the eastern states where the fragmented nature of land holdings has been a major factor in low development of groundwater usage.

24.3 How much Groundwater does India has?

The annually 'replenishable' groundwater resources of the country have been assessed as 432 billion cubic meter (BCM). Keeping aside a basic provision of 71 BCM per year for domestic and industrial use, 361 BCM per year is available for irrigation. The Ganga basin has the highest potential followed by Godavari basin and Brahmaputra basin. In fact the Indo-Gangetic alluvial plain with an area of around 25,000 km^3 is one of the largest groundwater reservoirs in the world. In addition to the 'replenishable' resources we have the 'in storage' groundwater resources. These resources which lie below the lowest level of groundwater fluctuation can be used in extreme situations. The in-storage groundwater resources up to depth of 450 meters in hard rock terrain have been estimated as 10812 BCM.

24.4 What Contaminates our Groundwater?

It was way back in mid eighties that a systematic study of groundwater quality and identification of 'problematic zones' was under taken by Central Pollution Control Board (CPCB). Over the years the problem areas that emerged witnessed excessive exploitation of groundwater for domestic and industrial uses. As pollution control enforcement activities gained momentum there were observed cases of indiscriminate waste disposal, subsurface discharge of effluent and inappropriate wastewater management by industries. Today, as a result of these malpractices there is a palpable stress on groundwater, in terms of quantitative imbalances as well as qualities deterioration.

24.5 Indiscriminate Disposal of Sewage and Garbage

With increasing urbanization, the groundwater pollution due to indiscriminate disposal of untreated sewage and garbage has also required alarming proportions. With 70 to 80 per cent of water supply getting converted into wastewater and limited facility of only 26 per cent for its treatment, further compound the problem, and in many instances led to outbreak of water borne diseases apart from microbial contamination of groundwater.

Protecting Groundwater

Mapping of vulnerable areas of groundwater depletion and pollution

☆ Notification of critical areas of groundwater

☆ Notification for banning commercial sale of groundwater

☆ Special studies on areas of high concentration of carcinogenic elements in groundwater

☆ Directives in industries/mining/commercial establishments for regulation over withdrawal of groundwater

☆ Environment impact study for groundwater

Campaigns to create public awareness for judicious use and conservation of groundwater.

24.6 Salt Water Intrusion

Along about 7000 km long Indian coast line coastal aquifers form a vital source of freshwater. On the other hand, the aquifers being in hydraulic contact with sea are equally vulnerable to contamination due to intrusion of salt water from sea. The intrusion in these areas is caused by concentrated withdrawal of groundwater and reversal of natural hydraulic gradient. The problem has been reported in areas of Saurashtra, TN, AP and WB. It must be clearly understood that less than even 2 per cent of sea water can diminishes water portability. The recommended remedial methods for salt water intrusion include modification of pumping pattern, artificial recharge, physical barrier and hydraulic barrier.

24.7 Action Plan for Water Conservation

24.7.1 Conservation of Surface Water Resources

A large number of dams have been constructed in the country to store rainwater. At the end of IX Plan, 4050 large dams creating live storage capacity of 213BCM have been constructed and 475 large projects are on ongoing, which will add another 76BCM on completion. Projects under consideration will add another 108BCM of storage. All efforts have to be made to fully utilize the monsoon run-off and store rainwater at all probable storage sites. In addition to creating new storages it is essential to renovate the existing tanks and water bodies by desilting and repairs. The revival of traditional water structures should also be given due priority.

24.7.2 Conservation of Groundwater Resources

Groundwater is an important component of hydrological cycle. It supports the springs in hilly regions and the river flow of all peninsular rivers during the non-monsoon period. For sustainability of groundwater resources it is necessary to arrest the groundwater outflows by

(a) Construction of sub-surface dams

(b) Watershed management

(c) Treatment of upstream areas for development of springs

(d) Skimming of freshwater outflows in coastal areas and islands

24.7.3 Rainwater Harvesting

Rainwater harvesting is the technique of collection and storage of rainwater at the surface or in subsurface aquifers, before it is lost as surface run-off. Groundwater augmentation through diversion of rainfall to subsurface reservoirs, by various artificial recharge techniques, has special relevance in India where due to terrain conditions most of the rainwater is lost as flash floods and local streams remain dry for most of the part of the year. Central Groundwater Board (CGWB) has identified an area of about 4.5lakhs sq km in the country, which shows a declining trend in groundwater levels and needs urgent attention to meet the growing needs for irrigation, industry and domestic purpose. It is estimated that in these identified areas of water scarcity, about 36.1BCM of surplus monsoon surface run-off is available which can be fruitfully utilized to augment the groundwater resources. A twin strategy of adopting simple artificial recharge in rural areas like percolation tanks, check dams, recharge shafts, dug well recharge and sub-surface dykes and adopting roof-top rainwater harvesting in urban areas, can go a long way in redeeming the worsening situation of groundwater.

About 2.25 lakhs artificial recharge structures in rural areas and about 37 lakhs Rooftop rainwater harvesting structures in the cities are feasible. The design and viability of various low cost structures have been demonstrated by CGWB by undertaking 174 schemes throughout the country during Ninth Five Year Plan under the Central Sector Scheme "Study of Groundwater Recharge". Rainwater harvesting is to be taken up in a big way to solve the crisis of water scarcity. Uncovered areas, particularly in urban and semi-urban localities, are continuously diminishing due to phenomenal pace of industrialization and urbanization and massive use of concrete all around in the country. This phenomenon is constantly causing reduced scope for percolation of rain waters to the ground during monsoon and thus perpetual reduction in groundwater recharge year after year. With a view to offset this loss in recharge of groundwater there is apparent need for making roof rainwater harvesting mandatory, either through legislation or by promulgating ordinance, for every public as well as private new and existing buildings in urban and semi-urban areas within specified time frame. Apart from this, harvesting of surface runoff in open areas, both public and private, may also need to be encouraged.

24.7.4 Protection of Water Quality

The rapid in density of human population in certain pockets of the country as a result of urbanization and industrialization is making adverse impact on the quality of both surface and groundwater. Demand for water is increasing on one hand and on the other hand the quantity of "utilizable eater resources" is decreasing due to human intervention in the form of pollution of freshwater. Thus the protection of existing water resources from pollution is a very vital aspect of water conservation.

24.7.5 Cleaning up of Polluted Rivers, Lakes and Water Bodies

Rivers, lakes, ponds and water bodies are the main sources of water on which a civilization grow and develop. Water bodies get polluted as a result of human interference and unplanned development activities. The main reason for pollution is discharge of untreated domestic and municipal waste and also the industrial waste. The wastewater encounters alternate aerobic and anaerobic microbial population which convert carbonaceous and to a lesser extent nitrogenous and phosphatic, contaminants in the water to less polluting materials. Further microorganisms can form biofilms around lower stems, which can then trap particles suspended in the wastewater by absorption. It is expected to achieve an effluent quality of 5-10mg/l of suspended solids. A total of 495 MGD of recycled water can be made available for irrigation, horticulture and industrial needs and for domestic non-drinking supply (http://www.delhijalboard.nic.in/djbdocs/consumer/conservation.htm).

The cleaning up of these water bodies is of utmost importance to provide water supply to the population on the one hand and on the other hand to maintain the environment to the desired level. The action points in this regards are as follows:

1. To control and check the flow of pollution to the rivers, lakes and ponds through appropriate measures/action.
2. Treatment of effluent up to the appropriate standard before discharging the river.
3. Proper maintenance and uninterrupted operation of the sewage treatment plant.
4. System of incentive and disincentive for discharging pollutants/untreated waste into the rivers.
5. Adopting remedial measures in the particular river stretch where the problem is acute.
6. Adopting appropriate technology for removal of pollution from lakes and reservoirs.
7. Declaring particular site/location as water heritage site and adoption by different organizations/departments for maintaining the same to the desired standard.

On account of continuous discharge of industrial effluents in water bodies like rivers, canals, lakes, ponds, etc. and contamination of groundwater aquifers with polluted waters, these water bodies at places have become polluted to an enormous extent and apparently huge financial resources are needed for decontaminating them. This suggests for taking stringent measures like imposition of huge penalty for abusing such water bodies, cancellation of license or permission for operation of water polluting industrial units. Pollution control boards at Central and State levels may be provided legal powers through legislation to deal with such delinquent agencies and industrial units. Sensitizing general public and involvement of nongovernmental organizations with requisite experience and interest in implementation of legislation for control of pollution of water bodies may also prove useful and effective. Media has also a very vital role to enact by way of highlighting lapses on the part of individuals and industrial units. Traditionally, in India, rivers are revered as Goddess. With time, such a feeling has started diluting. People particularly young generation, may be inculcated to bestow respect to rivers and other water bodies to strengthen this traditional belief of sacred status of rivers and streams and maintain their aesthetic values through mass awareness,

24.7.6 Groundwater Protection

Groundwater resources are getting polluted at an alarming pace due to lack of proper wastewater and sewerage disposal system in urban areas. The application of excessive fertilizers in agriculture sector and disposal of hazardous effluents from the industries are putting great strain on availability of freshwater. the action points to safeguard the water bodies may be as follows:

1. Use of organic fertilizers should be encouraged to protect groundwater from pollution due to excessive use of chemical fertilizers. Groundwater vulnerable zones may be identified by preparing vulnerability maps for physical, chemical and biological contaminants for the whole country.

2. Notification on banning industries, landfills and disposal sites of industrial effluents and sewerage, which are hazardous to groundwater aquifer systems.

3. Devising groundwater solute transport model for contaminants plum migration studies.

4. Research and development studies for corrective action techniques on polluted aquifers.

24.8 Recent Attempts for Water Conservation

(a) The MP Government implemented watershed development under Rajiv Ghandi Watershed Mission. In the initial phase beginning with 1994, the objective of the programme was to arrest degradation of resources that were critical to peoples' livelihood. The programme evolved over a period of time and culminated in the year 2001 as "Pani Roko Abhiyan". It was a peoples' movement, which was backed by financial commitment and technical support of the Government. The programme was so successful that 14 districts, which were not covered under the drought relief programme in 2001, were also enabled through the banking channels to take up "Pani Roko Abhiyan".

(b) PRADHAN, a NGO has adopted this strategy in Jharkhand. Under the Indo –German Bilateral Watershed Project, it seeks to promote livelihood improvement through water harvesting. It is an innovative and simple technique of collecting rainwater in a two-meter deep pit in 5 per cent of the total area of the plot. PRADHAN provided assistance and guidance to the villagers for construction of farm tanks. The farmers have been able to harvest two crops from the same land due to availability of the water in the field tank.

(c) The Government of Gujarat has adopted rainfall harvesting technique by constructing small dams in water deficit areas. Farmers are also forward to adopt this methodology. Other states may undertake similar water conservation measures.

24.9 Action Plan for Water Conservation

An important of component of water conservation involves minimizing water losses, prevention of water wastage and increasing efficiency in water use. "Resource saved is resource created" should be kept uppermost in mind. The action points towards water conservation in different sectors of water use are listed below:

Irrigation Sector

Important action points towards water conservation in the irrigation sector are as follows:

☆ Performance improvement of irrigation system and water utilization;

☆ Proper and timely system maintenance;

☆ Rehabilitation and restoration of damaged/and silted canal system to enable them to carry designed discharge;

☆ Rehabilitation and restoration of damaged/and silted canal systems to enable them to carry designed discharge;

☆ Selective lining of the canal and distribution systems, on technoeconomic consideration, to reduce seepage losses;

☆ Registration/provision of appropriate control structures in the canal system with efficient and reliable mechanism;

☆ Conjunctive use of surface and groundwater to be restored to, specially in the areas where there is threat to water logging;

☆ Adopting drip and sprinkler systems of irrigation foe crops, where such systems are suitable;

☆ Adopting low-cost innovative water saving technology;

☆ Renovation and modernisation of existing irrigation systems;

✰ Preparation of realistic and scientific system operation plan keeping in view the availability of water and crop water requirements;

✰ Execution of operation plan with reliable and adequate water measurement structures;

✰ Revision of cropping pattern in the event of change in water availability;

✰ Impairing training to farmers about consequences of using excess water for irrigation;

✰ Rationalization of water rate to make the system self-sustainable;

✰ Formation of Water Users Associations and transfer of management to them;

✰ Introducing night irrigation practice to minimize evaporation loss;

✰ In arid regions crops having longer root such as linseed, berseem, lucerne guar, gini grass, etc. may be grown as they can sustain in dry hot weather;

✰ Assuming timing an optimum irrigation for minimizing water loss and water logging;

✰ Modern effective and reliable communication systems may be installed at all strategic locations in the irrigation command and mobile communication systems may also be provided to personal involved with running and maintenance of systems. Such an arrangement will help in quick transmission of messages and this in turn will help in great deal in effecting saving of water by way of taking timely action in plugging canal breaches., undertaking repair of systems and also in canal operation particularly when water supply is needed to be stopped due to sudden adequate rainfall in the particular areas of the command;

✰ With a view to control over irrigation to the field on account of un-gated water delivery systems, all important outlets should be equipped with flow control mechanism to optimize irrigation water supply;

✰ As far as possible with a view to make best use of soil nutrients and water holding capacity of soils, mixed cropping such as cotton with groundnut, sugarcane with black gram or green gram or soya bean may be practiced;

✰ It has been experienced that with scientific use of mulching in irrigated agriculture, moisture retention capacity of soil can be increased to the extent of 50 per cent and this in turn may increase yield up to 75 per cent.

The concept of strip tillage on the contour with ripping to encourage infiltration was tested in Tanzania by Macartney *et al.* (1971), and in Botswana, Willcocks (1981) showed that precision strip tillage was more effective than other cultivation methods, particularly if the strips could be consistently aligned along the previous crop rows.

The practice of cultivating to leave ridges and furrows on the contour without any additional land shaping was discussed. It appears to be more common for maize and sorghum to be planted on the ridges, and small grains in the furrows. The practice of drilling wheat, barley, and rye into small furrows is reported from Poland by Wollen (1974) and is common in the wheat growing areas of Kazakhstan, USSR.

In low rainfall areas a large ratio of catchment to cropped area is required, but it is not easy to design a system which will give the best result for all variations of annual rainfall. In some experiments in North America the ratio of 33:1 was tried, and with 190 mm of rain gave 530 mm of run-on to the farmed area, which was more than required and more than could be absorbed. The next year the ratio was reduced to 15:1, but the rain was greater at 246 mm and with a higher run-off coefficient gave 390 mm of run-on, about the right amount. Retaining the ratio of 15:1, the next year had 140 mm rain, giving 220 mm run-on which was not sufficient. The authors (Morin and Matlock 1975) suggest that mathematical modelling is an appropriate tool to design the optimum ratio for varying conditions of soil and climate, but as we have discussed previously, any system is going to have widely varying degrees of success or failure from year to year. In a region of low winter rainfall, using saw-tooth ridges shaped by motor grader, Shanan and Tadmor (1979) found that the best ratio of catchment to planted strip could vary from 4-20. The very wide range suggests that trial and error may be as effective as mathematical modelling.

Domestic and Municipal Sector

Important action points for water conservation in domestic and municipal sector are as under:

☆ Management of supply through proper meter as per rational demand;

☆ Intermittent domestic water supply may be adopted to check its wasteful use;

☆ Realization of appropriate water charges so that the system can be sustainable and wastage is reduce;

☆ Evolving norms for water use for various activities and designing of optimum water supply system accordingly;

  (a) Modification in design of accessories such as flushing system, tap, etc. to reduce water requirement to optimal level;

  (b) Possibility for recycling and reuse of water for purpose like gardening, flushing to toilets, etc. may be explored;

☆ Optimum quantity of water required for waste disposal to be worked out;

☆ In public building the taps, etc. can be fitted with sensors to reduce water losses.

Industrial Sector

Important action points for water conservation in industrial sector are given below; Setting-up norms for water budgeting; Modernization of industrial process to reduce water requirement;

☆ Recycling water with a re-circulating cooling system can greatly reduce water use by using the same water to perform several cooling operations;

☆ Three cooling water conservation approaches are evaporative cooling, ozonation and air heat exchange. The ozonation cooling water approach can result in a five-fold reduction in blow down when compared to

Table 24.1: Typical Use of Water

Typical Use of Water	
Drinking	4 per cent
Cooking and other kitchen uses	8 per cent
Personal hygienic	29 per cent
Washing cloths	10 per cent
Toilet flushing	39 per cent
House cleaning/gardening, etc.	10 per cent

traditional chemical treatment and should be considered as an option for increasing water saving in a cooling tower;

☆ The use of de-ionized water in reusing can be reduced without affecting production quantity by eliminating some plenum flushes, converting from a continuous flow to an intermittent flow system and improving control of the use.

☆ Proper processing of effluents by industrial units to adhere to the norms for disposal.

☆ Rational pricing of industrial water requirement to ensure consciousness/action for adopting water saving technologies.

Table 24.2: Water Usage and Its Saving

What we Do?	What Should be Done?	Saving of Water (L)
Bathing with shower: 100 liter	Bathing with bucket: 18 liter	82
Bathing with running water:40 liter	Bathing with bucket: 18 liter	22
Using old style flush: 20 liter	Using new style flush: 6 liter	14
Shaving with running water: 10 liter	Shaving by use of a mug : 1 liter	9
Brushing teeth with running water: 10 liter	Brushing teeth by taking water in mug: 1 liter	9
Washing clothes with running water: 116	Washing clothes with bucket: 36 l	80
Washing car with running water: 100	Washing car with wet cloth: 18	82
Washing clothe with running water (15'x10') 50 liter	Washing floor with wet cloth: 10 liter	40
Washing hands with running water tap: 10 liter	Washing hands with mug: 0.5 liter	9.5

24.10 Regulatory Mechanism for Water Conservation

Groundwater is an unregulated resource in our country with no price tag. The cost of construction of a groundwater abstraction structure is the only investment. Unrestricted withdrawal in many areas has resulted in decline of groundwater levels. Supply side management of water resources is very important for conserving this vital resource for a balanced use. An effective way is through energy pricing restriction on supply and proving incentives to help in conservation of water. Action plan, in this regard, may include the following:

Tips for Conserving Water for Domestic Use

☆ Timely detection and repair of all leaks;

☆ Avoiding/minimizing use of shower/bath tub in bath room;

☆ Turning off faucets while soaping and rinsing clothes;

☆ Avoiding use of extra detergent in washing clothes; Using automatic washing machine only when it is fully loaded;

☆ Using overflow stop valve in the overhead tanks to check over flow of water;

☆ Turning off the main valve of water while going outdoor;

☆ Minimizing water used in cooling equipment by following manufacturer's recommendations;

☆ Watering of lawn or garden during the coolest part of the day (early morning or late evening hours) when temperature and wind speed are the lowest. This reduces losses from evaporation;

☆ Planting of native/or drought tolerant grasses, shrubs and trees and grouping of plants based on water needs while planting them;

☆ Setting sprinklers to water the lawn or garden only, not the street or sidewalk;

☆ Installation of high-pressure, low-volume nozzles on spray washers;

☆ Replacement of high-volume hoses with high-pressure, low volume cleaning systems;

☆ Washing vehicles less often, or using commercial car wash that recycles water.

In case of big establishments like hotels, large offices and industrial complexes, community centres, etc. dual piped water supply may be insisted upon. Under such an arrangement one supply may carry freshwater for drinking, bathing and other human consumptions whereas recycled supply from second line may be utilized for flushing out human solid wastes. Similarly, water harvesting through storming of water run-off including rainwater harvesting in all new buildings on plots of 100 sq m and above may be made mandatory.

24.11 Tips for Conserving Water for Industrial Use

☆ Using fogging nozzles to cool product;

☆ Installing in-line strainers on all spray headers; regular inspection of nozzles for clogging;

☆ Adjusting pump cooling and water flushing to the minimum required level;

☆ Choosing conveying systems that use water efficiently;

☆ Handling waste materials in a dry mode whenever possible;

☆ Replacing high-volume hoses with high-pressure, low volume cleaning systems;

☆ Equipping all hoses with spring loaded shutoff nozzles – it should be ensured that these nozzles are not removed;

☆ Turning off all flows during shutdowns unless flows are essential for cleanup – using solenoid valves to stop the flow of water when production stops (the valves could be activated by trying them to drive motor controls);

☆ Adjusting flows in sprays and other lines to meet minimum requirements;

☆ Making an inventory of all cleaning equipments, such as hoses in the plant – determining how often equipment are used and whether they are water-efficient;

☆ Driveways, loading docks, parking areas or sidewalks with water may be avoided – using sweepers and vacuum may be considered;

☆ Avoiding run-off and making sure that sprinklers cover just the lawn or garden, not sidewalks, driveways, or gutters;

It is imperative that users from all sectors of water use, stakeholders including state and central governments, agencies, institutions, organizations, NGOs, municipalities, and duties and responsibilities of individuals as well as of organizations and institutions towards judicious and optimal use of water.

24.12 Water Users' Association (WUA) and Legal Empowerment

Water Users' Association, though relatively a new concept in the country but is prevalent in some states in irrigation sector. It is considered that involvement of farmers in water management will facilitate equitable and judicious allocation of irrigation waters among farmers of head, middle and tail reach and improve collection of water charges, irrigation projects may not languish for maintenance for want of funds and in this way overall efficiency of irrigation system will improve. This will help saving of water and optimum utilization of water. Illegal tapping of water from supply lines or lifting water of canals are also prevalent at places. It has also been observed that inhabitants, in general, are less sensitive to leakage or water loss from the system. Similarly in case of industrial sector, it is not very uncommon to discharge untreated or partially treated industrial effluents in water bodies like rivers, lakes, ponds, canals, etc. including groundwater aquifers. WUA in domestic and industrial sectors of water use may address these issues and may help in conservation of water and control pollution of water bodies from industrial pollutants. WUA may be duly empowered through legislation or promulgation of ordinance to punish errant water users. (Source: Central Water Commission: http://cwc.gov.in/Acts_laws_rules_guidelines.htm)

References

http://cwc.gov.in/Acts_laws_rules_guidelines.htm

http://www.delhijalboard.nic.in/djbdocs/consumer/conservation.htm

http://edugreen.teri.res.in/explore/water/conser.htm

Macartney J.C., Northwood P.J., Dagg M. and Lawson R. 1971,The effect of different cultivation techniques on soil moisture conservation and the establishment and yield of maize at Kongwa, Central Tanzania. *Trop. Agric.* Trinidad 48: 9-23.

Morin G.C.A. and Matlock W.G.1975 Desert strip farming - computer simulation of an ancient water harvesting technique. US Agric. Res. Stn. Service, Western Region ARS-W Agric. Res. Serv. US Dep. Agric. Res. Serv. US Dep. Agric. 22, 141-150.

Shanan L. and Tadmor N.H. 1979. Micro-catchment systems for arid zone development. Centre Int. Agric. Cooperation, Min. of Agriculture, Rehovot, Israel.

Willcocks T.J. 1981The tillage of clod-farming sandy loam soils in the semi-arid climate of Botswana. *Soil Tillage Research* 1: 3223-350.

Wollen L. 1974 Furrow cultivation (sowing) of cereals in Poland. Akademia Rolnicza, Krakow, Poland. Postepy Nauk Rolniczych, 1974, No. 4: 17-26.

2015, Impact of Global Warming and Climate Change on
Human and Plant Health
Pages 258–267

Editors: **Dr. Arun Arya and Prof. V.S. Patel**
Published by: **DAYA PUBLISHING HOUSE, NEW DELHI**

Chapter 25

"Biodiesel": An Ecofriendly Option

Abhipsa R. Makwana*

Civil Engineering Department,
Faculty of Technology and Engineering,
The M.S. University of Baroda, Vadodara, Gujarat, India

ABSTRACT

Now a day the world is facing the crises of fuel depletion and environment degradation. As far as environmental degradation is concerned the burning problem is global warming and emission of greeenhouse gases. Vehicles and industries etc are responsible for emission of greeenhouse gases. Use of fossil fuel in vehicles are chief contributors of air environment degradation. Burning of fossil fuel generates CO, CO_2, SOx, NOx, unburnt or partially burnt hydro carbons and particulate emission.

This paper covers the concept of biofuel use which are eco friendly as far as air environment is concerned. Biodiesel when used in the place of conventional fossil fuel substantially reduces emission of unburnt hydrocarbon, CO and CO_2 and particulate matter. It has been reported that the global warming potential of biodiesel is about 23 per cent that of fossil fuel when used along with advanced engine technology and saturated biodiesel. This paper reviews the technology developed for Bio diesel towards eco friendliness.

Keywords: *Biodiesel, Fossil fuel, Particulate emission, Hydrocarbon.*

---

* E-mail: abhipsamakwana@gmail.com

25.1 Introduction

With increasing power consumption and an increase in number of transport vehicle the coal reserves are going to deplete very soon. The world at present is heavily dependent upon petroleum fuels for transportation and for operating agriculture machinery. Diesel engines dominate the field of transportation and agriculture machinery on account of its superior fuel efficiency, the consumption of fuel in India is several times higher than that of petrol consumption. Rough estimate of petrol and diesel consumption is 30 and 70 per cent, respectively. The diesel engine is a major contributor to air pollution especially within cities and along urban traffic routes. In addition to air pollution that causes ground level ozone and smog in the atmosphere, diesel exhaust also contains particulate and hydrocarbon toxic air contaminants (TAC). Now society has become more aware of harmful effects of the various exhaust emission coming out of the engines and there is tremendous pressure on researchers to reduce exhaust emissions. Various harmful effect of exhaust emission are already established and known to the society. Carbon monoxide, if inhaled, enters the blood stream and causes hypoxia, which leads to further health problems. Hydrocarbon emissions are irritant and odorants and some of them carcinogenic. Oxides of nitrogen are found to be responsible for many of the pulmonary diseases.

Diesel particulate matter (PM) is made up of very small particles that are inhaled deep into the human lung. Since there is no effective natural removal process from this area of the lung, the particles are increasingly urgent health concern. Therefore, it has become very essential to develop the technology of IC engines, which will reduce the consumption of petroleum fuels and exhaust gas emissions. Irrecoverable rapid depletion of petroleum reserves, high price fluctuations, uncertainty in supply to consuming nations, high expenditure on fuel import, harmful effects of various exhaust emission on the human being and environment forces to search for alternative fuels that they themselves can produce. These alternative fuels should be preferably available from renewable sources. Therefore, attention is mainly focused towards biomass- based fuels. Alternative considered are ethanol, methanol, biogas and vegetable oil, methyl or ethyl ester of vegetable oil (biodiesel). Biodiesel is a vegetable oil-based fuel that runs diesel engines. One can blend it with regular diesel or run 100 per cent biodiesel.

25.2 Biodiesel

Biodiesel is composed of long-chain fatty acids with an alcohol attached, often derived from vegetable oils. It is produced through the reaction of a vegetable oil with methyl alcohol or ethyl alcohol in the presence of a catalyst. Animal fats are another potential source. Commonly used catalysts are potassium hydroxide (KOH) or sodium hydroxide (NaOH). The chemical process is called transesterification which produces biodiesel and glycerine.

25.3 Basic Chemistry and Terminology

The Fuel Standards Regulations 2001 under the Fuel Quality Standards Act 2000 define biodiesel as "a diesel fuel obtained by esterification of oil derived from plants or animals." Esterification is the conversion of a compound into an ester. An ester is a compound formed by the reaction between an acid and an alcohol with the elimination of a molecule of water.

Production of methyl esters of long chain fatty acids requires the following:

1. A feedstock source of triglycerides. Vegetable oils (*e.g.* soyabean oil, rapeseed/canola oil, palm oil), animal fats (*e.g.* beef tallow), and waste cooking oils (*e.g.* reused frying oil) can all be used as feedstock in biodiesel production.

2. An alcohol. The most common alcohol used in biodiesel production is methanol, but other alcohols, typically ethanol, can be used.

3. A catalyst. Most biodiesel reactions are alkali catalyzed, with the most common alkali source as potassium hydroxide. Sodium hydroxide can also be used as a catalyst.

The major products of the esterification reaction are:

1. Methyl esters of long chain fatty acids (the biodiesel); and

2. Glycerol

Biodiesel is produced by a reaction of a vegetable oil or an animal fat with an alcohol. The alcohol is charged in excess to assist in quick conversion and recovered for reuse. The catalyst is usually sodium or potassium hydroxide which has already been mixed with the alcohol. The finished biodiesel derives some 10 per cent of its mass from the reacted alcohol. The alcohol used in this reaction may or may not come from renewable resources. If methanol is used the result is methyl esters and if ethanol is used the result is ethyl esters.

Depending on the feedstock and processes employed, by-products may include glycerin, fatty acids, fertilizer and oil seed meal (for grain feed stocks). Wastewater also results from biodiesel production. This may or may not be treated on site.

Biodiesel may have other terminology depending on the feedstock used to produce it, for example Fatty Acid Methyl Ester (FAME), Plant Methyl Ester (PME), Rapeseed Methyl Ester (RSME or RME), vegetable oil Methyl Ester (VOME) or Soybean oil Methyl Ester (SOME).

Biodiesel can be used in the pure form, or blended in any amount with diesel fuel for use in compression ignition engines. Figure 25.1 shows basic transesterification technology. The transesterification process of converting vegetable oils to biodiesel is shown in Figure 25.2. The "R" groups are the fatty acids, which are usually 12 to 22 carbons in length. The large vegetable oil molecule is reduced to about 1/3.

25.4 Emission Characteristics of Biodiesel

The global population of motor vehicle on the roads today is half a billion, which is more that ten times higher than what was in 1950s. Combustion of various fossil fuels leads to emission of several pollutants, which are categorized as regulated and unregulated pollutants. Regulated pollutants are the once, whose limits have been prescribed by environmental legislation. Whereas there are some pollutants for

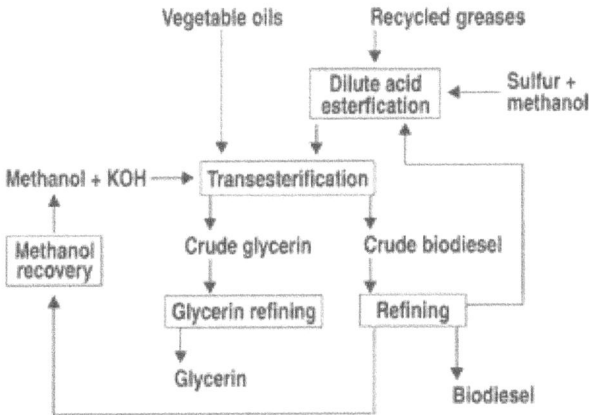

Figure 25.1: Basic Transesterification Technology.

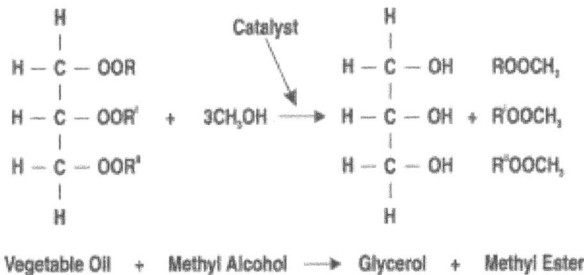

Vegetable Oil + Methyl Alcohol ⟶ Glycerol + Methyl Ester

Figure 25.2: Transesterification of Vegetable Oils.

which no legislative limits have been prescribed. These are categorized as unregulated pollutants. Regulated pollutants include NOx, CO, HC, Particulate matter (PM) and unregulated pollutant include formaldehyde, benzene, toluene, xylene (BTX), aldehydes, SO_2, CO_2 methane etc. these pollutants also contribute towards several regional and global environmental effects. Regional environmental effects such as summer smog are because of aldehydes, CO, NOx etc. Winter smog is because of particulate. Acidification is caused by NOx, SOx etc. Several global effects like ozone layer depletion, global warming are caused by CO_2, CO, methane, non methane hydrocarbon, NOx.

Biodiesel is the only alternative fuel to have a complete evaluation of emission results and potential health effects submitted to the U.S.EPA under the Clean Air Act Section 211(b). These programs include the most stringent emissions testing protocols ever required by EPA for certification of fuels in the U.S. Emission results for pure biodiesel (B100) and mixed biodiesel (B20-20 per cent biodiesel and 80 per cent petro diesel) compared to conventional diesel are given in Table 25.1.

The use of biodiesel in a conventional diesel engine results in substantial reduction of unburnt hydrocarbons, carbon monoxide and particulate matter. Emissions of nitrogen dioxides are either slightly reduced or slightly increased depending on the duty cycle or testing methods. Biodiesel decreases the solid carbon fraction of particulate matter (since the oxygen in the fuel enables more complete combustion to CO_2), eliminates the Sulphur fraction (as there is no Sulphur in the fuel), while the soluble or hydrogen fraction stays the same or is increased.

Table 25.1: Biodiesel Emissions Compared to Conventional Diesel

Emissions	B100	B20
Regulated Emissions		
Total Unburned Hydrocarbons	−93 per cent	−30 per cent
Carbon MoNOxide	−50 per cent	−20 per cent
Particulate Matter	−30 per cent	−22 per cent
NOx	+13 per cent	+2 per cent
Non-Regulated Emissions		
Sulphates	−100 per cent	−20 per cent *
Polycyclic Aromatic Hydrocarbons (PAH)**	−80 per cent	−13 per cent
NPAH (Nitrated PAHs)**	−90 per cent	−50 per cent ***
Ozone Potential of Speciated HC	−50 per cent	−10 per cent
Life-Cycle Emissions		
Carbon Dioxide (LCA)	−80 per cent	
Sulphur Dioxide (LCA)	−100 per cent	

Note. *Estimated from B100 results. **Average reduction across all compounds measured. ***2-nitroflourine results were within test method variability.

The life-cycle production and use of biodiesel produces approximately 80 per cent less carbon dioxide and almost 100 per cent less sulphur dioxide compared to conventional diesel. From Table-1 it is clear that biodiesel gives a distinct emission benefit almost for all regulated and non-regulated pollutants when compared to conventional diesel fuel but emissions of NOx appear to increase from biodiesel. NOx increases with the increase in concentration of biodiesel in the mixture of biodiesel and petrodiesel. This increase in NOx may be due to the high temperature generated in the fairly complete combustion process on account of adequate presence of oxygen in the fuel. This increase in NOx emissions may be neutralized by the efficient use of NOx control technologies, which fits better with almost nil sulphur biodiesel then conventional diesel containing sulphur. A comparative emission scenario with petrodiesel, biodiesel and biodiesel blends evolved from a real-life fleet study is presented in Figure 25.3.

25.5 Comparative Emissions from Petrodiesel and Biodiesel

NOx and Biodiesel

Opinion regarding emissions of nitrogen dioxides varies from one study to another study. Some fleet tests concluded NOx emissions to have increased with the use of biodiesel as fuel while other studies proved that emissions of NOx can be controlled, if not decreased, by adjustments like retarding the injection timing or by adding heavy alkylate replacing 20 per cent of the fuel of B20 per cent blend biodiesel. Conclusions from different scientific studies are compiled below.

☆ "Adjustment of injection timing and engine operation temperature will result in reduction of NOx emission levels from biodiesel below that of petrodiesel levels"- Dr. Kerr Walker, Scottish Agricultural College (Journal of Royal Society of England).

"Nitrous Oxides (NOx) are reported by several researchers to be increased with biodiesel. However, our own data shows a reduction in NOx, very consistently,

Figure 25.3: Comparative Emission Scenarios with Petrodiesel, Biodiesel and Biodiesel Blends.

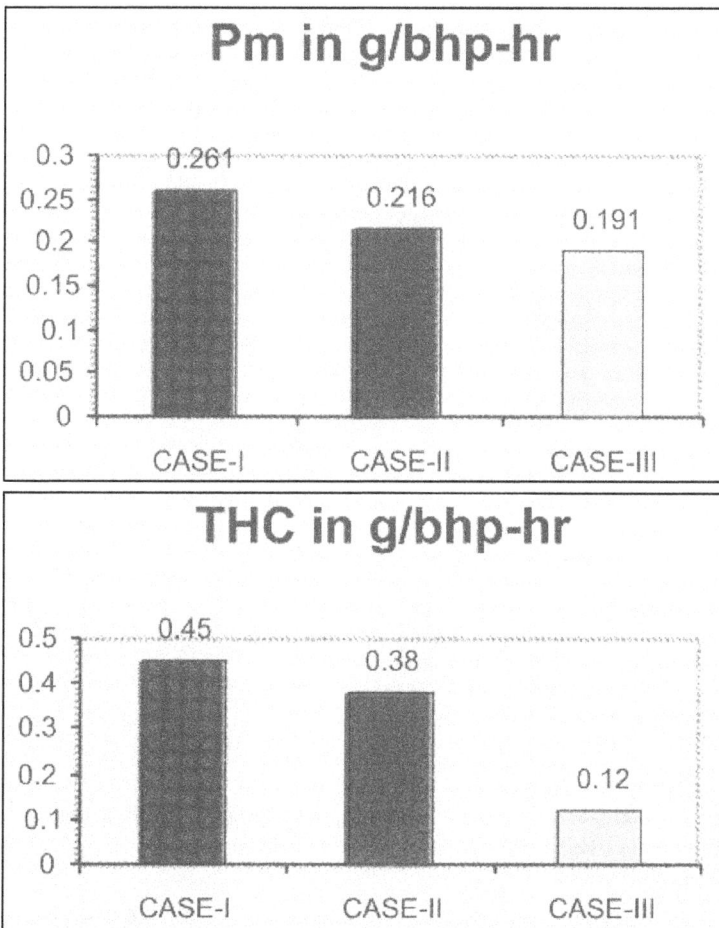

Contd...

Figure 25.3–*Contd...*

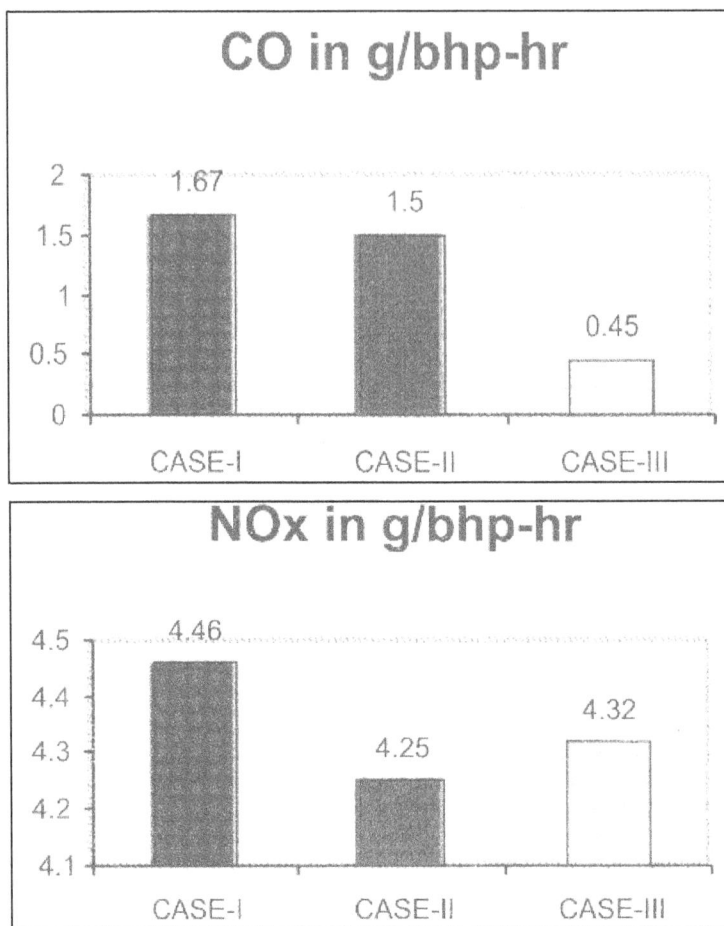

CO in g/bhp-hr

2
1.67
1.5
1.5
1
0.5
0.45
0
CASE-I   CASE-II   CASE-III

NOx in g/bhp-hr

4.5
4.46
4.4
4.32
4.3
4.25
4.2
4.1
CASE-I   CASE-II   CASE-III

CASE-I: Petrodiesel with 0.05 per cent sulphur. CASE-II: 20 per cent biodiesel with 3-degree injection timing adjustment. CASE-III: CASE-II + Catalytic Converter (Source: Twin Rivers Technologies, USA).

☆ Throughout all our dynamometer tests. NOx started at 6.2 gm/mile for diesel and goes down to around 5.6 gm/mile with 100 per cent biodiesel, with slightly more reduction with REE (rapeseed ethyl ester) than RME (rapeseed methyl ester)"- Toxicology, biodegradability and environmental benefits of Biodiesel; Charles L. Peterson and Daryl Reece, University of Idaho.

☆ "Fueling with biodiesel/diesel fuel blends reduced PM, THC and CO, while increasing NOx. Retarding the fuel injection timing reduced NOx emissions while maintaining the other emissions reductions"-DDC engine emission tests using methyl ester; L.G. Schumacher, W.G. Hires; University of Missouri.

☆ "As the concentration of biodiesel increased, NOx emissions decreased. The B20A20 fuel blend effectively reduced the oxides of nitrogen emissions below that of baseline diesel fuel. Retarding the timing was an effective way of reducing NOx. Moreover, NOx emissions from biodiesel can be successfully reduced below that of baseline diesel fuel either retarding injection timing or replacing 20 per cent of the baseline diesel fuel of B20 blend with heavy alkylate"-Engine exhaust emissions evaluation of a Cummins L10E with biodiesel; William Marshall, Leon G. Schumacher; Society of Automotive Engineers, SAE paper # 952363.

☆ "Nitrogen Oxides (NOx) emissions from biodiesel increase or decrease depending on the engine family and testing procedures. NOx from pure biodiesel (100 per cent) increased in this test by 13 per cent. However, biodiesel lack of sulphur allows use of NOx control technologies that cannot be used with conventional diesel. So, biodiesel NOx emissions can be effectively managed and efficiently eliminated as a concern of the fuel's use"- U.S. National Biodiesel Board, Biodiesel Report.

☆ "There are reliable proven methods for base lining or even reducing NOx produced when using biodiesel. I have certified emissions for the urban bus retrofit program with EPA using this technology. This package included use of an

25.6 Particulate Composition and Biodiesel

Table 25.2 gives a comparison of particulate emissions of all forms (insoluble, fuel soluble, and lube soluble and inorganic soluble) from petrodiesel and RSME (biodiesel from rapeseed methyl ester).

Table 25.2: Particulate Composition-Diesel vs. Biodiesel

Test	Fuel	Total PM (g/mile)	Insoluble (g/mile)	Fuel Soluble (g/mile)	Lube Soluble (g/mile)	Soluble Inorganic (g/mile)
Cold FTP	Diesel	0.311	0.259	0.021	0.031	17
Difference	RSME	0.258	0.118	0.104	0.036	54
Per cent		−17%	−54%	+49%	+16%	+318%
Hot FTP	Diesel	0.239	0.206	0.012	0.021	14
Difference	RSME	0.190	0.101	0.068	0.021	47
Per cent		−21%	−51%	+567%	0%	+335

The study on mechanism of soot formation from diesel as well a biodiesel (RSME) indicates reduction in total particulate matter. When the engine is operated on RSME, soot emissions (insoluble) are dramatically reduced, but the proportion of emissions composed of fuel derived hydrocarbons (fuel soluble), condensed on the soot, is much higher as can be seen from the Table given below. This implies that the RSME may not burn to completion as readily as diesel fuel. It should, however, be noted that gaseous HC emissions were reduced with RSME in the above tests. Since concern

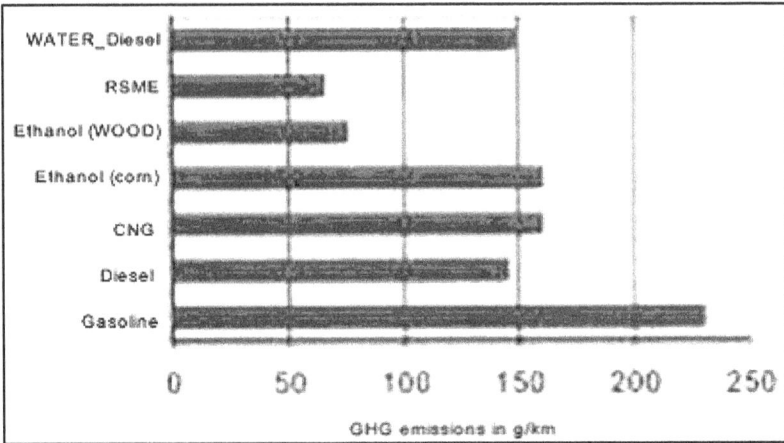

Figure 25.4: Life cycle Analysis – Greeenhouse Gases Emission of different Fuels.

over particulates arises partly from the potential harmful effects of the soluble fraction, it might be suspected that emissions from RSME would be more harmful however data shows no tendency for the mutagenic capacity of exhaust gas to increase for a vehicle running on 20 per cent RSME and 80 per cent diesel blends.

25.7 Greenhouse Gases and Biodiesel

Comparative emissions of greenhouse gases for diesel and biodiesel in various stages of life cycle are depicted in Table 25.3. Life cycle analysis for various fuels including biofuels is diagrammatically represented in Figure 25.4, which shows that biodiesel (RSME) has the lowest Greenhouse emissions followed by ethanol from wood. Emissions of greenhouse gases during the production of diesel are about 32 g/

Table 25.3: Emissions of Greenhouse Gases (g/km) from Diesel and Biodiesel

Diesel		Biodiesel	
Extraction	15.84	Fertilizer Production	15
Transport	2.74	Fertilizer Application	10
Refining	13.63	Agricultural Machinery*	25
Distribution	0.95	Oil Production	3
Vehicle Operation	245	Processing Straw**	1
		Processing Gas	17
		Transport	5
		Vehicle Operation	0
Total	278.16	Total (Straw Processing)	59
		Total (Gas Processing)	75

*Assumed mineral diesel oil used.

**Emissions of straw include those from transporting straw.

km (Figure 25.5). These are hardly a half of the emissions from producing biodiesel even when straw rather than electricity is used to fire the processing. However, this difference is far outweighed by the emissions of CO_2 during the combustion of the diesel itself (245 g/km).

25.8 Conclusions

1. Biodiesel is free from sulphur, halogens, and aerobatics so it reduces SOx

2. Biodiesel is considered as carbon neutral because all the carbon dioxide released consumption has been sequestered from the atmosphere for the growth of vegetable oil crop and hence biodiesel helps in reducing global warming.

3. Study has shown that the regulated emission such as CO, SOx, THC, PM etc. decreases while using biodiesel produced from the vegetable oil used for diesel engine, which is due to presence of its inbuilt oxygen content and higher cetane number which leads to complete combustion of the fuel.

4. The major studies have found sharp reduction in particulate matter as compared to the petro diesel. This reduction is mainly caused by reduced soot formation and soot oxidation. The oxygen content and absence of aerometric content have been pointed out as main reasons.

5. There is slight increase in NOx while using the biodiesel but this problem can be overcome by the implementation of certain engine advancement such as port injection of ethanol, exhaust heat recirculation system. Selection and modification of biodiesel (such as utilization of more saturated biodiesel and utilization of biodiesel in blended and fumigated mode) can also be utilized to overcome the problem of higher NOx emission

References

Agarwal D, Sinha S, Agarwal A K, Experimental Investigation of Control of NOx emission from Biodiesel.

USEPA, A comprehensive analysis of bio diesel impact on exhaust emission, EPA 420, 2002

Emission reduction with bio diesel, www.cytoculture.com

Deepak T., Proceeding of National Seminar on Environmental Pollution Control and Measures 2008 (EPCM 08)

CPCB guide Line.

DOE. Feb. 2002. "Biodiesel-Clean, Green Diesel Fuel". DOE/GO-102001-1449, National Renewable Energy Lab., US Department of Energy.

2015, **Impact of Global Warming and Climate Change on Human and Plant Health**

Pages 268–275

Editors: **Dr. Arun Arya and Prof. V.S. Patel**
Published by: **DAYA PUBLISHING HOUSE, NEW DELHI**

Chapter 26

Awareness of Carbon Trading in Gujarat

Hitesh A. Solanki* and Ruby Patel

*Department of Botany University School of Sciences,
Gujarat University, Ahmedabad, Gujarat, India*

ABSTRACT

The idea to have carbon market is one of the most effective policies for tackling climate change. It inspires operational excellence and incentivizes business investments in low-carbon technologies. Global warming has spawned a new form of commerce: the carbon trade. This new economic activity involves the buying and selling of "environmental services," Carbon is the common denominator in all-polluting gases that cause global warming. The idea behind carbon trading is quite similar to the trading of securities or commodities in a marketplace. The 1997 Kyoto Protocol of framework of UNFCCC is a key step towards the mitigation of climate change due to increased greenhouse gases accumulation in atmosphere. India signed and ratified the Protocol in August, 2002. Gujarat is the first state in India to sign such a MoU with the World Bank. GFL (Gujarat fluorochemicals limited) is amongst India's largest refrigerant gas manufacturing company and setting up a project for Greenhouse Gas Emission Reduction by Thermal Oxidation of HFC 23, at Gujarat in India.

Keywords: *Carbon trading, Greenhouse gases, Kyoto protocol, Gujarat.*

26.1 Introduction

The Earth has an atmosphere of the proper depth and chemical composition. About 30 per cent of incoming energy from the sun is reflected back to space while the

---

* *Corresponding author.* E-mail: husolanki@yahoo.com

rest reaches the earth, resulting in warming the air, oceans, and land, and maintaining an average surface temperature of about 15°C. The chemical composition of the atmosphere is also responsible for nurturing life on our planet. Most of it is nitrogen (78 per cent); about 21 per cent is oxygen, which all animals need to survive; and only a small percentage (0.036 per cent) is made up of carbon dioxide which plants require for photosynthesis. The atmosphere carries out the critical function of maintaining life-sustaining conditions on Earth, in the following way: each day, energy from the sun is absorbed by the land, seas, mountains, etc. If all this energy were to be absorbed completely, the earth would gradually become hotter and hotter. But actually, the earth both absorbs and, simultaneously releases it in the form of infra red waves (which cannot be seen by our eyes but can be felt as heat, for example the heat that you can feel with your hands over a heated car engine). All this rising heat is not lost to space, but is partly absorbed by some gases present in very small quantities in the atmosphere, called greenhouse gases (GHGs). Greenhouse gases (for example, carbon dioxide (CO_2), methane (CH_4), nitrous oxide (N_2O), water vapour), re-emit some of this heat to the earth's surface. If they did not perform this useful function, most of the heat energy would escape, leaving the earth cold (about minus 18°C) and unfit to support life (*http://www.tce.co.in/Downloads/bro pdf/papers/cdm_carbon_trading.pdf*).

However, ever since the Industrial Revolution began about 150 years ago, man-made activities have added significant quantities of GHGs to the atmosphere. The atmospheric concentrations of carbon dioxide, methane, and nitrous oxide have grown by about 31 per cent, 151 per cent and 17 per cent, respectively, between 1750 and 2000 (Intergovernmental Panel on Climate Change, IPCC 2001) (*www.ipcc.ch/*).

The carbon market is one of the most effective policies for tackling climate change. It inspires operational excellence and incentivizes business investments in low-carbon technologies. Not only is the market expected to save over two billion tons of CO_2 emissions by the end of 2012, but the development of the current global carbon market, now worth over US$140 billion, has catapulted climate change to the forefront of business decisions. But while it exhibits real environmental and economic impact, and helps achieve climate change goals, it remains vulnerable to external factors. In a period of downturn in economic growth, it is widely perceived to be positive that the carbon market has adjusted by a decrease in price. More problematic is a significant downturn in investment. *The State and Trends of the Carbon Market 2011* (World Bank) shows a 60 per cent decline in Clean Development Mechanism (CDM) investments because of the lack of a sufficiently ambitious emissions reduction agenda anywhere in the world (Derwent and Motty, 2008).

Global warming has spawned a new form of commerce: the carbon trade. This new economic activity involves the buying and selling of "environmental services," including the removal of greenhouse gases from the atmosphere, which are identified and purchased by eco-consulting firms and then sold to individual or corporate clients to "offset" their polluting emissions (Sada, 2007).

Carbon is the common denominator in all-polluting gases that cause global warming. Carbon dioxide is the gas most commonly thought of as a greenhouse gas; it is responsible for about half of the atmospheric heat retained by trace gasses. It is produced primarily by burning of fossils fuels and deforestation accompanied by burning and biodegradation of biomass. Analyses of gas trapped in polar ice samples indicate that pre-industrial levels of CO_2 in the atmosphere was approximately 260 ppm. Over the last 300 years, this level has increased to current value of around 375 ppm; most of the increase by far has taken place at an accelerating pace over the last 100 years.

The carbon trade is an idea that came about in response to the Kyoto Protocol. This is an agreement under which industrialized countries were supposed to reduce their greenhouse gas emissions between the years 2008 to 2012 to levels that were 5.2 per cent lower than those of 1990. The idea behind carbon trading is quite similar to the trading of securities or commodities in a marketplace. Carbon would be given an economic value, allowing people, companies or nations to trade it. If a nation bought carbon, it would be buying the rights to burn it, and a nation selling carbon would be giving up its rights to burn it. The value of the carbon would be based on the ability of the country owning the carbon to store it or to prevent it from being released into the atmosphere. A market would be created to facilitate the buying and selling of the rights to emit greenhouse gases. The market for carbon is possible because the goal of the Kyoto Protocol is to reduce emissions as a collective.

On the one hand, the idea of carbon trade seems like a win-win situation: greenhouse gas emissions may be reduced while some countries reap economic benefit. On the other hand, critics of the idea suspect that some countries will exploit the trading system and the consequences will be negative. While the proposal of carbon trade does have its merits, debate over this type of market is inevitable since it involves finding a compromise between profit, equality and ecological concerns (www.investopedia.com).

26.2 Kyoto Protocol

It is an agreement on global warming made under the United Nations Conference on Climate Change in Kyoto, Japan, in 1997. Kyoto Treaty was negotiated in December 1997, opened for signature on March 16, 1998, and closed on March 15, 1999. The agreement came into force on February 16, 2005 following ratification by Russia on November 18, 2004. As of September 2006, a total of 163 countries have ratified the agreement (representing over 61.6 per cent of emissions from Annex I countries). Notable exceptions include the United States and Australia. Other countries, like India and China, which have ratified the protocol, are not required to reduce carbon emissions under the present agreement.

The 1997 Kyoto Protocol of framework of UNFCCC is a key step towards the mitigation of climate change due to increased greenhouse gases accumulation in atmosphere. It was the first international agreement, which legally binds, developed nation to reduce worldwide emissions of greenhouse gases from these countries. The Kyoto Protocol, which was established for emission reduction target, states that the industrialized countries can achieve relatively inexpensive means of combating Climate change (Bruce, 1999). The Kyoto agreement encourages rich nations to cut greenhouse gas emissions by an average of 5.2 per cent below their 1990 levels over the next decade *i.e.*; by 2008-12. A country or company wishing to reduce or meet their

Carbon Trading emission targets can do so by investing in clean projects, which would contribute towards offsetting their GHG emissions, but would also earn the investor some "credits" which would go towards a net carbon reduction. A typical CDM project would be substituting fossil fuel-based power generation with renewable energy or a project that would improve existing energy efficiency levels. Or, as in India, by investing in forestation or community tree planting projects, called "carbon sinks". (http://unfccc.int/kyoto_protocol/items/2830.php)

26.3 The three Kyoto mechanisms are:

1. Joint Implementation,
2. Clean Development Mechanism and
3. Emissions Trading.

26.3.1. Joint Implementation

This is one of the so called 'flexibility mechanisms' are defined in Article 6 of the Protocol designed to help rich (Annex 1) countries meet their Kyoto commitment using methods other than directly via cuts in their own emissions. Under Joint Implementation, an Annex I Party (with a commitment inscribed in Annex B of the Kyoto Protocol) may implement an emission-reducing project or a project that enhances removals by sinks in the territory of another Annex I Party (with a commitment inscribed in Annex B of the Kyoto Protocol) and count the resulting emission reduction units (ERUs) towards meeting its own Kyoto target.

26.3.2. The Clean Development Mechanism

The clean development mechanism allows governments or private entities in rich countries to set up emission reduction projects in developing countries. They get credit for these reductions as 'certified emission reductions (CER's). This system is different from the Joint Implementation as it promotes sustainable development on developing countries. The Clean Development Mechanism (CDM) is the entry point for developing countries (non-Annex I) into the Kyoto Protocol on Climate Change. The mechanism was established under Article 12 of the Kyoto Protocol adopted by the Third Conference of the Parties to the Framework Convention on Climate Change on December 11, 1997.

26.3.3 Emissions Trading

Emissions' trading is one of the flexibility mechanisms allowed under the Kyoto Protocol to enable countries to meet their emissions reduction target. Countries/ companies with high internal emission reduction costs would be expected to buy certificates from countries/companies with low internal emission reduction costs. The latter entities would also be expected to maximize their production of low cost emission reduction so as to maximize their ability to sell certificates to high cost entities. The overall outcome is that the emission reduction target is met, but at a much lower cost than would be incurred by requiring each entity to achieve the emission reduction target on their own.

26.4 Carbon Credits: Some Studies

Following the literature survey, a study was undertaken using a structured questionnaire.

Currently, India takes part in an open-market carbon trade, where lots of consultants and brokers are ready to buy carbon from companies so that they can stock up on them and make good amount of money when the demand rises in another year's time in the global market. Sarkar and Manoharan (2009), in their research work gave an estimation of carbon in harvested wooden handicraft products. According to Bhayani (2009), India may be the second largest country after China in terms of carbon credit generation, but experts say Indian companies are unable to take full advantage of this. They end up selling almost a third of the credits generated to middlemen at very low rates. Many small and medium companies do not have adequate knowledge of how to register projects to be eligible for carbon emission certificates and where to sell them. That is where brokers step in; they help clients in completing formalities, but help themselves more by buying a third or even more of the credits generated at a predetermined rate. Companies get Certified Emission Reductions (CERs), popularly known as carbon credits, issued by the Clean Development Mechanism (CDM) Executive Board for emission reductions achieved under the rules of the Kyoto Protocol.

Das (2009), partner with Ernst and Young in advisory services, says some intermediaries represent the carbon credit buying community in the developed world. "Sometimes they misguide their clients, who start believing that getting their carbon emission certified and selling them is risky and the tedious and there is no proper return," he says, adding that many companies end up selling carbon credits at ridiculously low rates.

Emissions of carbon dioxide from the combustion of fossil fuels, which may contribute to long-term climate change, are projected through 2050 using reduced-form models estimated with national-level panel data for the period of 1950-1990. Using the same set of income and population growth assumptions as the IPCC, it is found that the IPCC's widely used emissions growth projections exhibit significant and substantial departures from the implications of historical experience. Sequestering C in agriculture requires a change in management practices, *i.e.* efficient use of pesticides, irrigation, and farm machinery. The C emissions associated with a change in practices have not traditionally been incorporated comprehensively into C sequestration analyses.

According to research conducted by Tami *et al.* (2004) the reductions in emission factors are partially offset by an increase in energy consumption since the base year of the previous study. The net effect is an overall decrease in emission estimates, especially those from fossil fuels.

Astha Projects (India) Ltd., Bithal Village, Himachal Pradesh Carbon trading is currently the central pillar of international climate change policy. In such cap and trade schemes, it is the level of the cap which determines how many emissions are allowed. It also determines what contribution those countries whose emissions have been capped will make towards the UNFCCC's stated aim of avoiding dangerous climate change and keeping global warming below 2°C. The level of the cap within countries or regions determines how much the largest polluting industries contribute to achieving these national or regional emission targets. IPCC recommended that greenhouse gas concentrations in the atmosphere would peak by 2015 and will be reduced by up to 85 per cent by 2050 to stabilize at 445-490 ppm CO_2e.

In 1976, the US Environmental Protection Agency (EPA) applied the concept of pollution trading to reduce the level of certain air pollutants. Companies were to be granted permission to build polluting factories in certain regions only if the company guaranteed to reduce pollution by a greater amount elsewhere (Reitze, 2001).

26.5 Conclusions

A system whereby countries or individual companies are set emission targets. Those that cannot meet their targets can buy credit from countries or companies that bear theirs. In economics, carbon trading is a form of emissions trading that allows a country to meet its carbon dioxide emissions reduction commitments, often to meet Kyoto Treaty requirements, in as low a cost as possible by utilizing the free market. It is a means of privatizing the public cost or societal cost of pollution by carbon dioxide.

People buy and sell such products because it is the most cost-effective way to achieve an overall reduction in the level of emissions, assuming that transaction costs involved in market participation are kept at reasonable levels. It is cost-effective because the entities that have achieved their own emission reduction target easily will be able to create emission reduction certificates "surplus" to their own requirements. These entities can sell those surpluses to other entities that would incur very high costs by seeking to achieve their emission reduction requirement within their own business. Similarly, sellers of carbon sequestration provide entities with another alternative, namely offsetting their emissions against carbon sequestered in biomass. (The Carbon Trade, BBC News, 20 April 2006).

26.5.1 Trading in Reducing Pollution

Suppose there are two companies, A and B, each emitting 100,000 tonnes of carbon dioxide a year. And, the government wants to cut their emissions by 5 per cent, so it gives each company allowances to emit only 95,000 tonnes. But now the government tells each company that if it doesn't want to cut its emissions by 5,000 tonnes each, it has another option. It can invest abroad in projects that 'reduce' emissions of carbon dioxide 5,000 tonnes 'below what would have happened otherwise'. Such projects might include growing crops to produce biofuels that can be used instead of oil; installing machinery at a chemical factory to destroy greenhouse gases; burning methane seeping out of a coal mine or waste dump so that it doesn't escape to the atmosphere; or building a wind power generator. The price of credits from such projects is only $ 4 per tonne, due to low labour costs, a plethora of 'dirty' factories, and government and World Bank subsidies covering part of the costs of building the projects and calculating how much carbon dioxide equivalent they save. In this situation, it makes sense for both company A and company B to buy credits from abroad rather than make reductions themselves. Company A saves $

5,000 by buying credits from projects abroad rather than cutting its own emissions. Company B meanwhile saves $ 55,000. The total saving for the domestic private sector is $ 60,000.Other names for project-based credit trading include 'baseline-and-credit' trading and 'off set' trading.

26.5.2 CDM Projects in India

India signed and ratified the Protocol in August, 2002. Since India is exempted from the framework of the treaty, it is expected to gain from the protocol in terms of transfer of technology and related foreign investments. At the G-8 meeting in June 2005, Indian Prime Dr. Singh pointed out that the per-capita emission rates of the developing countries are a tiny fraction of those in the developed world. Following the principle of *common but differentiated responsibility*, India maintains that the major responsibility of curbing emission rests with the developed countries, which have accumulated emissions over a long period of time (Sada, 2007).

"Problems with forest carbon arise when trees are being grown solely for their carbon. If there are other economic uses of the reforested land, such as producing fuel wood, rubber, fruits, and food crops, then the cost of carbon sequestration is lower. "Communities can use carbon payments to finance sustainable tree-growing investments that produce these non-carbon benefits."

As of 20 July 2005, 13 CDM projects had been registered by the Executive Board. Of these 12, three projects were from India, two from Honduras, and one each from Chile, China, Brazil, the Republic of Korea, Bhutan and Nepal. (The Indian Express, Jan, 28, 2007).

The following are some registered CDM projects registered from India:

1. GHG emission reduction by thermal oxidation of HFC-23
   Gujarat Fluorochemicals Ltd, Ranjitnagar, Gujarat
2. Biomass in Rajasthan: Electricity generation from mustard crop residue
   Kalpataru Power Transmission Ltd. (KPTL), Ganganagar, Rajasthan
3. 5 MW Dehar Grid-connected Small Hydroelectric Project (SHP) in Himachal Pradesh, India.

Gujarat is the first state in India to sign such a MoU with the World Bank. Under this agreement, Gujarat is planning to launch a campaign to reduce carbon emissions from the state. Emissions from industries and steps like safe handling of solid wastes will also be taken under this campaign. In return, the World Bank will provide financial incentives to the state.

GFL (Gujarat Fluorochemicals Limited) is amongst India's largest refrigerant gas manufacturing company. In the course of manufacture of HCFC22 (a coolant widely used in air-conditioning and refrigeration applications), HFC23 is generated as a waste product, which is a potent greenhouse gas, with a global warming potential equivalent to 11700 MT of CO_2. GFL is setting up a project for Greenhouse Gas Emission Reduction unit by Thermal Oxidation of HFC 23, at Gujarat in India. This project has been registered by the Executive Board of the CDM, established under the Kyoto

Protocol. Apart from being the largest project in India, it is also the first Indian and third in the world to be registered as a CDM. GFL expects to generate more than 3 million tons of CERs annually, which is expected to go up in the future as HCFC22 production grows. These CERs can be traded internationally and can be used as a compliance tool under the Kyoto Protocol as well as several other trading markets like the EU Emissions Trading Scheme. Trade in compliance grade emission reductions is expected to grow to € 10 billion per year by 2008, according to industry estimates (www.glf.co.in).

References

Anonymous, 2006 The Carbon Trade, BBC News, Thursday 20 April 2006.

Anonymous, 2007 The Indian Express, Jan, 28, 2007.

Bhayani, R., 2009 "Brokers Make Most of Carbon Credits" Mumbai, December 15, (2009). Business. Standard, July 1, 2010

Das, S., 2010 "India May Earn Rs 11K Crore Via Carbon Credits by Dec '12" New Delhi May 12, (2010), Business Standard, July 1, 2010

Derwent H. and Motty M, 2008. Crbon trading achievements key lessons and future forecasts.pdf, (www.climateactionprogramme.org) date assessed on 25/05/2012.

Reitze A W, 2001. *Air Pollution Control Law: Compliance and Enforcement.* Environmental Law Institute, pp. 79-80.

Sada R, 2007 Carbon trading, B.Sc. desertation, H.N.B. Garhwal University, Srinagar, Garhwal, Uttarakhand, India.

Sarkar A B and Manoharan T R, 2009. Benefits of Carbon Markets to Small and Medium Enterprises (SMEs) in Harvested Wood Products: A Case Study from Saharanpur, Uttar Pradesh, India. *African Journal of Environmental Science and Technology* 3 (9): 219-228.

Tami C B, David G S, Kristen F Y, Nelson S M, Jung-Hun Woo and Zbigniew K., 2004. "A Technology-Based Global Inventory of Black and Organic Carbon Emissions from Combustion" <www.agu.org/pubs/crossref/2004/2003jd003697.shtml>

http://www.tce.co.in/Downloads/bro_pdf/papers/cdm_carbon_trading.pdf, Date assessed on 25/05/2012. http://unfccc.int/kyoto_protocol/items/2830.php, *Date assessed on 24/05/2012.*

www.ipcc.ch/ *Date assessed on 24/05/2012.*

(www.investopedia.com) *Date assessed on 24/05/2012.*

(www.glf.co.in) *Date assessed on 24/05/2012.*

2015, Impact of Global Warming and Climate Change on
Human and Plant Health
Editors: Dr. Arun Arya and Prof. V.S. Patel
Published by: DAYA PUBLISHING HOUSE, NEW DELHI

Chapter 27

Suppression of Bio-diversity by *Lantana camara*: Simplest Solution to this Problem

Gunjan Motwani* and Hitesh Solanki**

*Department of Botany,
Gujarat University, Ahmedabad*

ABSTRACT

Every plant of the planet is a treasure to the human race. But invasive plants show negative impacts on the environment and global diversity. They are all around us- weeds defying the inhospitable city by growing out of sidewalks, grass and trees in city parks and suburban lots, forests and cultivated fields in the country side. Biotic invasion is one of the five main threats to the bio-diversity.

This study shows the status of such invasive plants in Gujarat. Also an allelopathic study regarding the mutual interaction of these invasive plants (*Lantana camara* and *Eucalyptus globulus*) was carried out. The results show that the reproductive growth of *Lantana* was totally curbed by the decrease in the concentration of phenols and sugars as both are essential components required for stimulation and initiation of flowering. Statistical analysis also supports these observations as phenols and sugars were found to be highly correlated with each other.

Keywords: *Invasive plants, Biodiversity, Allelopathy, Eucalyptus globulus.*

E-mail: *gunjan_motwani@yahoo.co.in; **husolanki@yahoo.com

27.1 Introduction

Invasive species are major concern of ecologists, biological conservationists and natural resource managers, the reason being their rapid spread, their threat to biodiversity and damage to ecosystems (Joshi *et al.,* 2004). The global extent and rapid increase in floral invasive species is recognized as a natural cause of global biodiversity loss and their effects include change in species composition and community structure (Wilcove and Chen, 1998). Drake *et al.* (1989) proposed bio-invasion as a significant component in global change and one of the major causes of species extinction. Some of the available literatures (Sivaramakrishan, 1976; Holm *et al.,* 1977; Huete *et al.,* 1985; Drake *et al.,* 1989; Thakur *et al.,* 1992; Wilcove and Chen, 1998; Rawat and Bhainsora, 1999; Lowe *et al.,* 2001; Murali and Setty, 2001; Clinton and Stuart., 2003; Joshi *et al.,* 2004 and Sharma *et al.,* 2005) on invasive species and exotic plants deal with the general problems associated.

Recently agronomists and botanists have started working on control of plant invasion using allelopathic interactions between plants. Several studies indicate that allelochemicals, particularly, phenolics affect the availability of soil nutrients (Dejong and Klinkhamber. 1985 and Appel, 1993). Plant species dominating the invasion in Gujarat are *Prosopis juliflora, Parthenium hysterophorus, Lantana camara, Eichhornia crassipes, Eucalyptus globulus, Leucaena leucocephala, Tamarix troupii, Zizyphus glabrata and Kapphaphycus aluarezii* (Algae). *Prosopis juliflora, Parthenium hysterophorus, Lantana camara and Eichhornia crassipes* are the major ones that have come up as a problem in Gujarat (Motwani and Solanki, 2009). Among these *Eucalyptus* and *Lantana* were selected for the study.

27.2 Methodology

Allelopathic interaction between the two weeds of Gujarat, *Eucalyptus globulus* and *Lantana camara* were studied, with an aim to reducing the growth of both the plants. The attempt aimed in only reducing and not totally destroying the plants as they are integral part of Indian bio-diversity and have various medicinal as well as commercial importance. Aqueous leaf extracts of *Eucalyptus* were used to treat *the Lantana.* Change in various bio-chemicals and metabolites of *Lantana* were studied using spectrophotometric quantification (Solanki *and* Motwani,. 2011).

27.3 Aqueous Extracts from *Eucalyptus* Leaves

Old leaves were collected from Gujarat University campus, washed and air dried for 2 weeks. Those leaves were then masticated to fine powder and used for extraction. 1 gm dried powder of plant material was soaked in 100 ml distilled water (D.W.) for preparation of 1 per cent aqueous extract. The mixture was heated up to 80°C and stirred well. The extraction was allowed to proceed for 48hrs. The extract was then filtered and used for the experiment. Various percentages (1 to 5) of extracts were prepared through this method.

The experiment involved treatment of this *Eucalyptus* extract on the *Lantana* through foliar spray. Freshly prepared 1-5 per cent aqueous extract was used for foliar spray. The plant was treated with this extract for 7 days every month.

27.4 Quantification Methods for Various Metabolites

Various physiological parameters, bio-chemicals and metabolites were studied are fresh weight, dry weight, shoot length, root length, IAA-Oxidase activity (Mahadevan, 1964), total phenols (Bray and Thorpe, 1954), polyphenol oxidase (PPO) activity (Kar and Mishra, 1976), starch (Chinoy, 1939), amylase activity (Summer and Howell, 1935), total Sugars (Nelson, 1944), invertase activity (Hatch and Glasziou, 1963), total proteins (Bradford, 1976) and protease activity (Cruz *et al.*, 1970).

Various laboratory tests involving carefully controlled chemical reaction were carried out to obtain their Optical Densities (O.D.) using spectrophotometer. Since the instrument is capable of only measuring the amount of light being allowed to pass through the cuvette, their readout devices display per cent of light transmitted or mathematically derived absorbance. Concentration from per cent transmittance or absorbance was evaluated through the use of a standard curve. For this purpose, standard curves with absorption on the Y axis, and increasing concentrations of the standard along the X axis were plotted. The above mentioned standard curve was constructed after obtaining the absorption readings from a number of solutions of known concentration (standards) used in a reaction. After the readings were obtained each was plotted on per cent transmittance against the corresponding concentration using the Golden Grapher software (version 4).

27.5 Results and Discussion

Aqueous extract of *Eucalyptus* leaves was used to treat *Lantana*. Post-treatment changes in the quantity of metabolites like total sugars, starch, proteins and phenols along with the post-treatment changes in the enzymes like invertase, Total Amylase, Protease, Polyphenol oxidase and IAA oxidase of *L. camara* were measured. The concentrations of these metabolites were obtained using standard curve experiments. The regression equation, coefficient of determination (r^2) and Root mean square error (RMSE) obtained as a results of these experiments are summarized in Table 27.1.

These concentrations obtained using these regression equation were then compared with the concentrations of metabolites in the controlled plant (Non-treated plants) and results along with its statistical analysis revealed the following interesting features:

27.6 Root Length, Shoot Length, Fresh Weight and Dry Weight

Comparison in variation of root length between the treated and controlled *Lantana* plant for 1-5 per cent treatment of aqueous extract from *Eucalyptus* leaves has been shown in graph 5.4(a). The results of various growth measurements showed that root length is increased 1.31 times more than the controlled plant when treated with 4 per cent aqueous extract as shown in Figure 27.1.

These results reflect that fresh and dry weight is higher at 4 per cent of aqueous extract treatment.

Shoot length is also increased 1.30 times more than the controlled plant when treated with 4 per cent aqueous. Both fresh and dry weights are increased when

Table 27.1: Regression Equation, Correlation Coefficient (r^2) and Rot Mean Square Error (RMSE) obtained through Standard Curve Experiments for Various Metabolites

Sl.No.	Metabolite	Regression Equation	Coefficient of Determination (r^2)	Root Mean Square Error (RMSE)
1.	IAA-Oxidase	O.D.=0.009* Conc.	1	1.62E-17
2.	Phenols and Polyphenol Oxidase	O.D. =0.0025* Conc.+0.0002	0.99	0.00013
3.	Starch and Total Amylase	O.D.=0.0014* Conc.+0.00029	0.99	0.0001
4.	Total sugars and Invertase	O.D. =0.0048* Conc.+0.0016	0.99	0.00064
5	Proteins and Protease	O.D. =0.0033* Conc.+0.0003	0.99	0.000221

Figure 27.1: Comparison between Root Length, Shoot Length, Fresh Weight and Dry Weight of Treated and Controlled *Lantana* Plants for 1-5 per cent Treatment of Aqueous Extract from *Eucalyptus* Leaves.

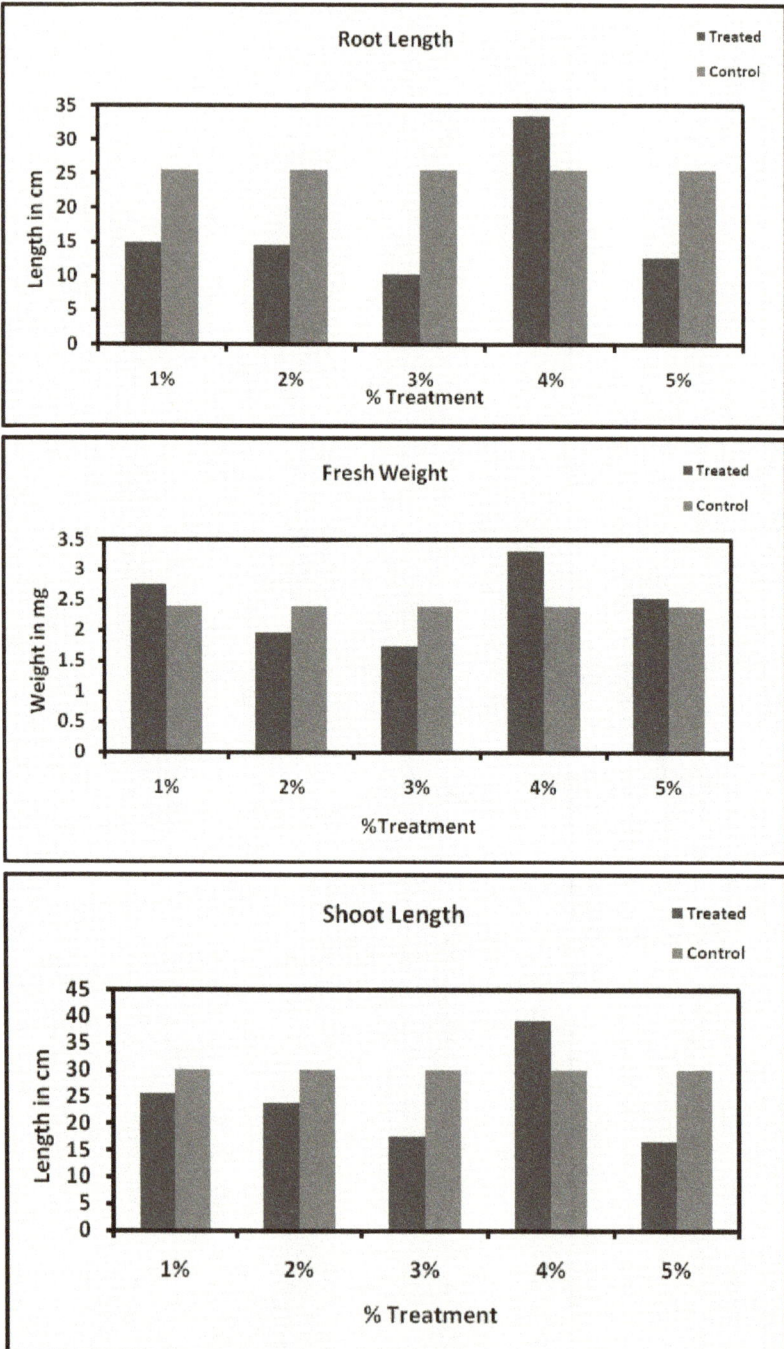

Contd...

Figure 27.1–*Contd...*

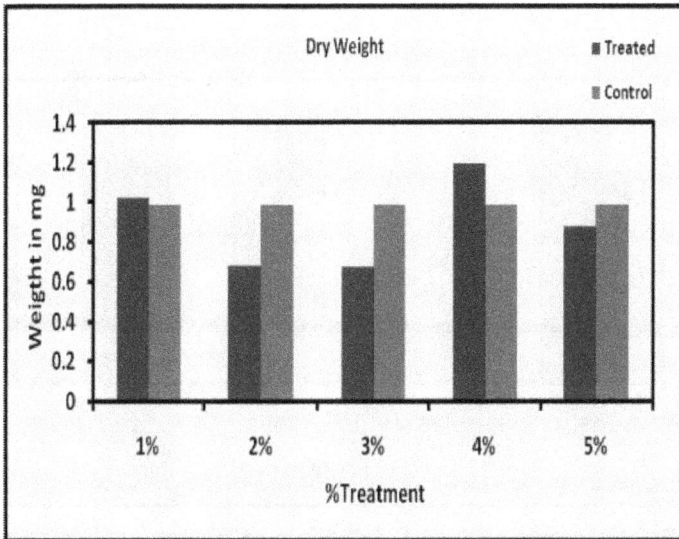

treated 4 per cent by 1.15 times and 1.38 times for 1 per cent extract concentration as shown in Figure 27.1.

The increased fresh weight may also be due to increased water content of the plant. Higher dry weight indicates addition of dry matter due to interaction of various enzymes. Growth parameters like root and shoot lengths also depend on water absorption or moisture content which would help in extended growth of the plant. It is also likely that higher dry weight is due to lesser utilization of dry matter. At 3 per cent aqueous extract treatment utilization of dry matter for continuous extension in growth of both fresh weight and dry weight are higher and show a direct correlation. Increased fresh and dry weight growth is combination and interaction of various factors like availability of water, interaction among various enzymes etc.

27.7 IAA-Oxidase

The treatment with *Eucalyptus* extract showed a decrease in the IAA Oxidase activity except for 3 per cent concentration of the treatment. Reduction in IAA-Oxidase activity corresponds to reduced oxidation of the growth hormone IAA.

The results of this IAA Oxidase quantification and its statistical analysis are shown in the Figure 27.2 which shows the comparison between the treated and controlled plant for variation in concentration of IAA Oxidase.

27.8 Phenols and Polyphenol Oxidase

Phenols were found to be at lowest concentration at 4 per cent treatment whereas the polyphenol oxidase is highest. Increase in the polyphenol oxidase content in the plant indicates increased degradation of the phenols. Decrease of phenolic content in the treated plants clearly coincides with the decrease in the reproductive growth of

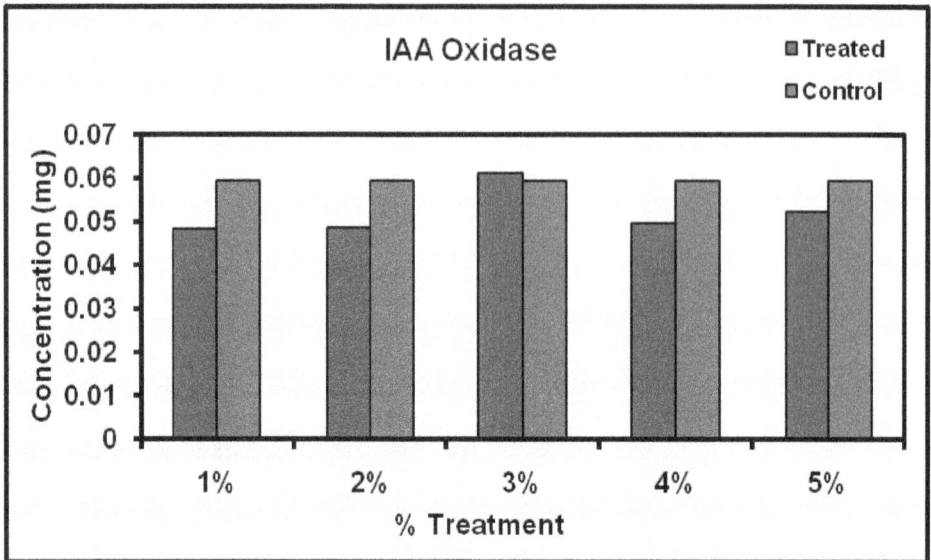

Figure 27.2: Comparison between IAA-Oxidase of Treated and Controlled *Lantana* Plants for 1-5 per cent Treatment of Aqueous Extract from *Eucalyptus* Leaves.

the treated plants. Flower initiation does not occur or is greatly reduced due to reduction in the phenols. Lesser amount of phenols also makes the plant more susceptible to infections and attacks by the predators. This can help in reducing the plant invasion as a secondary effect of the treatment. The variations in concentration of phenols and polyphenol oxidase, observed in the treated plant as compared to the controlled plant are summarized in Figure 27.3.

27.9 Starch and Total Amylase

With an exception of 3 per cent aqueous extract treatment from *Eucalyptus* leaves; the Total Amylase concentration is decreased in the *Lantana*; whereas the concentration of starch has increased in the plant. This increased starch is stored and when required utilized for the vegetative growth of the plant. The vegetative growth is enhanced in this case as there is no flowering (and so no fruiting). The variations in concentration of starch and total amylase, observed in the treated plant as compared to the controlled plant are shown in Figure 27.4.

27.10 Total Sugars and Invertase

Sugar content in all the treated plants decreased as compared to the controlled ones. This reduction in the sugar content causes floral inhibition in the plant. Reduction in the amount of phenols and sugars together result in total inhibition of flowering in the plant and thus check its further propagation through seeds. Increase in the invertase activity approximately coincides with the decrease in the concentration of sugar in the plant. The variations in concentration of Sugars and

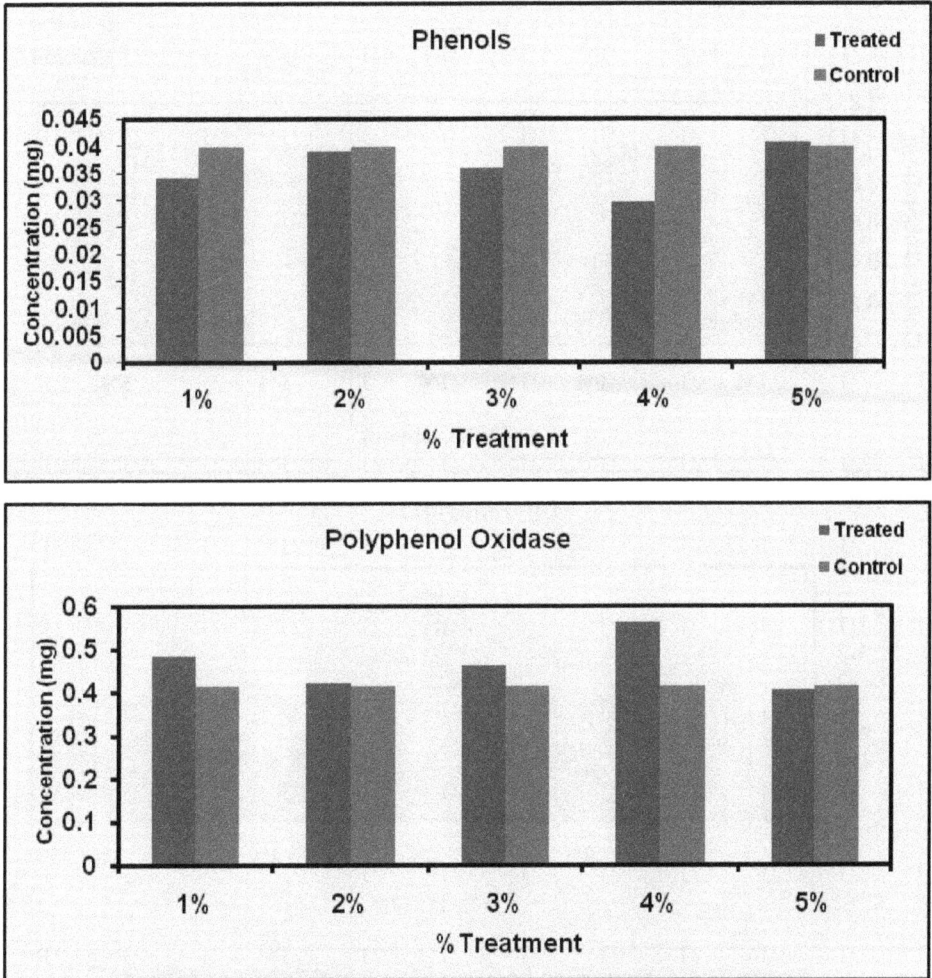

Figure 27.3: Comparison between Phenols and Polyphenol Oxidase of Treated and Controlled *Lantana* Plants for 1-5 per cent Treatment of Aqueous Extract from *Eucalyptus* Leaves.

invertase, observed in the treated plant as compared to the controlled plant are shown in Figure 27.5.

27.11 Proteins and Protease

With an exception of 3 per cent aqueous extract treatment from *Eucalyptus* leaves; the protease concentration is increased in *Lantana*. Increase in the enzyme content approximately coincides with the decrease in the concentration of proteins in the plant. Reduction in proteins concentration shows that the gene expression in the plant is affected corresponding to the change in the protein content. These variations in concentration of Proteins and protease, observed in the treated plant as compared to the controlled plant are shown in Figure 27.6.

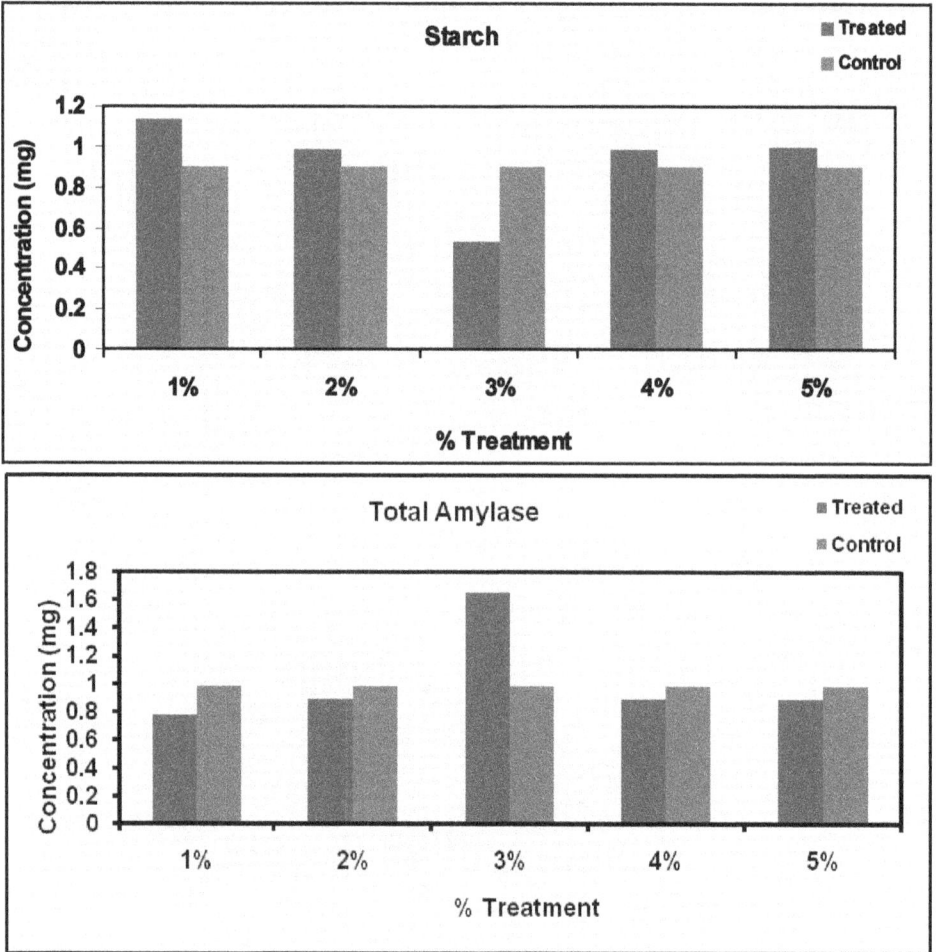

Figure 27.4: Comparison between Starch and Total Amylase of Treated and Controlled *Lantana* Plants for 1-5 per cent Treatment of Aqueous Extract from *Eucalyptus* Leaves.

27.12 Conclusions

Allelopathy is a strong and cost effective means of controlling these plants. Aqueous extract of *Eucalyptus* leaves increases the vegetative growth at 4 per cent concentration of the treatment. Other morphological changes as evident from the Figure 27.7 clearly show that flowering is inhibited or delayed with enhancement in the vegetative growth of the *Lantana* plant. It is well established that continuous vegetative growth will hamper or delay the inhibition of flowering as most of the metabolites will be utilized for vegetative growth. Flowering depends on interaction of many growth hormones like GA, IAA etc. and therefore a minimum or optimum concentration of various growth hormones, sugars and proteins have to be maintained to initiate flowering.

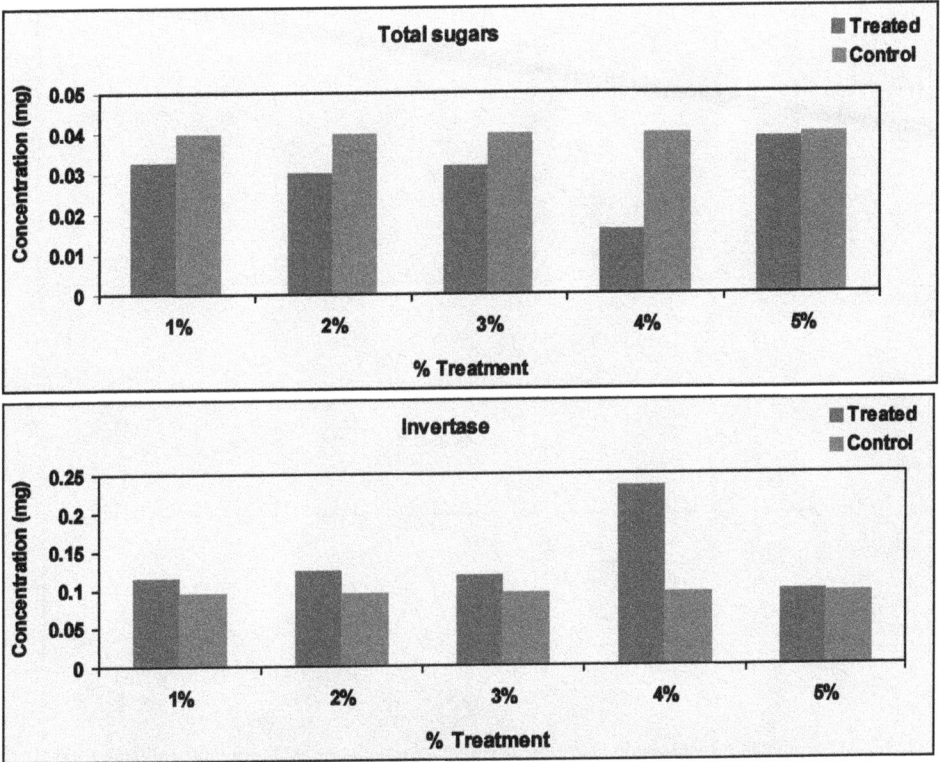

Figure 27.5: Comparison between Total Sugars and Invertase of Treated and Controlled *Lantana* Plants for 1-5 per cent Treatment of Aqueous Extract from *Eucalyptus* Leaves.

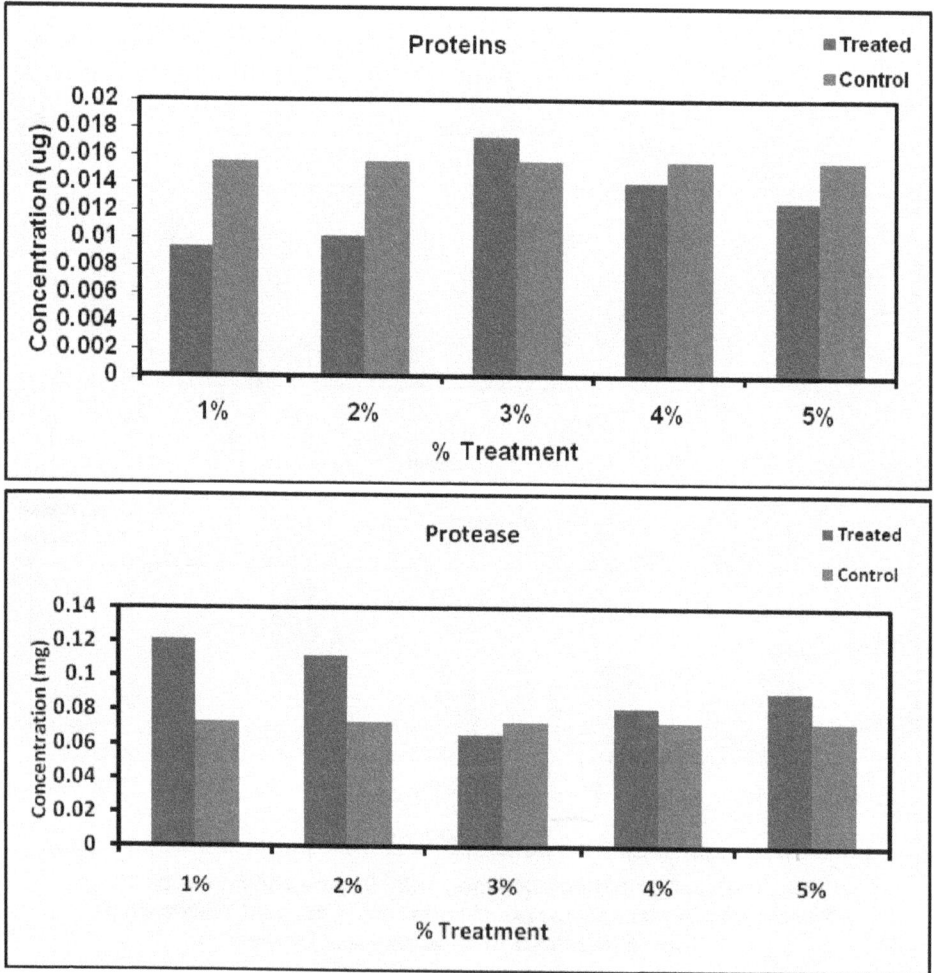

Figure 27.6: Comparison between Total Sugars and Invertase of Treated and Controlled *Lantana* Plants for 1-5 per cent Treatment of Aqueous Extract from *Eucalyptus* Leaves.

Figure 27.7: Pre-treatment and Post-treatment Photos of *Lantana camara*.

This study shows the potential of aqueous leaf leachate from fresh plants of *Eucalyptus* to reduce the growth of *L. camara*. Anderson and Loucks (1996) and Bell (1974) emphasized the importance of osmotic effects in allelopathic studies. But later studies of Achhireddy *et al.* (1985) showed that osmotic potential was not responsible for inhibition of growth. This growth inhibition occurred due to the presence of phyto-toxins in *Eucalyptus* and not due to the osmotic stress. Most of the water soluble allelochemicals are phenols. The results of this study also stand in support that phenolic extracts for *Eucalyptus* are responsible for inhibition of flowering in *Lantana*. The above observations of present study confirm that the controlled plant produced flowers whereas the plants treated with aqueous extract of *Eucalyptus* leaves failed to flower. With this method we curb the growth of plants without totally destroying it. Thus it causes no loss to the bio-diversity.

References

Achhireddy, N. R., Singh, M., Achhireddy, L. L., Nigg, H. N. and Nagy, S. 1985. Isolation and parial characterisation of phytotoxic compounds from Lantana (*Lantana camara* L.). *Journal of chemical ecology* 11:979-988.

Anderson, R. C. and Loucks, O. L. 1966. Osmotic pressure influence on germination tests for antibiosis. *Science* 152:771-773.

Appel H.M. 1993. Phenolics in ecological interactions: the importance of oxidation. *Journal of Chemical Ecology* 19:1521-1552.

Bell, D. T.1974. The influence of osmotic pressure infests for allelopathy III. *State Acad. Sci. Trans.* 67:312-317.

Bradford, M. M. 1976. A rapid and sensitive method for the quantitation of microgram quantities of proteins utilizing the principle of protein- dye binding. *Analytical Biochemistry* 72:248-254.

Bray, H. G. and Thorpe, W. V. T. 1954. Analysis of phenolic compounds of interest in metabolism. *Biochemical Analysis* 1: 2752.

Chinoy, J. J. 1939. A new colorimetric method for the determination of starch applied to soluble starch, natural starches and flour, part-I, colorimetric determination of soluble starch. *Mikrochemie* 26:132.

Clinton, N. J. and Stuart L. P. 2003. How big is the global weed patch? *Annals of Missouri Botanical Garden* 90:172-178.

Cruz, L. J., Cagampany, B. G. and Beinvenido, D. J. 1970. Biochemical factors affecting protein accumulation in rice grains. *Plant Physiology* 46: 743-747.

Dejong, T. J. and Klinkhamber, P. G. L. 1985. The negative effects of litter of parent plants of *Cirsium vulgare* to their offspring. Autotoxocity or immobilization. *Oecologia* 65:153-166.

Drake, J. A., Mooney H. A., Castri F. D., Groves R. H., Kruger, F. J., Aejmamek, M. and Williamson M 1989. *Biological invasions: a global perspective*. John Wiley and Sons: Chichester.

Hatch, M. D. and Glasziou, K. T. 1963. Sugar accumulation cycle in sugarcane, II. Relationship of invertase activity controlled environments. *Plant Physiology* 38:344-348.

Holm, L. G., Plucknett, D. L., Weldon, L. G. and Blackburn, R. D. 1977. The World's Worst Weeds: Distribution and biology, Univ. Press of Hawaii: Honolulu.

Huete, A. R., Jackson, R. D. and Post, D. F. 1985. Spectral response of a plant Canopy with different soil backgrounds. *Remote Sensing of Environment* 17:37-53.

Joshi, C., Leeuw, J. D. and Duren, ICV 2004. Remote sensing and GIS applications for mapping and spatial modelling of invasive species. In: ISPRS 2004: Proceedings of the XXth ISPRS congress: Geo imagery bridging continents, held at Istanbul, Turkey. Comm. VII, 669-677.

Kar, M. and Mishra, D. 1976. Catalase, Peroxidase and polyphenol oxidase activities during leaf senescence. *Plant Physiology* 57: 315-319.

Lowe, S., Browne, M. and Boudjelas, S. 2001. 100 of the world's worst invasive alien species. Invasive Species specialist Group (ISSG), IUCN/SSC GISP.

Mahadevan, S. 1964. Enzymes involved in the synthesis and breakdown of IAA. *Modern Methods of Plant Physiology* 7:233-259. (eds.) Linskons, M. F., Sanwar, B. D. and Tracey, M. V.

Motwani, G. S. and Solanki, H. A. 2009. Status of invasive plants in Gujarat. In: UGC sponsored State level seminar on *Environment Education and Disaster Management*, organized by M.G. Science Institute, Ahmedabad, Gujarat. 15.

Murali, K. S. and Setty, R. S. 2001. Effect of weeds *Lantana camara* and *Chromelina odorata* growth on the species diversity, regeneration and stem density of tree and shrub layer in BRT sanctuary. *Curr. Sci.* 80(5):675-678.

Nelson, N. 1944. A photometric adaption of the Somogyi's meyhod for the determination of glucose. *Journal of Biological Chemistry* 153: 375-380.

Rawat, G. S. and Bhainsora N. S. 1999. Woody vegetation of Shiwaliks and outer Himalaya in Himalaya in north western India. *Tropical Ecology* 40(1):119-128.

Sharma, G. P., Singh, J. S. and Raghubansi, A. S. 2005. Plant invasions: Emerging trends and future implications. *Curr. Sci.* 88(5):726-734.

Sivaramakrishan, V. R. 1979. Occurance of Lantana lace bug, *Teleonemia scrupulosa* Stal. (Hemiptera: Tinigdae) in South India. *Indian Forester* 102(9): 620-622.

Solanki, H. A. and Motwani, G. S. 2011. Allelopathic interaction between *Lantana camara* and *Eucalyptus globules*: a case study to control bio-invasion by *Lantana*. *Bioscience Guardian* 1(2):463-471.

Summer, J. B. and Howell, J. F. 1935. A method for determination of sacharase activity. *Journal of Biological Chemistry* 108: 51-54. Modified by Bernfeld, P. (1955). Methods in Enzymology. (eds.) Colowick, S. P. and Kaplan, N. O., Volume 1, Academic Press Inc. Publishers, New York.

Thakur, M. L., Ahmed, M. and Thank, R. K. 1992. Lantana weed (*Lantana camara* var. aculeate Linn.) and its possible management through natural insect pests in India. *Indian Forester* 118, 7.

Wilcove, D. S. and Chen, L. Y. 1998. Management costs for endangered species. *Conservation Biology* 12:1405-1407.

2015, Impact of Global Warming and Climate Change on   *Pages **291–303***
Human and Plant Health
Editors: **Dr. Arun Arya and Prof. V.S. Patel**
Published by: **DAYA PUBLISHING HOUSE, NEW DELHI**

Chapter 28

Problems of Groundwater Pollution and Their Management Strategies

Bina Rani[1]*, Upma Singh[2]** and Raaz Maheshwari[3]***

[1]*Department of Engineering Chemistry and Environmental Engineering, PCE,
Sitapura, Jaipur, Rajasthan, India*
[2]*School of Applied Sciences, Gautam Buddha University,
Greater Noida, U.P., India*
[3]*Department of Chemistry, SBRMGC, Nagaur, Rajasthan, India*

ABSTRACT

To cope up with global water scarcity, UN's General Assembly in 1992, declared March 22nd as World Water Day to create awareness of the dilemma amongst individuals and communities. Access to safe water should be a basic human right and it can be ensured by adopting various measures like watershed management through rehabilitation of existing systems like tanks, construction of check dams and Rainwater Harvesting (RWH). Among the different water management techniques, RWH is gaining favour but it's yet to become an essential part of day to day living for which mass awareness and group action necessary can be promoted by different committees, societies, NGOs, self help groups etc. Environmental arsenic exposure mainly occurs from arsenic-contaminated drinking water. Arsenic in drinking water is often from natural sources. Environmental exposure to arsenic also occurs from burning of coal containing naturally high levels of arsenic, and perhaps from wood treated with arsenical

E-mail: *binaraj_2005@rediffmail.com; **drupmasingh@gmail.com;
***drraazecoethics151260@gmaiil.com

preservatives. While, symptoms of acute intoxication in humans include fever, anorexia, hepatomegaly, melanosis, cardiac arrhythmia and, in fatal cases, eventual cardiac failure, the skin is a major target organ in chronic inorganic arsenic exposure. In this review, efforts are made at pointing the attention of the public to the danger of water pollutants and the need to reduce to the barest minimum the level of this contaminant in the environment. Therefore, the challenge is not only to generate new information, but to better use what we already know in order to make the best possible decisions and policies for our families and future generations. It's true that if each of us uses the water judiciously and augment the water resources by becoming the custodian rather than the absolute owner, the water resources can well be protected for the future mankind.

Keywords: *RWH, Artificial recharge, Erosion, Urbanization, Health hazards, Sustainable development, Check dams, Percolation tanks, Nadi-Talab-Johad-Baori-Kund.*

28.1 Introduction

Water is unarguably the most significant source of life. Of all the natural resources, it is most significant source for the urban and rural population, it helps in sustaining many wetland ecosystems (Ameetha *et al.,* 2002).

Life began in water and is nurtured with water. There are organisms, such as anaerobes, which can survive without Oxygen. But no organism can survive for any length of time without water. It is a universal solvent and as a solvent it provides the ionic balance and nutrients, which support all forms of life (Biswajit, 2001). The amount of water on the planet earth is estimated to be approximately 1388 million billion cubic meters (Sampat, 2001). Of this total amount, major part 1348 million billion cubic meters (97.3 per cent) is constituted by the salt water in the oceans. Only 37.5 billion cubic meter (2.7 per cent) water occurs in the form of freshwater. Of this total freshwater, 28200 thousand billion cubic meter (0.61 per cent) as groundwater and 127 thousand billion cubic meters in in the form of lakes, rivers etc.(CGWM, 1990).The daily demand of drinking water of a man is normally 7 per cent of his body weight.

The importance of water as a vital resource to the life system and an essential component of societal development cannot be overemphasized. Over the years rising populations, growing industrialisation, and expanding agriculture have pushed up the demand for water. Efforts have been made to collect water by building dams and reservoirs and digging wells. Water conservation has become the need of the day. The idea of groundwater recharging by harvesting rainwater is gaining importance in many cities.

Rainwater harvesting improves the quality of groundwater through the dilution of fluoride, nitrate, and salinity The most important step in the direction of finding solutions to issues of water and environmental conservation is to change people's attitudes and habits – this includes each one of us. Conserve water because it is the right thing to do. We can follow some of the simple things that have been listed below and contribute to water conservation.

28.2 Contamination of Groundwater

It was way back in mid eighties that a systematic study of groundwater quality and identification of 'problematic zones' was under taken by Central Pollution Control Board (CPCB). The usual and most neglected source of water is uncontrolled dumping of Municipal Solid Waste (MSW). Infiltration of water by rainfall, or water generated by biodegradation, cause the leachate to leave the dumping ground laterally or vertically and find its way into the groundwater thereby causing contamination. Over the years the problem areas that emerged witnessed excessive exploitation of groundwater for domestic and industrial uses. Studies carried out indicate that high concentration of cynide (2mg/L), hexagonal chromium(12.8 mg/L) have been observed in groundwater of Ludhiana, Punjab, chromium of 2mg/L in groundwater of Varanasi, 21mg/L in Kanpur, U.P. has also been observed. It is reported that 80µg of arsenic and40 µg of lead per 100 g of blood cause poisoning in adult and children (Panigrahi, 2014). Cary (1982) described action of pollutant chromium on animals. As pollution control enforcement activities gained momentum there were observed cases of indiscriminate waste disposal, subsurface discharge of effluent and inappropriate wastewater management by industries. Today, as a result of these malpractices there is a palpable stress on groundwater, in terms of quantitative imbalances as well as qualities deterioration.

28.3 Contamination due to Fluoride

In the world, around 200 million people from 25 nations have great health risks, with high fluoride in the drinking water. in the country (India), almost 60-65 million people drink fluoride contaminated groundwater and the number affected by fluorosis is estimated at 2.5 to 3 million in many states, especially, Rajasthan, AP, Punjab, TN and UP. In India, safe limit of fluoride in potable water is between 0.6 and 1.2ppm (mg/l)[BIS 2003]. Lower limit of fluoride (<0.6ppm) than that of the prescribed limit (0.6ppm) causes dental caries, while higher limit of fluoride (>1.2ppm) than those of the recommended limit (1.2ppm) results in fluorosis. The source of fluoride in groundwater is not only because of a wide spread occurrence of fluoride rich soil in India but also because of the excessive use of phosphatic fertilizer, mining (copper and iron) and allied industries. Fluorides are released into the environment naturally through the weathering and dissolution of minerals, in emissions from volcanoes and in marine aerosols. Fluorides are also released into the environment via coal combustion and process waters and wastes from various industrial processes, including steel manufacture, primary aluminium, copper and nickel production and use, glass, brick and ceramic manufacturing, and glue and adhesive production. The use of fluoride containing pesticides as well as the controlled fluoridation of drinking-water supplies also contributes to the release of fluoride from anthropogenic sources. Based on available data, phosphate ore production and use as well as aluminium manufacture are the major industrial sources of fluoride release into the environment. Other minerals containing fluoride are sellaite (MgF_2), villianmite (NaF), fluorite or fluorospar (CaF_2), cryolite (Na_3AlF_6), bastnaesite [(Ce, La)(CO_3)F] and fluorapatite [$Ca_3(PO_4)_2F$].

Fluoride ingested with water goes on accumulating in bones up to age of 55 years. At high doses fluoride can interfere with carbohydrate, lipid, protein, vitamins, enzymes and mineral metabolism. Long term consumption of water containing 1ppm of fluoride leads to dental fluorosis. White and yellow glistening patches on the teeth are seen which may eventually turn brown. The yellow and white, patches when turned brown present itself has horizontal streaks. The brown streaks may turn black and affect the whole tooth and may get pitted, perforated and chipped off at the final stage. Skeletal problems: Fluoride can also damage the foetus-if the mother consumes water and food with a high concentration of fluoride during pregnancy/breast feeding, infant mortality due to calcification of blood vessels can also occur. The others are: (1) Severe pain in the backbone, hip region and in the joints, (2) Stiffness of the backbone, (3) Immobile/stiff joints, (4) Increased density of bones, besides calcification of ligaments, (5) Construction of vertebral canal and vertebral foramen-pressure on nerves, and (6) Paralysis. Non-skeletal problems: (1) Neurological manifestations: Nervousness, depression, tingling sensation in fingers and toes, excessive thrust. Tendency to urinate frequently (polydypsia and poly urea are controlled by brain-appears to be adversely affected). (2) Muscular manifestations: Muscle weakness, stiffness, pain the muscle and loss of muscle power. (3) Allergic manifestations: Very painful skin rashes, which are perivascular inflammation – present in women and children, pinkish, red or bluish red spots on the skin that fade and clear up in 7-10 days, they are round or oval shape. (4) Gastero intestinal problems, acute abdominal pain, diarrhea, constipation, blood in stools, bloated feeling (gas) tenderness in stomach, feeling of nausea and mouth sores (5) Headache and (6) Loss of teeth (edentate) at an early age.

Treatment of high fluoride groundwater: treatment with lime and alum is the most common method practised both at the community and the household level but, more recently, activated alumina has come into use. In situ treatment has received much less attention. Alkaline soils can be remediated through the application of gypsum, pyrite or sulphuric acid. On a long-term basis, the planting of trees like *Acacia nilotica, Prosopis juliflora, Albizia lebbek* and *Polpulus deltoids* may alleviate sodicity in soils. Treatment of alkaline soils is usually initiated as a result of problems with soil structure and permeability. As is indicated by the relationship between soil, pH and fluoride in groundwater, lowering alkalinity may also decrease the mobility of fluoride. Gypsum treatment is the classical method of alleviating soil alkalinity. It has advantages in being cheap as gypsum is abundant in India, even in hard rock

GROUNDWATER IN INDIAN CITIES

With twenty five per cent of India's population living in urban areas, Indian cities are posed with the problem of over-exploitation of groundwater. The poor urban infrastructure has no systematic provision of sewage or solid waste management. Unplanned growth, unorganised land-use and poor drainage system further compound the groundwater quality concerns. In fact the very process of urbanization in Indian cities, has led to phenomenal decrease of natural (groundwater) recharge due to paved roads and soil compaction, thus promoting imbalance in the overall groundwater budget.

areas. The gypsum treatment will give harder water. this may be an advantage as that means a higher intake of Ca which, as mentioned earlier, mitigates the effect of fluoride.

28.4 Contamination due to Fluoride

Arsenic contamination of groundwater is a natural occurring high concentration of arsenic in deeper levels of groundwater, which became a high-profile problem in recent years due to the use of deep tubewells for water supply in Ganges Delta, causing serious arsenic poisoning in large number of people. A 2007 study found that over 137 million people in more than 70 countries are probably affected by arsenic poisoning of drinking water. Arsenic contamination of groundwater is found in many countries throughout the world, including the USA. Approximately 20 incidents of groundwater arsenic contamination have been reported from all over the world. Of these, four major incidents were in Asia, including locations in Thailand, Taiwan, and Mainland China. South American countries like Argentina and Chile have also been affected. There are also many locations in the United States where the groundwater contains arsenic concentrations in excess of the Environmental Protection agency standard 10ppb adopted in 2001. According o a recent film funded by the US Superfund, "In Small Doses", millions of private wells have unknown arsenic levels, and in some areas of the US, over 20 per cent of wells may contain levels that are not safe.

Arsenic is a carcinogen which causes many cancers including skin, lung, and bladder as well as cardiovascular disease. Some research concludes that even at the lower concentrations, there is still a risk of arsenic contamination leading to major causes of death. A study conducted in a contiguous six-country area of southeastern Michigan investigated the relationship between moderate arsenic levels and 23 selected disease outcomes. Disease outcomes included several types of cancer, diseases of the circulatory and respiratory system, diabetes mellitus, and kidney and liver diseases. Elevated mortality rates were observed for all diseases of the circulatory system. The researchers acknowledged a need to replicate their findings. A preliminary study shows a relationship between arsenic exposure measured in urine and type II diabetes the resulted supported the hypothesis that low levels of exposure to inorganic arsenic in drinking water may play a role in diabetes prevalence. Arsenic in Drinking water may also compromise function "Scientist link influenza A (H_1N_1) susceptibility to common levels of arsenic exposure".

The story of arsenic contamination of groundwater in Bangladesh is a tragic one. Many people have died from this contamination. Diarrheal diseases have long plagued the developing world as a major cause of death, especially in children. Prior to the 1970s, Bangladesh had one of the highest infant mortality rates in the world. Ineffective water purification and sewage systems as well as periodic monsoons and flooding exacerbated these problems. As a solution UNICEF and the World Bank advocated the use of wells to tap into deeper groundwater for a quick and inexpensive solution. Millions of wells were constructed as a result. Because of this action, infant mortality and diarrheal illness were reduced by fifty per cent. However, with over 8 million wells constructed, it has been found the last two decades that approximately one in five of these wells is now contaminated with arsenic above government's water standard.

In the Ganges Delta, the affected wells are typically more than 20 m and less than 100 m deep. Groundwater closer to the surface typically has spent a shorter time in the ground, therefore likely absorbing a lower concentration of arsenic; water deeper than 100 m is exposed to much older sediments which have already been depleted of arsenic. Dipankar Chakraborty from West Bengal brought the crises to international attention in 1995. Beginning his investigation in west Bengal in 1988, he eventually published, in 2000, the results of a study conducted in Bangladesh, which involved the analysis of thousands of water samples as well as hair, nail and urine samples. They found 900 villages with arsenic above the government limit. Chakraborty has criticized aid agencies, saying that they denied the problem during the 1990s while millions of tube wells were sunk. The aid agencies later hired foreign experts, who recommended treatment plants which were not appropriate to the conditions, were regularly breaking down, or were removing the arsenic.

Chakraborty says that the arsenic situation in Bangladesh and West Bengal is due to the negligence. He also adds that in West Bengal water is mostly supplied from rivers. Groundwater comes from deep tubewells, which are few in number in the state. Because of the low quantity of deep tubewells, the risk of arsenic patients in West Bengal is comparatively less. According to the WHO, "In Bangladesh, West Bengal (India) and some other areas, most drinking-water used to be collected from open dug wells and ponds with little or no arsenic, but with contaminated water transmitting diseases such as diarrhoea, dysentery, typhoid, cholera and hepatitis. Programmes to provide 'safe' drinking water over the past 30 years have helped to control these diseases, but in some areas they have had the unexpected side-effect of exposing the population to another health problem – arsenic." The acceptable level as defined by WHO for maximum concentrations of arsenic in safe drinking water is 0.01 mg/l. the Bangladesh government's standard is at a slightly higher rate, at 0.05 mg/l being considered safe. WHO has defined the ares under threat: Seven of the nineteen districts of West Bengal have been reported to have groundwater arsenic concentrations above 0.05 mg/l. the local population in these seven districts is over 34 million, with the number using arsenic-rich water is more than 1 million (0.05 mg/l). The number increases to 1.3 million when the concentration is above 0.01 mg/l. according to British Geological survey (BGS) study in 1998 on shallow tube-wells in 61 of the 64 districts in Bangladesh, 46 per cent of the samples were above 0.01 mg/l and 27 per cent were above 0.05 mg/l. When combined with the estimated 1999 population, it was estimated that the number of people exposed to arsenic concentrations above 0.05 mg/l and the number of those exposed to more than 0.01 mg/l is 46-57 million (BGS 2000).

Throughout Bangladesh, as tubewells get tested for concentrations of arsenic, ones which are found to have arsenic concentrations over the amount considered safe are painted red to warn residents that the water is not safe for drink. The solution according to Chakraborty is "By using surface water and instituting effective withdrawal regulation West Bengal and Bangladesh are flooded with surface water.

We should first regulate proper watershed management. Treat and use available surface water, rain-water and others. The way we're doing at present is not advisable."

There are many locations across the United States where groundwater contains naturally high concentrations of arsenic. Cases of groundwater-caused acute arsenic toxicity, such as those found in Bangladesh, are unknown in the United States where the concern has focussed on the role of arsenic as a carcinogen. The problem of high arsenic concentrations has been subject to greater scrutiny in recent years because of changing government standards for arsenic in drinking water. Some locations in the United States, such as Fallon, Nevada, have long been known to have groundwater with relatively high arsenic concentrations (in excess of 0.08 mg/l). Even some surface waters, such as Verde River in Arizona, sometimes exceed 0.01 mg/l arsenic, especially during low-flow periods when the river is dominated by groundwater discharge. A drinking water standard of 0.05 mg/l (equal to 50ppb) arsenic was originally established in the United States by the Public Health Service in 1942. The Environmental Protection Agency (EPA) studied the pros and cons of lowering the arsenic Maximum Contaminant Level (MCL) for years in the late 1980s and 1990s. A study of private water wells in the Appalachian mountains found that 6 per cent of the wells had arsenic above the US MCL of 0.010 mg/l.

In Nepal there is a serious problem with arsenic contamination particularly in Terai region, worst being near Nawakparasi District, where 26 per cent of shallow wells failed to meet WHO standard of 10ppb. A study by Japan International Cooperation Agency and the Environment in the Kathmandu Valley, particularly in deep wells, of which 71.6 per cent failed to meet the WHO standard, and 11.9 per cent failed to meet the Nepali standard of ppb.

The first signs of arsenic contamination in India were detected in West Bengal as early as 1988. Today symptoms of arsenicosis are being observed in more and more states. Recently, the groundwater in UP's Ballia district was found to be contaminated with arsenic. The UN's estimate is that currently 35 million people in Bangladesh and India are in danger of drinking arsenic contaminated water. In January 2004, as many as 2,404 samples of water drawn from handpumps in 55 villages in Ballia were tested. More than half the samples had arsenic level above the Indian guideline of 10μg per litre; eight per cent had arsenic levels above 500. The samples of water were tested and analysed under the guidance of Dr Dipankar Chakraborty, Director, School of Environmental Studies (SOES), Jadhavpur University, Kolkata, Dr Chakraborty is often offered as the 'arsenic hunter'.

The report on arsenic contamination in Ballia district was released by Sunita Narain, Director of Delhi-based Centre for Science and Environment (CSE), at a conference in Delhi recently. Narain said the local administration had promised to send Dr Chakraborty's report to the Industrial Toxicology Research Centre (ITRC) in Lucknow. ITRC had reported in July 2004 that the samples of water they tested from the district were safe, and so the villagers in the area had gone back to drinking water drawn up by handpumps. Having seen obvious symptoms of arsenicosis in the villagers of Ballia, CSE sponsored their own tests on handpump water, as well as blood, nail and hair samples from the area in August 2004. Analysed at the Shri Ram

Institute for Industrial Research Laboratory (Delhi), the water samples were found to contain levels of arsenic ranging from 15 to 129 μg per litre – dangerously high considering what the Indian guideline level suggests as safe. Arsenic in groundwater can enter the body by drinking arsenic laced water or by eating food cooked in the water. arsenic does not evaporate in to the air and is not easily absorbed through the skin. Most foods, including vegetables, fish and seafood also contain some arsenic. Studies have shown that drinking water containing elevated levels of arsenic can cause the following health effects:

☆ Thickening and discolouration of the skin, sometimes leading to skin cancer, which may be curable at an early stage.

☆ Digestive problems such as stomach pain, nausea, vomiting and diarrhoea.

☆ Numbness in the hands and feet.

Arsenic laced water is mostly found at an intermediate water depth of 20-100 meter below ground level. Occurrence of 'arseno-pyrite' in the region and the change of geo-chemical environment due to over-exploitation of groundwater and excessive fluctuations of groundwater table, has introduced this dreaded substance into freshwater.

Small scale water treatment: A simpler and less expensive form of arsenic removal is known as the Sono arsenic filter, using 3 pitchers containing cast iron turnings and sand in the fist pitcher and wood activated carbon and sand in the second. It is claimed that thousands of these systems are in use can last for years while avoiding the toxic waste disposal problem inherent to conventional arsenic removal plants. Although novel, this filter has not been certified by any sanitary standards such as NSF, ANSI, and WQA, and does not avoid toxic waste disposal similar to any other iron removal process. In the United S6ates small "under the sink" units have been used to remove arsenic from drinking water. this option is called "point of use" treatment. The most common types of domestic treatment use the technologies of adsorption (using media such as Bayoxide E33, GFH, or titanium dioxide) or reverse osmosis. Ion exchange and activated alumina have been considered but not commonly used.

Coagulation/filtration removes arsenic by coprecipitation and adsorption using iron coagulants. Coagulation/filtration using alum is already by some utilities to remove suspended solids and may be adjusted to remove arsenic. But problem of this type of filtration system is that it gets clogged very easily, mostly within two or three months. The toxic arsenic sludge are disposed of by concrete stabilization, but there is no guarantee that they won/t leach out in future.

Iron oxide adsorption filters the water through a granular medium containing ferric oxide, which has a high affinity for adsorbing dissolved arsenic. The iron oxide medium eventually becomes saturated, and must be replaced. The sludge disposal is a problem here too. Activated alumina is another filter medium known to effectively remove dissolved arsenic. It has also been used to remove undesirably high concentration of fluoride.

Ion exchange has long been used as a water-softening process, although usually on a single-home basis. It can also be effective in removing arsenic with a net ionic charge. (Note that arsenic oxide, As_2O_3, is a common form of arsenic in groundwater that is soluble, but has no net charge.) But the main advantage is that, the media is pretty much expensive.

Both Reverse osmosis and electrodialysis (also called *electrodialysis reversal*) can remove arsenic with a net ionic charge. Some utilities presently use one of these methods to reduce total dissolved solids and therefore improve taste. A problem with both methods is the production of high-salinity wastewater, called brine, or concentrate, which then must be disposed of.

28.5 Subterranean Arsenic Removal (SAR) Technology

In SAR, aerated groundwater is recharged back into the aquifer to create an oxidation zone which can coprecipitate iron and arsenic. The oxidation zone created by aerated water boosts the activity of the arsenic-oxidizing microorganisms which can oxidize arsenic from +3 to +5 state SAR technology. no chemicals are used and almost no sludge is produced during operational stage since iron and arsenic are trapped under the earth. Thus toxic waste disposal and risk of its future mobolizatuion is prevented by this technology. Also, it has very long operational life similar to the long lasting tube wells drawing water from the shallow aquifers. The first community water treatment plant based on SAR technology was set up at Kashinpore near Kolkata in 2004 by a team of European and Indian engineers. Researchers from Bangladesh and the united kingdom have recently claimed that dietary intake of arsenic adds a significant amount of total intake, where contaminated water is used for irrigation.

Biologists have identified genes in plants that control arsenic accumulation. The study promises to prevent the metalloid from entering the food web and also clean contaminated sites through bioengineered plants. Researchers had been looking for the genes for 25 years. "Identifying them is a crucial step in keeping arsenic from accumulating in the edible parts of the plants, like rice grains and fruits," stated Julian Shroeder, biologist at Columbia University, USA. As the plant absorbs water, arsenic binds with a plant molecule, phytochelatin, which transports it to vacuoles, a storage structure in plant cells. Eating plant parts that contain arsenic filled vacuoles is harmful. Shroeder's team analysed the genetic material of a yeast, *Schizosaccharomyces pombe,* and found its ABC2 gene controls the activity of the phytochelatin and plays a role in arsenic accumulation in the yeast, the findings were reported in the November, 10, issue of *Journal of Biological Chemistry.*

Shroeder later teamed up with researchers of University of Zurich in Switzerland and found two genes in *Arabidopsis* plant that control transport and storage. Knocking them off will prevent the uptake and accumulation of arsenic in plants. Researchers at the US Geological survey, stated the findings suggest non-food plants can be bioengineered to absorb arsenic from contaminated sites. Rhay can be later burnt to eliminate the metalloid.

28.5 Contamination due to Nitrite

Nitrate is a naturally occurring form of nitrogen in soil essential to all forms of life. Most crop plants require large quantities to sustain high yields. The formation of nitrates is an integral part of the nitrogen cycle in our environment. in moderate amounts, nitrate is harmless constituents of food and water. due to its high mobility, nitrate also can leach into groundwater. Although nitrates occurs naturally in some groundwater, in most cases higher levels are thought to result from human activities. Common sources of nitrate include:

☆ Fertilizer and manure,

☆ Animal feed,

☆ Municipal wastewater and sludge,

☆ Septic tanks, and

☆ N-fixation from atmosphere by legumes, bacteria and lightning

Health effects of nitrates: High nitrate levels in water can cause methemoglobinemia or blue baby syndrome, a condition found especially in infants less than six months. The stomach acid of an infant is not as strong as in older children and adults. This causes an increase in intestinal bacteria that can readily convert nitrate (NO_3^-) to nitrite (NO_2^-). Nitrate is absorbed in the blood, and haemoglobin (the oxygen carrying component of blood) is converted to methemoglobin, which does not carry oxygen efficiently. This results in a reduced oxygen supply to vital tissues such as brain. Methemoglobin in infant blood cannot change back to haemoglobin, which normally occurs in adults. Severe methemoglobinemia can result in brain damage and death. Pregnant women, adults with reduced stomach acidity, and people deficient in the enzyme that changes methemoglobin back to normal haemoglobin are all susceptible to nitrite-induced methemoglobinemia. The most obvious symptom of methemoglobinemia is a bluish colour of the skin, particularly around the eyes and mouth. Other symptoms include headache, dizziness, weakness or difficulty in breathing.

Healthy adults can consume fairly large amounts of nitrate with few known health effects. In fact, most of nitrate we consume is from our diets, particularly from raw or cooked vegetables. This nitrate is readily absorbed and excreted in the urine. However, prolonged intake of high levels of nitrate are linked to gastric problems due to the formation of nitrosamines. N-nitrosamine compounds have been shown to cause cancer in test animals. Clinical infant methemoglobinemia was first recognized in 1945. About 2,000 cases were reported in North America and Europe by 1971.fatality rates were reported to be approximately 7 to 8 per cent.

28.6 Clean Drinking Water

28.6.1 Protecting our Drinking Water

Protecting our drinking water supply from contamination is important for health and to protect property values and minimise potential liability. High nitrate levels often are associated with poorly constructed or improperly located wells. Locate new wells uphill and at least 100 feet away from septic systems, barn yards and chemical storage facilities. Properly seal or cap abandoned wells. Manage non-point sources of water pollution (fields, lawns) to limit the loss of excess water and plant nutrients.

Match fertilizer and irrigation applications to precise crop uptake needs in order to minimise groundwater contamination. A recent report by the Central Groundwater Board (CGWB) has revealed nitrate contamination in excess of the admissible limit (45ppm) in water in various parts of country. The report says that nitrate content in excess of the desirable limit from groundwater sources has been noticed in several localised pockets of the country. Actions have been initiated to provide safe water in the affected areas, with allocation of funds for the purpose.

28.6.2 Purification of Contaminated Water

While it may be technically possible to treat contaminated groundwater, it can be difficult, expensive and not totally effective. For this reason, prevention is the best way to ensure clean water. water treatment include distillation, reverse osmosis, ion exchange or blending.

☆ Distillation boils the water, catches the resulting steam, and condenses the steam on a cold surface (a condenser). Nitrates and other minerals remain in the boiling tank.

☆ Reverse osmosis forces water under pressure through a membrane that filters out minerals and nitrates. One-half to two-thirds of the water remains behind the membrane as reject water. Higher-yield systems use higher water pressures.

☆ Ion-exchange takes another substance, such as chloride, and trades places with nitrate. An ion exchange unit is filled with special resin beads that are charged with chloride. As water passes over the beads, the resin takes up nitrate in exchange for chloride. As more water pass over the resin, all the chloride is exchanged for nitrate. The resin is recharged by back washing with sodium chloride solution. The back wash solution, which is high in nitrate, must be properly disposed of.

☆ Blending is another method to reduce nitrates in drinking water. mix contaminated water with clean water from another source to lower overall nitrate concentration. Blended water is not safe for infants but id=s acceptable fpr livestock and healthy adults.

☆ Total removal of heavy metal from mixed plating rinse wastewater was described by Sapari *et al.* (1996).

☆ Charcoal filters and water softeners do not adequately remove nitrates from water. Boiling nitrate-contaminated water does not make it safe to drink and actually increases the concentration of nitrates. In many cases, the most effective alternative is to use bottled water for drinking and cooking.

28.7 Conclusions

Providing clean drinking water is a key challenge, which requires public participation. Mass awareness on the need for water conservation and providing common tips to effectively participate in this important mission is need of the time. The simple domestic purpose and how to save water under this sector, as given below, may help in creating awareness. Programmes related to water conservation may be telecasted periodically from regional television channels to create awareness among countrymen particularly people living in rural areas.

It is imperative that users from all sectors of water use, stakeholders including state and central governments, agencies, institutions, organizations, NGOs, municipalities, village panchayats, public sector undertakings and other such bodies directly or indirectly involved in planning, development and maintenance of water resources projects and providing services to the users, may need to be involved for making integrated and continuous efforts for creating mass awareness towards importance of saving and conservation of water, and duties and responsibilities of individuals as well as of organizations and institutions towards judicious and optimal use of water.

It has been observed that adverse situation of water supply to domestic sector, when supply is not adequate to meet demand, some residents use water pumps in water supply lines to boost supplies in their dwellings and thereby causing hardship to other residents of the locality. Illegal tapping of water from supply lines or lifting water of canals are also prevalent at places. It has also been observed that inhabitants, in general, are less sensitive to leakage or water loss from the system.

Similarly in case of industrial sector, it is not very uncommon to discharge untreated or partially treated industrial effluents in water bodies like rivers, lakes, ponds, canals, etc. including groundwater aquifers, this should be prevented.

References

Ameetha R. Baptiste, K. A., 2002. " *Water chemistry and heavy metals in some freshwater ponds of Palavakkam, Chennai, Tamil Nadu*" *Jr. Aqua. Biol.*, 23-25.

Biswajit R., 2001. " *Water quality and corrosivity of groundwater of northwestern part of Bankura District, West Bengal*." *Jr.Environ. Poll*, 8(4): 329-332.

C.G.W.M. 1990. " *About groundwater protection*"Wright State University, Dayton, Ohio.

Cary E.E. 1982 Chromium in air, soil and natural waters. In: Langard S. (ed) Biological and environmental aspects of chromium. Elsevier, Amsterdam, 48-64

Iqbal M.A. and Gupta S.G. 2009 Studies on Heavy Metal Ion Pollution of Groundwater Sources as an Effect of Municipal Solid Waste Dumping. *African Journal of Basic and Applied Sciences* 1 (5-6) : 117-122

Panigrahi A. 2012 Environmental Pollution: In Aquatic life: Problem and Prevention, Ekush Satak, Calcutta, pp225

Panigrahi A. 2014 Heavy Metal Toxicity, *Everyman's Science* 48 (6): 418-423.

Sampat, P., 2001. "The hidden threat of groundwater pollution" *USA Today*, 130.

Sapari, N. Idris, A. And Hisham N. 1996. Total removal of heavy metal from mixed plating rinse wastewater. *Desalination*, 106: 1-3

Smedley PL, Kinniburgh DG (2002) "A review of the source, behaviour and distribution of arsenic in natural waters". *Appl. Geochem.* 17 (5): 517–568.

Smith A.H, Lingas EO and Rahman M (2000) Contamination of drinking-water by arsenic in Bangladesh: a public health emergency. *Bullet. World Hlth. Org.* 78 (9): 1092-1103.

2015, Impact of Global Warming and Climate Change on
    Human and Plant Health
Editors: Dr. Arun Arya and Prof. V.S. Patel
Published by: DAYA PUBLISHING HOUSE, NEW DELHI

Pages 304–310

Chapter 29

Dust Pollution Prevention by Tree Leaves: Micromorphological and Phyllosphere Studies

Prakash Thanki, Prachi Mehta and Arun Arya*

*Environment Studies, Faculty of Science,
The Maharaja Sayajirao University of Baroda, Vadodara, Gujarat, India*

ABSTRACT

Use of plants in protection of environment is well known. Plants act as pollutant sink they reduce the CO_2 and absorb pollutant gases. In urban areas on both sides of road massive plantation is practiced. In wider roads and on high ways the plantation known as green belt is suggested. Many of the plants accumulate dust on their leaves and act as dust filters. An effort was made to find out plants with high pollution scavenging potential. Parameters like canopy cover, leaf area, dry weight ratio, stomatal conductance, presence of trichomes/hairs, presence of chlorophyll, ascorbic acid are usually taken into consideration to assess their potential to resist pollution. Plant surface is a natural habitat, which represents a heterogeneous population of microbes comprising of both pathogens and non-pathogens. Qualitative assessment of plant surface mycoflora has indicated that these fungi comprise of a wide range of organisms including yeasts and filamentous fungi.

Keywords: *Environmental pollution, Dust, Micromorphological studies, Azadirachta indica, Fungi.*

---

* *Corresponding author.* E-mail: aryaarunarya@rediffmail.com

29.1 Introduction

In metropolitan cities, exhaust from automobiles and gases from the industries are major sources of air pollution. In India over 20 million small scale industries and about 2000 industrial estates are working. So it is important to take adequate steps to control the pollution. Role of Plants as pollution sink is well recognized. Plants help us to reduce many types of pollution like air pollution soil pollution and erosion, noise pollution, water pollution and water harvesting. Plants are good scavengers they take in the pollutants within themselves converting it into useful products. The epidermis is multilayered in xerophytic plant *Nerium odorum*. It may have guard cell covering the stomata and in old stem, it forms cork layers called periderm.

Different plant organs represent specific ecological niches depending upon the nature of their substrates and the microorganisms harboured by them. This is the reason why a particular pathogen under favourable conditions for growth, only attacks a specific part (of definite age and maturity) of a plant.

Epidermis

The epidermis is the outermost protective cell layer of the primary plant body. It is the dermal tissue system of leaves, stems, roots, flowers, fruits, and seeds; it is usually transparent. Epidermal cells are tightly linked to each other and provide mechanical strength and protection to the plant. The walls of the epidermal cells of the above ground part of plants contain cutin and are covered with a cuticle. The cuticle reduces water loss to the atmosphere. The cuticle tends to be a thicker mostly on the top of the leaf.

Plant surface is a natural habitat, which represents a heterogeneous population of microbes comprising of both pathogens and non-pathogens. Qualitative assessment of plant surface microbes has indicated that these microbes comprise a wide range of organisms including yeasts, filamentous fungi, bacteria, actinomycetes, blue green algae, lichens and even pigmy ferns.

Trichomes

Trichomes are the hair like structures present on the epidermal cells. Dense covering of trichomes control the rate of transpiration and reduce the heating effect of sunlight. In plants like *Nerium* they are present in large number around the sunken stomata and make the micro environment around stomata cooler, thus transpiration is reduced. They aid in protection of plant body from outer harmful agencies. Active secretory cells of glandular trichomes have dense protoplast and elaborate various substances such as volatile oils, resins, mucilage and gums, etc. They are very near to each other so dust can easily t rap in them and pathogens also trap along with dust. If the plant have trichomes on both surface and no of trichomes are more than the dust capturing capacity is good of that leaf.

Pathogen

The agent that by its persistent association causes the disease is a pathogen. It may be living (infectious). For example: Bacteria, Fungi, Viruses and Actinomycetes.

Puccinia graminis tritici is a fungus which causes black stem rust of wheat so *P. graminis tritici* is a pathogen. *Streptomyces scabies* is an actinomycetes which causes soft rot of potato tubers so *S. Scabies* is a pathogen. **Biotrophs**: Parasitic microorganisms like bacteria and fungi which strictly grow on living host tissues. If the plant is dead the growth of biotrophs would also stop. These are host specific. *Puccinia graminis tritici* attacks wheat leaf. **Necrotrophs**: These are unspecialized microbes. These cause diseases in plants and would continue to grow even if the host is dead. *Alternaria* and *Curvularia* are common necrotrophs. **Residents**: The microbes that are adapted to phylloplane habitat *i.e.* they are capable of multiplying and giving rise to epiphytic flora are termed as Residents. **Casuals**: The microbes that fail to grow on leaf surface or fail to establish are called casuals.

A large number of plant pathogenic fungal species develop specialized cells called appresoria, that are able to breach the outer cuticle of Plants and thereby gain entry to epidermal cells (Wilson and Talbot, 2009). The plant cells are not ruptured in this process, but instead the fungus is able to invaginate the plasma membrane and grow within the apoplast- space between the plasma membrane and plant cell wall. This allows the fungus to occupy intact, living plant cells and set up a specialized interface to allow sequestration (extraction) of nutrients directly from host cells.

Effects Produced by Microorganisms

1. Competition for nutrients
2. Antagonism caused due to production of metabolites
3. Mechanical obstruction due to dense cell growth of inhibitor
4. Growth of pathogens may initiate production of antibiotic comp./ **Phytoalexins**
5. Saprophytes may induce systemic resistance similar to immunization in animals.

29.2 Materials and Methods

Twenty leaves were collected from each plant growing on both sides of VIP road. They were divided into two sets of 10 leaves in each group. Now take 500 ml of water and take a set of 10 leaves. All the 10 leaves were washed with clean water. Collect the supernatant. Now take initial weight of filter paper and filter the collected water after washing the leaves through it. Keep the filter papers for 24 hours in oven to dry it and then take final reading. Take the leaf area and do the calculations. The area of leaves was calculated by Laser powered leaf area meter. Phyllosphere mycoflora was studied by placing the cut pieces of leaves on PDA petriplates and observing the growth of fungi after 7 days.

29.3 Results and Discussion

Table 29.1 shows the presence of fungi on leaf surface of trees growing on both sides of road in Vadodara. Leaves of 15 trees showed spores of 12 different fungi. Both pathogenic and saprophytic fungi were encountered on leaf surface.

**Table 29.1: Showing Presence of different Fungal Spores on
Tree Leaves Collected with Dust**

Trees	Type Leaf	Fungal Organisms	
Azadirachta indica A.Juss.	C	*Aspergillus niger*	Saprophyte
		Alternaria alternaria	
Bambusa arundinacea (Retz.)Willd.	S	*Aspergillus fumigatus*	Saprophyte
Polyalthia longifolia Benth. Hook. F.	S	*Chaetomium globosum*	Pathogen
		Cladosporium herbarum	Pathogen
Ficus benghalensis L.	S	*Alternaria alternaria*	Pathogen
Mimusops elengi L.	S	*Curvularia prasadii*	Pathogen
Bauhinia racemosa L.	S	*Fusarium oxysporum*	Pathogen
Bougainvillea spectabilis Willd.	S	*Aspergillus niger*	Saprophyte
Plumeria alba Tourn. ex L.	S	*Fusarium pallidorosium*	Pathogen
Cassia sp.	C	*Cladosporium herbarum*	Pathogen
Peltophorum ferrugianum	C	*Aspergillus niger*	Saprophyte
		Monilia sitophila	Pathogen
Ficus religiosa L.	S	*Alternaria alternaria*	Pathogen
Melia azadirach L.	C	*Cladosporium herbarum*	Pathogen
Callistemon lanecolatus R.Br.	S	*Penicillium citrinum*	Saprophyte
		Gliocladium virence	Saprophyte
Eucalyptus gobulus Labill.	S	*Alternaria alternaria*	Pathogen
Syzygium cumini (L.) Skeels		*Cladosporium herbarum*	Pathogen

Note: S: Simple and C: Compound leaf.

Growth of Phylloplane Microflora is Influenced by

i) Maturity of the Leaf

Sharma and Sinha (1971) found that exudates from the young leaves of different sorghum varieties inhibited the germination and growth of *Collectotrichum graminicola*. With the maturity of leaf this effect was reduced.

ii) Effect of Pollen Grains

Pollen grains are airborne bio particles. These settle on leaf surfaces by impaction. The wall of pollen grains is made up of sporopollenin a protein. The carbohydrates present in them serve as a good source of nutrition. For example presence of pollen grains on rye leaves enhances the growth of pathogens like *Cladosporium herbarum, Helminthosporium sativum*. On sugar beet leaves the presence of pollen enhanced the growth of *Phoma betae*.

iii) Inhibitory Effect of Bacteria

Metabolites produced by bacteria may inhibit the growth of fungi. Bacteria present on leaf of Chrysanthemum inhibited the growth of leaf spot pathogen *Mycospherella*

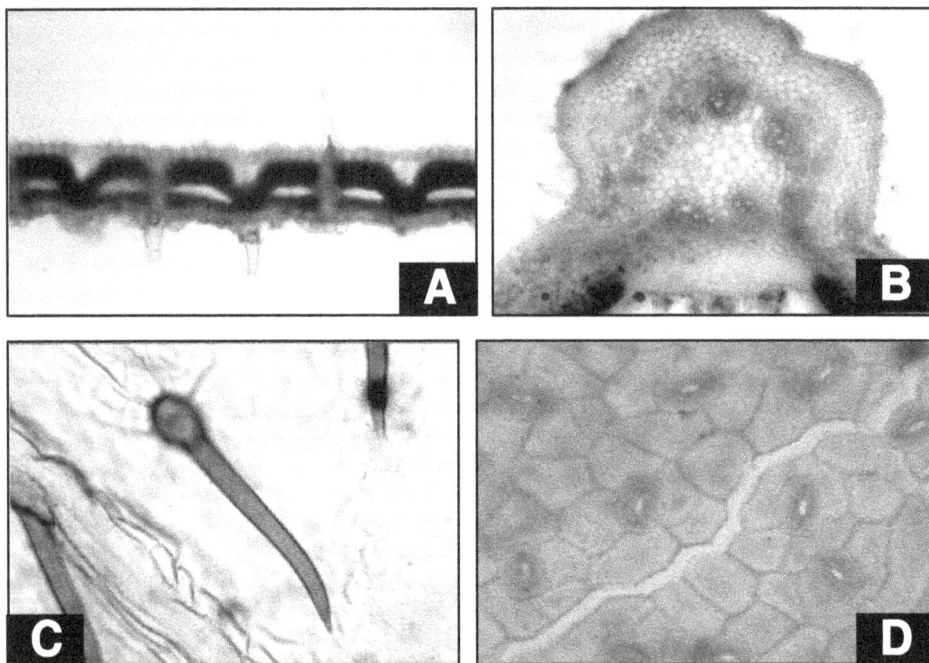

Figure 29.1: A: T.S. Leaf of *Nerium* Showing, B: Section Passing through Mid Rib (Photo on right); C: Epithelial Layer Showing Glandular Hairs, which Hold the Dust Particles, D: Epidermis Showing Enlarged Stomata.

liquicola. It was found that *Alternaria* blight of chillies could be controlled by the presence of bacteria. Sprays of *Bacillus mycoides, B. thruingiensis* and *Pseudomonas cepcia* (grown in nutrient dextrose broth) successfully controlled *Alternaria* leaf spot of tobacco and *Cercospora* leaf spot of peanut (Tikka disease).

iv) Effect of Leaf Exudates

Leaf exudates affect the growth of saprophytes and parasites. The chemicals diffusing from inside the leaf on its surface provide nutrition to *Phylloplane microflora*. The nature of these metabolites may change with maturity of leaf or environmental conditions. Presence of carbohydrates like sucrose and fructose has been detected from rose petals. Growth hormones are produced by yeasts and bacteria. *Cladosporium* and *Aureobasidium* can produce auxins.

v) Antagonism

An antagonism includes (a) competition for some nutrient or other commodity in limited supply but needed by the pathogen, (b) antibiosis, resulting from the liberation of a chemical harmful to the pathogen, and (c) predation, hyperparasitism, mycoparasitism or other form of direct exploitation of a pathogen by another organism

An antagonist should have (a) a high rate of sporulation, (b) fast growth or multiplication (c) better survival under different environmental conditions, and (d) very aggressive or antagonistic nature.

vi) Production of Phytoalexins

In response to pathogens various plants produce defense chemicals called phytoalexins. Like pea produces **Pisatin, Shakariun (*Ipomoea batata*)** produces **Ipomeameron** and potato produces ***Rishitin.*** If a pathogen can detoxify or breakdown these chemicals it can infect the leaf. Gossipol in cotton and Catechol in onion may be present from the beginning and can prevent the infection in host plants. Smudge disease in Red scaled onion caused by *Colletotrichum circinans* can be prevented by the presence of Catechol and not in white scaled ones.

vii) Induced Resistance or Immunization

Microbes may induce this effect. When tobacco leaves were injected with heat killed bacterium ***Pseudomonas*** tobacco host became resistant to super infection by living cell. Prof. Kuc has explained this as SAR reaction.

vii) Hypersensetive Response

A necrotic defense reaction caused by the entry of biotrophic organisms. With this reaction challenged host cells die and prevent further spread of pathogen in host plant cells.

29.4 Estimation of Dust Pollution

The results were obtained for 15 plants, type of leaves simple or compound, leaf area and amount of dust collected in mg are listed in Table 29.2.

Table 29.2: Showing Leaf Area and Dust Collected from different Plants Present on Road Side

Plants	Type of Leaf	Leaf Area (cm^2)	Dust Collected (mg)	Dust (mg/cm^2)
Azadirachta indica A.Juss.	C	7.9	6.9	0.87
Bambusa arundinacea (Retz.)Willd.	S	82.9	7.5	0.09
Polyalthia longifolia Benth. Hook. F.	S	62.4	8.2	0.13
Ficus benghalensis L.	S	88.1	9.8	0.11
Nerium odorum L.	S	63.1	5.8	0.15
Bauhinia racemosa L.	S	82.9	7.5	0.09
Bougainvillea spectabilis Willd.	S	25.0	1.5	0.06
Plumeria alba Tourn. ex L.	S	159.92	116.6	0.72
Cassia sp.	C	162.36	4.25	0.26
Peltophorum ferrugianum	C	409.42	23.55	0.057
Ficus religiosa L.	S	30.0	4.81	0.16
Melia azadirach L.	C	150	1.85	0.01
Callistemon lanecolatus R.Br.	S	2.7	1.22	0.45
Eucalyptus gobulus Labill.	S	26.5	1.35	0.05
Syzygium cumini (L.) Skeels	S	63.7	4.18	0.065

Note: S: Simple and C: Compound leaf.

The results of leaves of 15 plants, depicted in Table 29.2 revealed that *Peltophorum ferrugianum* has maximum leaf area followed by *Cassia* sp. and *Plumeria alba*. More amount of dust was accumulated by *Azadirachta indica* A.Juss and *Plumeria alba* Tourn. ex L. Similar type of studies have been conducted by Arya (2010) and Miroslavov (1998). They have found that epicuticular structure like trichromes, hair and arrangement of glands on leaves help in accumulation of large particles of dust.

The leaves of *Melia azadirach* L. *Eucalyptus gobulus* Labill. and *Bougainvillea spectabilis* Willd were poor dust collectors. Small *Bougainvillea* leaves are smooth and shiny and thus failed to hold dust particles. This plant is preferred on road dividers due to its beautiful flowers and better pollution tolerance capacity. Planting of more trees is suggested to reduce the pollution load.

References

Arya A. 2010. Plants recommended as potential pollutant sink in urban areas.

97th Indian Science Congress Association Thiruvanthpuram

Miroslavov E A Vozesenkaya E.V and Koteeva N. K. 1998. Comparative description of leaf anatomy of Arctic and Boreal Plants *Botanicheskil Zhumal* 83 21-27

Sharma J.K. and Sinha S. 1971. Role of humidity on sporulation pattern and infection in Sorghum caused by *Colletotricum graminicola*.www.new1.dli.ernet.in › data1 › upload › insa › INSA_1

Wilson RA, *Talbot* NJ. 2009. Under pressure: investigating the biology of plant infection. *2009* Mar;7(3):185-95. doi: 10.1038/nrmicro2032. Under pressure: investigating the biology of plant infection by *Magnaporthe oryzae. www.ncbi.nlm.nih.gov › pubmed*

2015, **Impact of Global Warming and Climate Change on Human and Plant Health**

Pages **311–321**

Editors: **Dr. Arun Arya and Prof. V.S. Patel**

Published by: **DAYA PUBLISHING HOUSE, NEW DELHI**

Chapter 30

Diversity of Fungal Endophytes in the Leaves of Five *Terminalia* spp.

Pradyut Dhar* and Arun Arya**

Department of Botany, Faculty of Science,
The Maharaja Sayajirao University of Baroda, Vadodara, Gujarat, India

ABSTRACT

Fungal endophytes both bacteria and fungi may reside in internal tissues of living plants without causing any immediate damage to the plant. Grass endophytes are known to provide protection against herbivory and pathogens that increases their fitness. Endophytes belonging to the members of Xylariaceae family frequently produce compounds of high biological activity. Such increase in secondary metabolites has led to screen a large number of plants for the occurrence of fungal endophytes. These organisms are used to produce pharmaceutically active compounds and discovery of novel drug compounds. From bark and twig of *T. arjuna,* species of *Pestalotiopsis, Myrothecium* and *Trichoderma* were isolated from Srirang patna by mycologist. The host specificity among endophytes is expressed at the family level. However, some endophytic genera such as *Phomopsis* and *Phyllosticta* occur in a wide host range. A study was conducted to isolate various endophytes from the leaves of 5 different *Terminalia* spp. The observations revealed the presence of 11, 10, 7, 15 and 12 endophytes from leaves, petiole and leaf tip respectively.

Keywords*: Fungal endophytes, Enzymes, Terminalia, Phomopsis.*

E-mail: *pradyutdhar@gmail.com; **aryaarunarya@rediffmail.com

30.1 Introduction

At the most basic level, endophyte simply means the location of an organism, with "endo" means "inside" and "phyte" means "plants". Therefore, endophyte refers to organisms that live within plants (Wilson, 1995). Fungi and bacteria are the most common organisms associated with the term endophyte. Fungi colonize foliar surface and on twigs as epiphytes. Inside plants they enter as biotrophs and necrotrophs. Biotrophic pathogens are parasites that have evolved the means to grow within living plant cells without stimulating plant defence mechanisms. The term endophyte has been used to denote a particular type of systemic, nonpathogenic symbiosis. The grass endophytes provide their hosts with a number of benefits, such as protection against herbivory and pathogens, that increase their fitness (Clay, 1990, Saikkonen *et al.,* 1998). Fungal endophytes reside in healthy tissues of all terrestrial plant taxa studied to date and are diverse and abundant in tropical woody angiosperms. Endophytic fungi were studied from *T. arjuna*, an important ethnopharmacological plant extensively used for heart diseases. Tejasvi *et al.* (2005) reported 22 fungal genera from 5 plants growing in 3 different locations. *Pestalotiopsis, Chaetomium* and *Myrothecium were* most common fungi.

Pathogenic fungi capable of symptomless occupation of their hosts during a portion of the infection cycle "quiescent infections" (Williamson, 1994), and strains with impaired virulence can be considered endophytes (Schardle *et al.,* 1994). Most biologists agree that species composition of the internal mycobiota is distinct for various hosts, organs and tissues.

Endophytic fungi exhibit a complex web of interactions with host plants and have been extensively studied over the last several years as prolific sources of new bioactive natural products. Fungal enzymes are one of them which are used in food, beverages, confectionaries, textiles and leather industries to simplify the processing of raw materials. They are often more stable than enzymes derived from other sources. Enzymes of the endophytes are degraders of the polysaccharides available in the host plants. The use of simpler solid media permits the rapid screening of large populations of fungi for the presence or absence of specific enzymes.

30.2 Bioprospecting for Natural Products

Natural products are naturally derived metabolites and/or by-products from microorganisms, plants, or animals (Baker *et al.,* 2000). These products have been exploited for human use for thousands of years, and plants have been the chief source of compounds used for medicine. Even today the largest users of traditional medicines are the Chinese, with more than 5,000 plants and plant products in their pharmacopoeia (Bensky and Gamble, 1993). In fact, the world's best known and most universally used medicinal is aspirin (salicylic acid), which has its natural origins from the glycoside salicin which is found in many species of the plant genera *Salix* and *Populus*. Examples abound of natural-product use, especially in small native populations in a myriad of remote locations on Earth. For instance, certain tribal groups in the Amazon basin, the highland peoples of Papua New Guinea, and the Aborigines of Australia each has identified certain plants to provide relief of symptoms varying from head colds to massive wounds and intestinal ailments (Isaacs,

2002). History also shows that now-extinct civilizations had also discovered the benefits of medicinal plants. In fact, nearly 3,000 years ago, the Mayans used fungi grown on roasted green corn to treat intestinal ailments (Bush and Hayes, 2000). More recently, the Benedictine monks (800 AD) began to apply *Papaver somniferum* as an anesthetic and pain reliever as the Greeks had done for years before (Grabley and Thieriecke, 1999). Many people, in past times, realized that leaf, root, and stem concoctions had the potential to help them. These plant products, in general, enhanced the quality of life, reduced pain and suffering, and provided relief, even though an understanding of the chemical nature of bioactive compounds in these complex mixtures and how they functioned remained a mystery.

It was not until Pasteur discovered that fermentation is caused by living cells that people seriously began to investigate microbes as a source for bioactive natural products. Then, scientific serendipity and the power of observation provided the impetus to Fleming to usher in the antibiotic era via the discovery of penicillin from the fungus *Penicillium notatum*. Since then, people have been engaged in the discovery and application of microbial metabolites with activity against both plant and human pathogens. Furthermore, the discovery of a plethora of microbes for applications that span a broad spectrum of utility in medicine (*e.g.*, anticancer and immunosuppressant functions), agriculture and industry is now practical because of the development of novel and sophisticated screening processes in both medicine and agriculture. These processes use individual organisms, cells, enzymes, and site-directed techniques, many times in automated arrays, resulting in the rapid detection of promising leads for product development.

Even with untold centuries of human experience behind us and a movement into a modern era of chemistry and automation, natural-product-based compounds have had an immense impact on modern medicine since about 40 per cent of prescription drugs are based on them. Furthermore, 49 per cent of the new chemical products registered by the U.S. Food and Drug Administration are natural products or derivatives thereof (Brewer, 2000). Excluding biologics, between 1989 and 1995, 60 per cent of approved drugs and pre-new drug application candidates were of natural origin (Grabley and Thieriecke, 1999). From 1983 to 1994, over 60 per cent of all approved cancer drugs and cancer drugs at the pre-new drug application stage were of natural origin, as were 78 per cent of all newly approved antibacterial agents. In fact, the world's first billion-dollar anticancer drug, paclitaxel (Taxol), is a natural product derived from the yew tree (Strobel, *et al.* 1999). Many other examples abound that illustrate the value and importance of natural products in modern civilizations.

Recently, however, natural-product research efforts have lost popularity in many major drug companies and, in some cases, have been replaced entirely by combinatorial chemistry, which is the automated synthesis of structurally related small molecules (Bills *et al.*, 2002). In addition, many drug companies have developed interests in making products that have a larger potential profit base than anti-infectious drugs. These include compounds that provide social benefits, that reduce the symptoms of allergies and arthritis, or that can soothe the stomach. It appears that this loss of interest can be attributed to the enormous effort and expense that is required to pick and choose a biological source and then to isolate active natural products, decipher their structures, and begin the long road to product development (Grabley and Thieriecke, 1999). It is also apparent that combinatorial chemistry and other synthetic chemistries revolving around certain basic chemical structures are now serving as a never-ending source of products to feed the screening robots of the drug industry. Within many large pharmaceutical companies, progress of professionals is primarily based upon numbers of compounds that can be produced and sent to the screening machines. This tends to work against the numerous steps needed even to find one compound in natural-product discovery. It is important to realize that the primary purpose of combinatorial chemistry should be to complement and assist the efforts of natural-product drug discovery and development, not to supersede it (Grabley and Thieriecke, 1999). The natural product often serves as a lead molecule whose activity can be enhanced by manipulation through combinatorial and synthetic chemistry. Natural products have been the traditional pathfinder compounds, offering an untold diversity of chemical structures unparalleled by even the largest combinatorial databases.

30.3 Fungal Endophytes

Endophytes are microbes that live inside plants and other microbes and alter their physiology without causing any disease symptom. While combinatorial synthesis produces compounds at random, secondary metabolites, defined as low-molecular-weight compounds not required for growth in pure culture, are produced as an adaptation for specific functions in nature (Demain, 1981). Schutz (2001) noted that certain microbial metabolites seem to be characteristic of certain biotopes, on both an environmental as well as organismal level. Accordingly, it appears that the search for novel secondary metabolites should center on organisms that inhabit unique biotopes. Thus, it behaves the investigator to carefully study and select the biological source before proceeding, rather than to have a totally random approach in the biological source material. Careful study also indicates that organisms and their biotopes that are subjected to constant metabolic and environmental interactions should produce even more secondary metabolites ((Schutz, 2001). Endophytes are microbes that inhabit such biotopes, namely, higher plants, which is why they are currently considered to be a wellspring of novel secondary metabolites offering the potential for medical, agricultural, and/or industrial exploitation. Currently, endophytes are viewed as an outstanding source of bioactive natural products because there are so many of them occupying literally millions of unique biological niches (higher plants) growing in so many unusual environments. Thus, it appears that these biotypical factors can be important in plant selection, since they may govern the novelty and biological activity of the products associated with endophytic microbes.

Since the discovery of endophytes in Darnel, Germany, in 1904 (Tan and Zou, 2001), various investigators have defined endophytes in different ways, which is usually dependent on the perspective from which the endophytes were being isolated and subsequently examined. Bacon and White give an inclusive and widely accepted definition of endophytes–"microbes that colonize living, internal tissues of plants without causing any immediate, overt negative effects" (Bacon and White,2000). While the symptomless nature of endophyte occupation in plant tissue has prompted focus on symbiotic or mutualistic relationships between endophytes and their hosts, the observed biodiversity of endophytes suggests they can also be aggressive saprophytes or opportunistic pathogens. Both fungi and bacteria are the most common microbes existing as endophytes. It seems that other microbial forms, *e.g.*, mycoplasmas and archaebacteria, most certainly exist in plants as endophytes, but no evidence for them has yet been presented. The most frequently isolated endophytes are the fungi. It turns out that the vast majority of plants have not been studied for their endophytes. Thus, enormous opportunities exist for the recovery of novel fungal forms, taxa, and biotypes. Hawksworth and Rossman estimated there may be as many as 1 million different fungal species, yet only about 100,000 have been described (Hawksworth and Rossman. 1987). As more evidence accumulates, estimates keep rising as to the actual number of fungal species. For instance, Dreyfuss and Chapela estimate there may be at least 1 million species of endophytic fungi alone (Dreyfuss and Chapela, 1994). It seems obvious that endophytes are a rich and reliable source of genetic diversity and novel, undescribed species. Finally, in our experience, novel microbes usually have associated with them novel natural products. This fact alone helps eliminate the problems of dereplication in compound discovery.

30.4 Selection of Plants for the Study of Endophytes

It is important to understand the methods and rationale used to provide the best opportunities to isolate novel endophytic microorganisms as well as ones making novel bioactive products. Thus, since the number of plant species in the world is so great, creative and imaginative strategies must be used to quickly narrow the search for endophytes displaying bioactivity.

A specific rationale for the collection of each plant for endophyte isolation and natural-product discovery is used. Several reasonable hypotheses govern this plant selection strategy and these are as follows. (i) Plants from unique environmental settings, especially those with an unusual biology, and possessing novel strategies for survival are seriously considered for study. (ii) Plants that have an ethnobotanical history (use by indigenous peoples) that are related to the specific uses or applications of interest are selected for study. These plants are chosen either by direct contact with local peoples or via local literature. Ultimately, it may be learned that the healing powers of the botanical source, in fact, may have nothing to do with the natural products of the plant, but of the endophyte (inhabiting the plant). (iii) Plants that are endemic, that have an unusual longevity, or that have occupied a certain ancient land mass, such as Gonwanaland, are also more likely to lodge endophytes with active natural products than other plants. (iv) Plants growing in areas of great biodiversity also have the prospect of housing endophytes with great biodiversity.

Just as plants from a distinct environmental setting are considered to be a promising source of novel endophytes and their compounds, so too are plants with an unconventional biology. For example, an aquatic plant, *Rhyncholacis penicillata*, was collected from a river system in Southwest Venezuela where the harsh aquatic environment subjected the plant to constant beating by virtue of rushing waters, debris, and tumbling rocks and pebbles (Strobel *et al.,* 1999). This created many portals through which common phytopathogenic oomycetes could enter the plant. Still, the plant population appeared to be healthy, possibly due to protection from an endophytic product. This was the environmental biological clue used to pick this plant for a comprehensive study of its endophytes. Eventually, a potent antifungal strain of *Serratia marcescens* was recovered from *R. penicillata* and was shown to produce oocydin A, a novel antioomycetous compound having the properties of a chlorinated macrocyclic lactone. It is conceivable that the production of oocydin A by *S. marcescens* is directly related to the endophyte's relationship with its higher plant host. Currently, oocydin A is being considered for agriculture use to control the ever-threatening presence of oomyceteous fungi such as *Pythium* and *Phytophthora*.

Plants with ethnobotanical history, as mentioned above, also are likely candidates for study, since the medical uses to which the plant may have been selected relates more to its population of endophytes than to the plant biochemistry itself. For example, a sample of the snakevine, *Kennedia nigriscans*, from the Northern Territory of Australia was selected for study since its sap has traditionally been used as bush medicine for many years. In fact, this area was selected for plant sampling since it has been home to one of the world's long-standing civilizations–the Australian Aborigines. The snakevine is harvested, crushed, and heated in an aqueous brew by local Aborigines in southwest Arnhemland to treat cuts, wounds, and infections. As it turned out, the plant contained a novel endophyte, *Streptomyces* sp. strain NRRL 30562, that produces wide-spectrum novel peptide antibiotics called munumbicins. It is reasonable to assume that the healing processes, as discovered by indigenous peoples, might be facilitated by compounds produced by one or more specific plant-associated endophytes as well as the plant products themselves.

In addition, it is noteworthy that some plants generating bioactive natural products have associated endophytes that produce the same natural products. Such is the case with paclitaxel, a highly functionalized diterpenoid and famed anticancer agent that is found in each of the world's yew tree species (*Taxus* spp.) (Suffness, 1995). In 1993, a novel paclitaxel-producing fungus, *Taxomyces andreanae*, from the yew *Taxus brevifolia* was isolated and characterized (Strobel *et al.,* 1993).

30.5 Materials and Methods

Study Site

Study site was selected at Gora Forest Range, Narmada (21°49'40.85"N, 73°44'18.87"E, Elev. 779 ft), Sagai Forest Range (21°43'41.46"N, 73°38'57.89"E, Elev. 1297 ft) and Arboretum, The M.S. University of Baroda, Vadodara (22°19'15.93"N, 73°10'45.96"E, Elev. 129 ft).

30.5.1 Different Culture Media Listed below were Used for Isolation of Endophytes from *Terminalia* Plants

a. Modified Asthana and Hawker's Medium 'A'

b. Potato Dextrose Agar Medium

c. Czapek's Medium

Table 30.1: Showing Presence of Endophytes from Lamina of Leaves of Five *Terminalia* sp.

Fungal Organisms	Name of the Plant				
	T. bellerica	T. chebula	T. crenulata	T. catappa	T. arjuna
Alternaria alternata	+	+	+	+	+
Aspergillus flavus	+	+	+	−	+
Aspergillus niger	+	+	+	+	+
Chaetomium globosum	−	−	+	+	+
Curvularia prasadii	+	+	+	−	+
Emericella sp.	−	−	+	−	−
Fusarium oxysporum	−	−	−	+	−
Fusarium roseum	−	+	−	−	−
Nigrospora oryzae	−	−	−	−	+
Phomopsis phyllanthi	+	+	+	+	+
Rhizopus stolonifer	+	−	−	−	−
Trichoderma viride	−	−	+	−	−

+: Present; −: Absent.

Table 30.2: Showing Presence of Endophytes from Petiole of the Leaves of Five *Terminalia* sp.

Fungal Organisms	Name of the Plant				
	T. bellerica	T. chebula	T. crenulata	T. catappa	T. arjuna
Alternaria alternata	+	+	+	−	+
Aspergillus flavus	+	+	+	+	+
Aspergillus niger	+	+	+	+	+
Colletotrichum gleosporoides	−	−	−	−	+
Chaetomium fusisporale	−	−	+	−	−
Cladosporium sp.	−	−	−	+	−
Curvularia sp.	−	−	+	+	+
Drechslera rostrata	−	−	−	−	+
Emericella sp.	−	−	+	−	−
Fusarium oxysporum	−	−	−	+	−
Fusarium roseum	−	+	−	−	−
Gloeosporium sp.	−	+	−	−	−
Lasiodiplodia theobromae	−	−	−	+	−
Nigrospora sphaerica	+	−	−	−	+
Pestalotiopsis sp.	+	−	−	−	+
Phomopsis sp.	+	+	+	−	−
Trichoderma viride	−	−	+	−	−

+: Present; −: Absent.

30.5.2 Isolation of Endophytic fungi

Leaf samples were collected from survey area. First the leaves were washed in running tap water for 10 minutes to drain off the dust particles present on leaf surface. The leaves were cut into 0.5×0.5 cm^2 and surface sterilized in NaOCl solution (1 per cent available Cl$_2$) for 2 min. The leaves were washed in sterile distilled water for 2 minutes. Then the leaves were inoculated in suitable culture media. The growth of the organisms was observed after 7 and 21 days of inoculation.

30.6 Results

It was observed that 6, 6, 8, 5 and 7 endophytes were present in leaf lamina of *T. belerica, T. chebula, T. crenulata, T. catappa* and *T. arjuna* respectively (Table 30.1) whereas 6, 6, 8, 6 and 8 fungal endophytes were present in petiole of the same (Table 30.2).

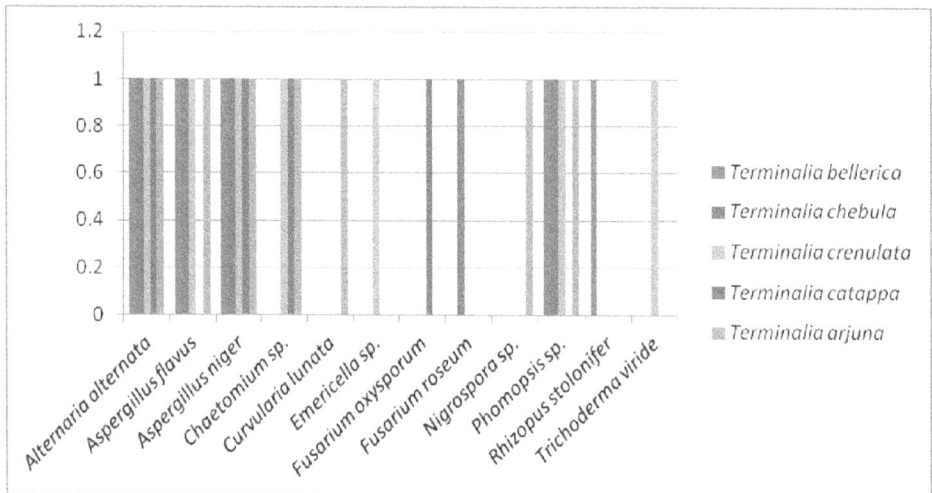

Figure 30.1: Presence of Fungal Endophytes on Leaf Tip of 5 different *Terminalia* sp.
1: Presence of endophytes; 0: Absence of endophytes.

Of all the fungal endophytes which were isolated *Aspergillus, Alternaria, Pestalotiopsis* sp., *Chaetomium* and *Phomopsis* sp. were present in promising numbers.

30.7 Conclusions

Fungi which colonize foliar surface and inhabit inside any plant are termed as biotrophs and necrotrophs. Biotrophic pathogens are parasites that have evolved the means to grow within living plant cells without stimulating plant defense mechanisms.

The term endophyte has been used to denote a particular type of systemic, nonpathogenic symbiosis. The plants used for the study of endophytes are chosen either by direct contact with local peoples or via local literature.

Table 30.3: Colonization Frequency (per cent) in Leaves of 5 different *Terminalia* Species

Fungal Organisms	Colonization Frequency (per cent) in Leaves				
	T1	T2	T3	T4	T5
Alternaria alternata	63.8	35.23	44.76	26.66	29.52
Aspergillus niger	69.52	63.8	66.66	46.66	75.23
Chaetomium globosum	34.28	0	39.04	34.28	0
Curvularia prasadii	24.76	35.2	21.9	0	8.57
Phomopsis sp.	12.1	29.52	41.9	10.47	17.14

T1: *Terminalia arjuna*, T2: *T. belerica*, T3: *T. crenulata*, T4: *T. catappa*, T5: *T. chebula*.

Total isolates- 105.

The healing powers of the botanical source, in fact, may have nothing to do with the natural products of the plant, but the endophyte,which inhabit the plant may produce the compounds or increase their concentration.

Most biologists agree that species composition of the internal mycobiota is distinct for various hosts, organs and tissues.

Endophytic fungi exhibit a complex web of interactions with host plants and have been extensively studied over the last several years as prolific sources of new bioactive natural products.

Fungal enzyme are often more stable than enzymes derived from other sources. Enzymes of the endophytes are degraders of the polysaccharides available in the host plants.

Acknowledgements

We are grateful to University Grants Commission (UGC) for providing financial assistance in the form of Major Research Project 41-458/2012 (SR) Dated 16/07/2012.

References

Bacon, C. W., and J. F. White. 2000. Microbial endophytes. Marcel Dekker Inc., New York, N.Y.

Baker, D., U. Mocek, and C. Garr. 2000. Natural products vs. combinatorials: a case study, p. 66-72. *In* S. K. Wrigley, M. A. Hayes, R. Thomas, E. J. T. Chrystal, and N. Nicholson (ed.), Biodiversity: new leads for pharmaceutical and agrochemical industries. The Royal Society of Chemistry, Cambridge, United Kingdom.

Bensky, D., and A. Gamble. 1993. Chinese herbal medicine. Materia medica, new ed. Eastland Press Inc., Seattle, Wash.

Bills, G., A. Dombrowski, F. Pelaez, J. Polishook, and Z. An. 2002. Recent and future discoveries of pharmacologically active metabolites from tropical fungi, p. 165-

194. *In* R. Watling, J. C. Frankland, A. M. Ainsworth, S. Issac, and C. H. Robinson. (ed.), Tropical mycology: micromycetes, vol. 2. CABI Publishing, New York, N.Y.

Brewer, S. 2000. The relationship between natural products and synthetic chemistry in the discovery process, p. 59-65. *In* S. K. Wrigley, M. A. Hayes, R. Thomas, E. J. T. Chrystal, and N. Nicholson (ed.), Biodiversity: new leads for pharmaceutical and agrochemical industries. The Royal Society of Chemistry, Cambridge, United Kingdom.

Buss, T., and M. A. Hayes. 2000. Mushrooms, microbes and medicines, p. 75-85. *In* S. K. Wrigley, M. A. Hayes, R. Thomas, E. J. T. Chrystal, and N. Nicholson (ed.), Biodiversity: new leads for pharmaceutical and agrochemical industries. The Royal Society of Chemistry, Cambridge, United Kingdom.

Clay, K. (1990). Fungal Endophytes of Grasses. *Ann. Rev. of Ecology and Systematics,* 21: 275-297.

Demain, A. L. 1981. Industrial microbiology. Science 214:987-994.

Dreyfuss, M. M., and I. H. Chapela. 1994. Potential of fungi in the discovery of novel, low-molecular weight pharmaceuticals, p. 49-80. *In* V. P. Gullo (ed.), The discovery of natural products with therapeutic potential. Butterworth-Heinemann, London, United Kingdom.

Grabley, S., and R. Thiericke (ed.). 1999. Drug discovery from nature, p. 3-33. Springer-Verlag. Berlin, Germany.

Hawksworth, D. L., and A. Y. Rossman. 1987. Where are the undescribed fungi? *Phytopathology* 87:888-891.

Isaacs, J. 2002. Aboriginal food and herbal medicine. New Holland Press, Sydney, Australia.

Saikkonen, K., Faeth, S.H., Helander, M. and Sullivan, T.J. (1998). Fungal endophytes: a continuum of interactions with host plants. *Ann. Rev. of Ecology and Systematics,* 29: 319-343.

Saikkonen, K., Lehtonen, P., Helander, M., Koricheva, J. and Faeth, S.H. (2006). Model systems in ecology: dissecting the endophyte-grass literature. *Trends in Plant Sci.* 11: 428-433.

Schardl, C.L., Leuchtmann, A.H., Tsai, F., Collett, M.A., Watt, D.M. and Scott, D.B. (1994). Origin of a fungal symbiont of perennial rye-grass by interspecific hybridization of a mutualist with the ryegrass choke pathogen, *Epichloe typhina. Genetics,* 136: 1307-1317.

Schutz, B. 2001. Endophytic fungi: a source of novel biologically active secondary metabolites, p. 20. *In* British Mycological Society, international symposium proceedings: bioactive fungal metabolites–impact and exploitation. University of Wales, Swansea, Wales.

Strobel, G. A., A. Stierle, D. Stierle, and W. M. Hess. 1993. *Taxomyces andreanae* a proposed new taxon for a bulbilliferous hyphomycete associated with Pacific yew. Mycotaxon 47:71-78.

Strobel, G. A., J. Y. Li, F. Sugawara, H. Koshino, J. Harper, and W. M. Hess. 1999. Oocydin A, a chlorinated macrocyclic lactone with potent anti-oomycete activity from *Serratia marcescens. Microbiology145*:3557-3564

Suffness, M. 1995. Taxol, science and applications. CRC Press, Boca Raton, Fla.

Tan, R. X., and W. X. Zou. 2001. Endophytes: a rich source of functional metabolites. *Nat. Prod. Rep.*18:448-459.

Tejasvi, V., Basavanna, M., Monnanda, S.N., Harishchandra, S.P., Kukkundoor, R.K., Subbiah, V. and Hunthrike, S.S. (2005). Endophytic fungal assemblages from inner bark and twig of *Terminalia arjuna* W. and A. (Combretaceae). *World Journal of Microbiology and Biotechnology.* 21: 1535-1540.

Williamson, P.R. (1994). Biochemical and molecular characterization of the diphenol oxidase of *Cryptococcus neoformans: identification* as a laccase. *J Bacteriol,* 176: 656–664.

Wilson, D (1995). Endophyte- the evolution of the term, a clarification of its use and definition, *Oikos,* 73: 274-276.

2015, Impact of Global Warming and Climate Change on
Human and Plant Health

Pages 322–340

Editors: **Dr. Arun Arya and Prof. V.S. Patel**
Published by: **DAYA PUBLISHING HOUSE, NEW DELHI**

Chapter 31
Carbon Sequestration by Mangroves of Gujarat, India

Richa Pandey[1]* and C. N. Pandey[2]**

[1]Department of Forests (Development and Management),
[2]Office of Add. Principal Chief Conservator of Forests
(Development and Management),
3rd Floor, B Wing, Aranya Bhavan, Opp. St. Xeivar's School,
Near Cha 3, Sector 10, Gandhinagar – 382 010, Gujarat, India

ABSTRACT

Gujarat has the second largest mangrove cover (1058 sq km) of country. Mangrove being the major woody habitats forms the important carbon sinks in the coastal regions. Present work has examined the carbon sequestration by mangroves of Gujarat. Tree harvesting methods was used for estimation of total biomass (above and below ground) by stratified random sampling. Further, carbon sequestered in the soil (up to 30 cm depth) was also examined. A total of 8.116 million ton carbon has been sequestered by mangroves of Gujarat. Mangrove soils contribute more than mangrove plants in the overall carbon sequestration.

Keywords: *Gujarat, Mangrove, Carbon sequestration, Climate change, Soil organic carbon.*

31.1 Introduction

Tropical forests in general are an important part of the global carbon cycle (Clark, 2001). Mangrove ecosystems are large and dynamic resorvior of carbon, which is an

E-mail: *richa81181@gmail.com; **cnpandey@gmail.com

important part of global carbon cycle and a potential sink of atmospheric carbon dioxide (Matsui, *et al.,* 2010). Although the area covered by mangrove forests represent only a small fraction of the tropical forests, their position at the terrestrial-ocean interface and potential exchange with coastal ocean waters suggest that these forests make a unique contribution to total carbon cycle in coastal ocean (Twilley, 1992). The potential impact of mangroves on the coastal zone carbon dynamics has been a topic of debate in past decades (Steven, *et al.,* 2008). The contribution of coastal and marine ecosystems to mitigate climate change through carbon sequestration and storage indicate that these ecosystems can rival their terrestrial counterparts (Yee., 2010). Blue carbon sinks include the open oceans, kelp forests, salt marshes, sea grass beds, coral reefs and mangroves. Management of these blue carbon sinks are currently not accounted for in climate change policies and are excluded from national carbon inventories and international carbon payment schemes (Lasco, 2004).

The biomass assessment is very significant (Zhang, *et al.,* 2006) and it endeavors to have two fundamental sets of information *i.e.* resource availability and environment management. In forest biomass studies, two biomass units are used, fresh weight (Araujo, *et al.,* 1999) and dry weight (Aboal, *et al.,* 2005; Montagu, *et al.,* 2005). Lu, *et al.* (2004) has mentioned three methods for biomass estimation *viz.* field measurement, remote sensing and GIS-based approach amongst them the field measurement has been considered to be accurate. Further, in remote sensing and GIS-based approach also field based data is required for validation. All the three methods *i.e.* harvest method (field measurement), mean tree method (remote sensing) and the allometric method (GIS method) have been used to estimate the biomass of different mangrove forests of the world. The mean tree method may be utilized only in forests with homogenous tree size and distribution such as plantation areas. In the case of allometric method, whole or partial weight of a tree is estimated with the help of measurable tree dimensions such as trunk diameter and height by using allometric equations. The allometric method is based on the assumption that one part of the organism is proportional to that of another. Therefore, the trunk diameter is correlated with the trunk weight. However, these allometric equations have been reported to vary with the species and site as well (Cloug, *et al.,* 1997; Smith III and Whelan, 2006; Komiyana, *et al.,* 2008).

Most of the allometric equations of mangrove have been developed for single stemmed mangrove species. However, some allometric equations have also been developed for multi-stemmed mangrove species such as *Rhizophora, Avicennia* and *Excoecaria* (Cloug, *et al.,* 1997). Saenger (2002) has sited 43 allometric equations for the above ground biomass of mangroves. The species specific allometric equations (reported by different scientists) are significantly different for the same mangrove species (Tam, *et al.,* 1995; Saenger, 2002; Ong, *et al.,* 2004; Comley, *et al.,* 2005). Further, the question arises as to whether a species specific allometric equation may be applicable to different mangrove forests of the world. Ong. *et al.* (2004) have found similar equation for two sites for *Rhizophora apiculata.* Clough, *et al.* (1997) have reported different relationships at different mangrove sites. Therefore, allometric equations need to be developed for each site and species. Common allometric equations for mangroves have also been proposed by various scientists (Chave, *et al.,* 2005;

Komiyama, *et al.,* 2005). These common allometric equations have been examined for various mangrove species (Komiyana, *et al.,* 2008) and were reported to show significant relative error with reference to the site specific and species specific equations.

Majority of the biomass estimation studies are focused on above-ground forest biomass (Brown, 1997; Kraenzel, *et al.,* 2003; Laclau, 2003; Losi, *et al.,* 2003; Aboal, *et al.,* 2005) because it accounts for majority of the total accumulated biomass in the forest ecosystem. Using this method, the amount of carbon sequestered in the biomass is generally taken as 50 per cent of the dry biomass (Losi, *et al.,* 2003; Montagu, *et al.,* 2005). Komiyana *et al.* (2008) have examined 23 publications of mangroves of last 50 years and found that only nine of them dealt with both above ground and below ground biomass.

Considering these facts, although destructive and very laborious, the present work has selected harvesting method for estimation of biomass (above and below ground) for understanding the carbon sequestered by mangroves of Gujarat.

31.2 Study Area

The state of Gujarat is located on the west coast of India. The average annual rainfall ranges from 2000mm (South Gujarat) to less than 400mm (North west-Kachchh). Its mangrove cover has been distributed over four regions *i.e.* Kachchh, Gulf of Kachchh, Saurashtra and South Gujarat. The state has the longest coastline of the country hosting the second largest (1048 sq km) mangrove cover (FSI, 2011). However, for the present work is not based on this estimation. Instead of this, the mangrove cover estimated by GEER has been taken for estimation of carbon sequestered in the mangrove forests of the state for the present work (Table 31.2). A total of 15 mangrove species have been reported from Gujarat (Table 31.1). However, the mangrove populations are largely represented by a single species *i.e. A. marina* and the distribution of mangrove species over four regions is not uniform (Pandey and Pandey 2009, 2010). Species diversity of mangroves is maximum in South Gujarat and minimum in Saurashtra (Table 31.1). It may be attributed to the hydrological and edaphic conditions.

31.3 Materials and Methods

For estimation of carbon sequestration by mangroves of Gujarat, the present work has considered two carbon pools *i.e.* vegetation biomass and soil. Further, the mangrove cover estimated by GEER - 912.47 sq km (unpublished) - has been taken (Table 31.2) for the assessment of mangrove biomass. The carbon sequestered in the litter has not been examined owing to the fact that the litter is generated by trees which is either decomposed or mineralized and mixed in soil transported by the tidal currents. Since, the carbon content of mangrove plants and soil have been estimated, the carbon content of litter has not been examined. Stratified random sampling method was used for assessment of mangrove populations. Mangrove forests of all the four regions were classified into three density classes *i.e.* dense, moderate and sparse. Subsequently, a sample comprising 0.03 per cent area of the state mangrove cover was taken for estimating the population values of each mangrove species. For this,

Table 31.1: Occurrence of Mangrove Species in Gujarat

Sl.No.	Species	Relative Occurrence in the State	Reported Location	Selected for Uprooting
1.	*Avicennia alba* Blume	Rare	South Gujarat	No
2.	*Avicennia marina* (Forsk.) Vierh.	Dominates the mangrove of Gujarat	All the four mangrove regions	Yes
3.	*Avicennia officinalis* L.	Rare	South Gujarat	Yes
4.	*Rhizophora apiculata* Bl.	Extremely rare	South Gujarat	No
5.	*Rhizophora mucronata* Lamk	Common in the Gulf of Kachchh, rare in other areas	Gulf of Kachchh and South Gujarat	Yes
6.	*Acanthus ilicifolius* L.	Rare	South Gujarat	Yes
7.	*Ceriops decandra* (Griff.) Ding Hou	Extremely rare (Only a few trees)	South Gujarat.	No
8.	*Ceriops tagal* (Perr.) C. B. Robinson	Common in the Gulf of Kachchh, rare in other areas	Kachchh, Gulf of Kachchh, South Gujarat	Yes
9.	*Bruguiera cylindrica* (L.) Bl.	Rare (More than 100 reproductively matured trees)	South Gujarat	Yes
10.	*Bruguiera gymnorrhiza* (L.) Lamk	Rare (less than 50 trees)	South Gujarat	No
11.	*Sonneratia apetala* Buch. –Ham.	Common in South Gujarat	South Gujarat	Yes
12.	*Excoecaria agallocha* L.	Rare (less than 100 trees)	South Gujarat	No
13.	*Aegiceras corniculatum* (L.) Blanco	Rare	Gulf of Kachchh, South Gujarat	Yes
14.	*Kandelia candel* (L.) Druce	Rare (Only one tree)	South Gujarat	No
15.	*Lumnitzera racemosa* Willd	Rare (About 200 trees)	South Gujarat	Yes

10X100 m plots were laid. A total of 316 such plots were laid for estimation of mangrove plants and recruits in the state (Table 31.3). Mangrove recruits and trees were categorized into five height classes (<1.5 m, 1.5 to 3m, 3 to 4 m, 4 to 5 m and > 5 m) and documented. The plants of less than 1.5m were treated as recruits.

Table 31.2: Distribution of Mangrove Cover (ha) in the Four Regions of Gujarat

Regions	Dense	Moderate	Sparse	Total	Per cent Distribution
Kachchh	25798	14370	15836	56004	61 per cent
Gulf of Kachchh	9275	4541	3595	17411	19 per cent
Saurashtra	2106	823	523	3452	4 per cent
South Gujarat	8474.5	0	5906.2	14380.7	16 per cent
Total	**45653.5**	**19734**	**25860**	**91247.7**	**100 per cent**

Table 31.3: Details of No. of Plots Laid for Mangrove Population Estimation and No. of Soil Samples Collected for SOC Estimation

Mangrove Regions	No. of Plots Laid for Mangrove Population Estimation	No. of Soil Samples Collected
Kachchh	120	76
Gulf of Kachchh	101	0
Saurashtra	39	24
South Gujarat	56	36
Total	**316**	**136**

For biomass estimation, nine out of the 15 mangrove species of the state were selected for uprooting (Table 31.1). Owing to the relatively rare occurrence of rest six mangrove species, they were not selected for harvesting (Table 31.1). Further, one representative of each height class (if available) was uprooted for the nine mangrove species and both the biomass (shoot and root) was reported. Shoot biomass comprised of stem, branches and leaves while root biomass represented by below ground and aerial roots. After uprooting each recruit/tree, it was segregated into trunk, branches, leaves, aerial roots and roots and the green weights of each part were taken immediately after uprooting. Dry weights of uprooted plants were taken for estimation of carbon content. For estimation of dry weight of uprooted tree, 500 gm sample of each part (trunk, branches, leaves, aerial roots and roots) were oven dried at 70°C for 48 hrs and respective dry weight were calculated. Further, these oven dried samples were powdered and analyzed using CHNS elemental analyzer (using combustion method at 600°C) at SICART laboratory, Vallabh Vidya Nagar, Anand, Gujarat which gave the percent carbon content of each part (by dry weight). The carbon sequestration potential (amount of carbon sequestered per recruit/tree) calculated for various height classes of the nine mangrove species. These values were multiplied by the respective no. of recruit/trees of different height classes of nine mangrove species to give total carbon sequestration by mangrove plants. These values were converted to area level estimates by using the values of recruit and tree density of mangroves in various density classes of four mangrove regions.

For examining the carbon pool of mangrove soil, the soil samples were collected from 68 locations spread over Kachchh, Gulf of Kachchh and South Gujarat. Since climatic conditions of Saurashtra and Gulf of Kachchh are similar, the soil samples have been collected from only one region *i.e.* Gulf of Kachchh to represent both. For each location two set of samples corresponding to the depths 1-15 cm and 16-30 cm have been collected. Therefore, a total of 136 soil samples, have been collected (Table 31.3). These soil samples were analyzed for soil density, bulk density, carbon content as per the standard method (Pansu, 2006) at laboratory of GEER Foundation.

The tree level estimates of carbon sequestration by mangrove trees were converted into area level estimates by using the data about mangrove tree and recruit density in different types of mangrove habitats. This has been aggregated with the area level estimation of soil carbon to give an estimate of the total carbon sequestered by the mangrove habitat.

31.4 Results

A total of 13644 lakhs recruits were estimated for mangroves of Gujarat of which 6658 lakh (49 per cent), 678 lakhs (5 per cent), 297 lakh (2 per cent) and 6009 (44 per cent) are from Kachchh, Gulf of Kachchh, Saurashtra and South Gujarat respectively (Table 31.4). For Kachchh, the recruit density is 16705 pa ha, 11788 per ha and 4139 per ha for the dense, moderate and sparse mangroves respectively. It was 5727 per ha, 2522 per ha and 914 per ha for the dense, moderate and sparse mangroves of Gulf of Kachchh respectively. In the case of Saurashtra, it was 10555 per ha, 7803 per ha and 2143 per ha for dense, moderate and sparse mangroves respectively. Interestingly the recruit density of the dense and sparse mangroves of South Gujarat was 58845 per ha and 16567 per ha respectively. Moderate type of mangrove forest is not found in South Gujarat region. Considering all the three density classes together, the recruit density comes out to be 11890 per ha, 3839 per ha, 8625 per ha and 41482 per ha for Kachchh, Gulf of Kachchh, Saurashtra and South Gujarat. Considering all the four regions and the three density classes together, it is 14905 for mangroves of Gujarat.

Table 31.4: Estimated No. of Mangrove Recruits in Gujarat

Mangrove Regions	Dense	Moderate	Sparse	Total	Per cent Distribution
Kachchh	430950430	169393560	65548371	665892362	49 per cent
Gulf of Kachchh	53122069	11453411	3287566	67863046	5 per cent
Saurashtra	22229708	6421645	1120528	29771880	2 per cent
South Gujarat	502789374	NA	98150275	600939648	44 per cent
Total	1009091581	187268616	168106739	1364466935	100 per cent

A total of 5377 lakh mangrove trees/bushes were estimated for Gujarat of which 4612 lakh, 566 lakh, 94 lakh and 104 lakh are in Kachchh, Gulf of Kachchh, Saurashtra

Table 31.5: Species-wise details About No. of Mangrove Trees in Four Mangrove Regions

Species	Mangrove Regions	151-300 cm	301-400 cm	401-500 cm	>500 cm	Total	Per cent of Sp.	Per cent of all Sp.
A. marina	Kachchh	356092990	92438016	12368662	0	460899668	88	**97.62**
	GOK	43403787	1443696	157872	175633	45180988	9	
	Saurashtra	9477983	0	0	0	9477983	2	
	South Gujarat	5113035	2178295	1267216	794046	9352592	1	
	Total	**414087795**	**96060007**	**13793750**	**969679**	**524911231**	**100**	
A. officinalis	Kachchh	0	0	0	0	0	0	**0.03**
	GOK	0	0	0	0	0	0	
	Saurashtra	0	0	0	0	0	0	
	South Gujarat	72581	32070	18567	11815	135033	100	
	Total	**72581**	**32070**	**18567**	**11815**	**135033**	**100**	
A. ilicifolius	Kachchh	0	0	0	0	0	0	**0.04**
	GOK	0	0	0	0	0	0	
	Saurashtra	0	0	0	0	0	0	
	South Gujarat	226540	0	0	0	226540	100	
	Total	**226540**	**0**	**0**	**0**	**226540**	**100**	
A. corniculatum	Kachchh	0	0	0	0	0	0	**0.22**
	GOK	1172671	0	0	0	1172671	99	
	Saurashtra	0	0	0	0	0	0	
	South Gujarat	17529	0	0	0	17529	1	
	Total	**1190200**	**0**	**0**	**0**	**1190200**	**100**	

Contd...

Table 31.5–Contd...

Species	Mangrove Regions	151-300 cm	301-400 cm	401-500 cm	>500 cm	Total	Per cent of Sp.	Per cent of all Sp.
B. cylindrica	Kachchh	0	0	0	0	0	0	**0.00**
	GOK	0	0	0	0	0	0	
	Saurashtra	0	0	0	0	0	0	
	South Gujarat	477	0	0	0	477	100	
	Total	**477**	**0**	**0**	**0**	**477**	**100**	
C. tagal	Kachchh	329389	0	0	0	329389	4	**1.41**
	GOK	7201579	45388	0	0	7246967	96	
	Saurashtra	0	0	0	0	0	0	
	South Gujarat	170	0	0	0	170	0	
	Total	**7531138**	**45388**	**0**	**0**	**7576526**	**100**	
R. mucronata	Kachchh	0	0	0	0	0	0	**0.56**
	GOK	3000935	23681	0	0	3024616	100	
	Saurashtra	0	0	0	0	0	0	
	South Gujarat	1	0	0	0	1	0	
	Total	**3000936**	**23681**	**0**	**0**	**3024617**	**100**	
L. racemosa	Kachchh	0	0	0	0	0	0	**0.00**
	GOK	0	0	0	0	0	0	
	Saurashtra	0	0	0	0	0	0	
	South Gujarat	0	200	0	0	200	100	
	Total	**0**	**200**	**0**	**0**	**200**	**100**	

Contd...

Table 31.5– Contd...

Species	Mangrove Regions	151-300 cm	301-400 cm	401-500 cm	>500 cm	Total	Per cent of Sp.	Per cent of all Sp.
S. apetala	Kachchh	0	0	0	0	0	0	**0.12**
	GOK	0	0	0	0	0	0	
	Saurashtra	0	0	0	0	0	0	
	South Gujarat	346720	154763	126225	41217	668925	100	
Total		**346720**	**154763**	**126225**	**41217**	**668925**	**100**	
Total mangrove trees		**426456387**	**96316109**	**13938542**	**1022711**	**5377733749**	**NA**	**NA**
Per cent distribution over different height classes		**79.3 per cent**	**17.9**	**2.6 per cent**	**0.2**	**100.0**	**NA**	**NA**

and South Gujarat respectively. In terms of density classes, 3687 lakhs, 1009 lakhs and 654 lakhs are in the dense, moderate and sparse mangrove areas respectively (Table 31.5). The tree density of Kachchh is 11737 per ha, 6578 per ha and 4036 per ha for the dense, moderate and sparse mangroves respectively. In the case of Gulf of Kachchh, it is 5383 per ha, 1304 per ha and 216 per ha for the dense, moderate and sparse forests respectively. For the mangroves of Saurashtra, it was 4216 per ha, 576 per ha and 238 per ha for the dense, moderate and sparse mangroves respectively. In the case of South Gujarat it was 1139 per ha and 127 per ha for the dense and sparse mangroves respectively. Considering all the three density classes together, the tree density of Kachchh, Gulf of Kachchh, Saurashtra and South Gujarat comes out to be 8236 per, 3252 per ha, 2746 per ha and 723 per ha respectively. Considering four mangrove regions together, the tree density of mangroves in Gujarat is 8132 per ha, 5114 per ha and 2535 per ha for the dense, moderate and sparse mangroves respectively giving an average tree density of mangroves in the state to the tune of 5893 per ha.

Of the total mangrove trees, 97.62 per cent belongs to a single species *i.e. A. marina*. The rest 2.38 per cent is collectively represented by other mangrove species (Table 31.5). Considering the height classes, 79.3 per cent trees belong to the height class 1.5 to 3 m (Table 31.5). Further, 17.9 per cent, 2.6 per cent and 0.2 per cent trees belong to the height class 3 to 4 m, 4 to 5 m and more than 5 m respectively.

The carbon sequestration potentials of different height classes of the nine mangrove species are given at Table 31.6. In case of recruits and plants up to 3 m height, it was maximum for *R. mucronata* and minimum for *A. officinalis* (Table 31.6). The recruits of *L. racemosa* were not reported from the state. For the height class 3 to 4 m, carbon sequestration potential was maximum for *R. mucronata* while minimum for *L. racemosa* (Table 31.6).

Table 31.6: Height-class-wise Details of Carbon Sequestration Potential (kg) of Nine Mangroves Species

Species	1-150 cm	151-300 cm	301-400 cm	401-500 cm	> 500 cm
A. marina	0.26	1.91	7.39	16.56	55.84
A. officinalis	0.01	0.28	3.72	6.17	65.66
S. apetala	0.18	0.90	8.09	14.61	146.25
A. ilicifolius	0.61	0.88	NA	NA	NA
A. corniculatum	0.23	1.98	NA	NA	NA
B. cylindrica	0.22	0.29	NA	NA	NA
C. tagal	1.00	4.27	9.97	NA	NA
R. mucronata	1.60	14.69	46.32	NA	NA
L. racemosa	NA	NA	0.96	NA	NA

The mangrove plants (including recruits) have sequestered 2.24 million ton of carbon. Considering the height classes, 0.36 million ton (16.40 per cent), 0.87 million ton (38.83 per cent), 0.71 million ton (31.76 per cent), 0.23 million ton (10.26 per cent) and 0.06 million ton (2.7 per cent) carbon has been sequestered by up to 1.5 m, 1.5 to

3 m, 3 to 4 m, 4 to 5 m and more than 5 m (Table 31.7). Further, 1.7 million ton (78 per cent), 0.21 million ton (10 per cent), 0.02 million ton (1 per cent) and 0.25 million ton (11 per cent) carbon has been sequestered by mangroves of Kachchh, Gulf of Kachchh, Saurashtra and South Gujarat respectively (Table 31.8).

Table 31.7: Carbon Sequestration (million ton) by Mangrove Plants of different Height Classes of Four Mangrove Regions

Mangrove Region	1-150 cm	151-200 cm	301-400 cm	401-500 cm	>500 cm	Total	Per cent Distribution
Kachchh	0.1753	0.6820	0.6824	0.2047	0.0000	1.7444	**78**
GOK	0.0297	0.1601	0.0122	0.0026	0.0098	0.2144	**10**
Saurashtra	0.0077	0.0181	0.0000	0.0000	0.0000	0.0258	**1**
South Gujarat	0.1551	0.0103	0.0175	0.0229	0.0512	0.2570	**11**
Total	0.3678	0.8705	0.7121	0.2302	0.0610	2.2416	**100**

Table 31.8: Carbon Sequestered in Mangrove Forests (million ton)

Regions	Particulars	Dense	Moderate	Sparse	All Density Classes
Kachchh	Soils	1.179	0.430	0.440	2.049
	Plants	1.180	0.344	0.220	1.744
	Total	**2.360**	**0.774**	**0.660**	**3.793**
Gulf of Kachchh	Soils	0.912	0.250	0.066	1.229
	Plants	0.190	0.021	0.005	0.215
	Total	**1.102**	**0.271**	**0.071**	**1.443**
Saurashtra	Soils	0.207	0.045	0.010	0.262
	Plants	0.023	0.003	0.001	0.026
	Total	**0.230**	**0.048**	**0.010**	**0.288**
South Gujarat	Soils	1.709	NA	0.626	2.335
	Plants	0.228	NA	0.029	0.257
	Total	**1.937**	**0.000**	**0.655**	**2.592**
All regions	Soils	**4.008**	**0.725**	**1.141**	**5.874**
	Plants	**1.621**	**0.367**	**0.254**	**2.242**
	Total	**5.629**	**1.092**	**1.395**	**8.116**

In terms of carbon sequestration by mangrove plants (including recruits) per unit area, it is 31 ton per ha, 12 ton per ha, 7 ton per ha and 18 ton per ha for Kachchh, Gulf of Kachchh, Saurashtra and South Gujarat respectively. Considering the density classes, it comes out to be 36 ton per ha, 19 ton per ha and 10 ton per ha for dense moderate and sparse mangroves.

In the case of mangrove soil, carbon sequestration was found more in the lower layer (16 to 30 cm depth) as compared to the upper layer (up to 15 cm depth).

Considering the density classes of four mangrove regions, the carbon sequestered by mangrove soils is 4.01 million ton, 0.73 million ton and 1.14 million ton (Table 31.8) by the dense, moderate and sparse mangroves. Hence, a total of 5.87 million ton carbon has been sequestered by mangrove soils (up to 30 cm depth) of Gujarat (Table 31.8). Considering the four mangrove regions, 2.05 million ton (35 per cent), 1.23 million ton (21 per cent), 0.26 million to (4 per cent and 2.33 million ton (40 per cent) carbon has been sequestered by mangroves of Kachchh, Gulf of Kachchh, Saurashtra and South Gujarat respectively (Table 31.8).

The mangrove soil and plants (including recruits) together has sequestered 8.116 million ton carbon in Gujarat (Table 31.8). Out of this 3.793 million ton, 1.443 million ton, 0.288 million ton and 2.592 million ton has been sequestered by Kachchh, Gulf of Kachchh, Saurashtra and South Gujarat respectively (Table 31.8). Considering the three density classes of the four mangrove regions, 5.629 million ton, 1.092 million ton and 1.395 million ton carbon has been sequestered by dense, moderate and sparse mangroves of the state (Table 31.8). Further, the carbon sequestration per unit area comes out to be 67.73 ton per ha, 82.90 ton per ha, 83.42 ton per ha and 180.24 ton per ha for Kachchh, Gulf of Kachchh, Saurashtra and South Gujarat respectively. Considering the four mangrove regions together, it comes out to be 88.95 ton per ha. In terms of three density classes, the carbon sequestration per unit area is 123.30 ton per ha, 55.35 ton per ha and 53.95 ton per ha for the dense, moderate and sparse mangroves respectively.

In terms of carbon dioxide, 29.76 million ton has been sequestered by mangrove plants (including recruits) and soil (up to 30 cm depth) together. In terms of carbon dioxide sequestered per unit area, it has been found to be 248.3 t ha^{-1}, 304.0 t ha^{-1}, 305.9 t ha^{-1} and 660.9 t ha^{-1} by Kachchh, Gulf of Kachchh, Saurashtra and South Gujarat mangroves respectively. Considering the density classes of all the four mangrove zones, the carbon dioxide sequestered per unit areas has been estimated as 452.1 t ha^{-1}, 203.0 t ha^{-1} and 197.8 t ha^{-1} by the dense, moderate and sparse mangrove respectively (Table 31.9).

Table 31.9: CO$_2$ Sequestered per Unit Area (ton per ha)

| Regions | CO$_2$ Sequestered per Unit Area (ton per ha) | | | |
	Dense	Moderate	Sparse	Average of all Density Classes
Kachchh	335.4	197.4	152.7	248.3
Gulf of Kachchh	435.7	218.5	72.1	304.0
Saurashtra	400.3	213.4	71.1	305.9
South Gujarat	838.2	NA	406.5	660.9
Avg. of all zones	452.1	203.0	197.8	326.1

31.5 Discussion and Conclusion

The recruit density of mangroves of Philippines has been estimated to range from 2,533±1,139 to 1,96,000±8,461 per ha (Antonio, *et al.*, 1999). The recruit density of the dwarf and fringe mangrove forests of Biscayne National Park, Florida (U.S.A.) has been estimated as 6,08,200±1,22,100 per ha and 1,07,200±42,900 per ha, respectively (Michael, *et al.,* 2001). (Nayar, 2008) has found that for Kerala mangroves the recruit density is highest for *Aegiceras corniculatum* (1,35,229 per ha) which is followed by *Bruguiera cylindrical* (42,823 per ha), *Acanthus ilicifolius* (14,414 per ha), *Avicennia officinalis* (12,629 per ha), *Excoecaria agallocha* (7,143 per ha), *Avicennia marina* (5,714 per ha), *Rhizophora mucronata* (1,886 per ha), *Kandelia candel* (943 per ha) and *R. apiculata* (471 per ha). The recruit density of mangrove estimated by the present work is within the range of available information.

The tree density of mangroves of Philippines has been estimated as 3,300±656 trees per ha (Antonio, *et al.,* 1999). For the dwarf and fringe mangrove forests of Biscayne National Park, Florida (U.S.A.), it has been estimated as 8,800±8,800 trees per ha and 57,800±9,000 trees per ha respectively (Michael, *et al.,* 2001). The tree density of Bahamas, Jamanica, Puerto Rico, Belize Venezuela and Bon Accord Lagoon mangrove forests of Tobago (of *R.* mangle) has been estimated as 1140 per ha, 4950 per ha, 2370 per ha, 4520 per ha, 2630 per ha and 4620 per ha respectively. Further, the tree density of mangroves of Bertioga and Guaratiba, Southeast Brazil estimated as 4100 trees per ha and 2560 trees per ha (average tree height 7 m) where the average annual rainfall was 1032mm and 2240mm respectively (Soares and Yara, 2005). The average annual rain fall ranges from 2000 mm (South Gujarat) to less than 400 mm (Kachchh) in the state. The present work has estimated the tree density for Kachchh, Gulf of Kachchh, Saurashtra and South Gujarat as 8,236 per ha, 3252 per ha, 2,746 per ha and 723 per ha respectively. However, considering all the four mangrove regions together, it is 5893 per ha. Further, in terms of three density classes, the present work has found the tree density as 8132 per ha, 5114 per ha and 2535 per ha for dense, moderate and sparse mangroves respectively. The tree density of *A. marina* for the mangroves of French Guiana and Northern Australia has been reported to be 1,06,000 per ha (0.6m height), 8,831 per ha (2.7 m height) and 7,500 per ha (3.4 m height) (Christophe, *et al.,* 2003).

The biomass carbon content values range from 25 t CO_2 ha^{-1} to 2,254 t CO_2 ha^{-1} and most estimates fall between 300 and 1,000 t CO_2 ha^{-1} (Samantha, *et al.,* 2011). Considering the vegetation carbon pool, the CO_2 sequestered per unit area in Kachchh, Gulf of Kachchh, Saurashtra and South Gujarat has been found to be 308 t ha^{-1}, 95.3 t ha^{-1}, 55.0 tons ha^{-1} and 117.3 t ha^{-1} respectively. All the mangrove regions put together, the CO_2 sequestered per unit area has been found to be 143.9 t ha^{-1} for the mangroves of Gujarat. In terms of the three density classes, the carbon dioxide sequestered by mangrove plants (including recruits) is 95.3 t ha^{-1}, 39.1t ha^{-1} and 19.3 t ha^{-1} for the dense, moderate and sparse mangroves of Gujarat.

Tropical wetland forests such as mangroves are important reservoirs of organic C as the soil depth is more (Page, *et al.,* 2001; Page, 2002; Murdiyarso, 2010). High carbon organic matter (as high as 75 per cent carbon concentration) can be found as deep as 8m belowground (Chmura, *et al.,* 2003). Carbon accumulation rates in mangroves are 10 times the rate for temperate forests and up to 50 times the rate for tropical forests (Laffoley, *et al.,* 2009). In other words, 1 km^2 of mangrove area results in the equivalent long term sequestration in 50 km^2 of tropical forests (Yee., 2010).

Peatlands received significant attention since 1997 (Duke, 2007), when peat fires associated with land clearing in Indonesia increased atmospheric CO_2 enrichment by 13 to 40 per cent over global annual fossil fuel emissions (Page, 2002). It has resulted in incorporation of peatlands in the international climate change mitigation strategies (Murdiyarso, *et al.,* 2010). However, mangroves were not given due importance in such strategies although it occur in 118 countries, adding 30 to 35 per cent to the global area of tropical wetland forest over peat swamps alone (FAO, 2007; Page, *et al.,* 2001).

Though mangroves are well known for high C assimilation and flux rates (Twilley, *et al.,* 1992; Chmura, *et al.,* 2003; Alongi, *et al.,* 2004; Komiyama, *et al.,* 2008; Kristensen, *et al.,* 2008), data are surprisingly lacking on whole-ecosystem carbon storage - the amount which stands to be released with land-use conversion. Limited components of carbon storage have been reported, most notably tree biomass (Twilley, *et al.,* 1992; Kristensen, *et al.,* 2008), but evidence of deep organic-rich soils (Golley, 1962; Matsui, 1998; Fujimoto, 1999) suggests these estimates miss the vast majority of total ecosystem carbon. These facts have been re-established by the present work which has found that of the total carbon sequestered by mangroves of Gujarat, 72 per cent is by soil and 28 per cent is by mangrove vegetation.

In terrestrial ecosystems the soil organic carbon shows a decreasing trend with the depth of soil (Yang, *et al.,* 2010). However, this is not the case with mangroves. Mangrove soils consist of a variably thick, tidally submerged suboxic layer (variously called 'peat' or 'muck') supporting anaerobic decomposition pathways and having moderate to high C concentration (Kristensen, *et al.,* 2008). The present work has also found that the soil organic carbon is more in the lower layer (16 to 30 cm) than the upper one (up to 15 cm depth).

Daniel, (2011) has examined 25 mangrove sites (n =10 estuarine, n = 15 oceanic) across the Indo-Pacific (8°S-22°N, 90°-163°E) and found high carbon concentration (per cent dry mass) throughout the top 1 m of the soil profile, with a decline below 1m. They have reported that mangroves are among the most carbon dense forests in the tropics (mean 1,023 t C ha^{-1}), and exceptionally high compared to mean carbon storage of the world's major forest domains (Boreal- ~350 t C ha^{-1} Temperate ~350 t C ha^{-1} upland~300 t C ha^{-1}). Estuarine sites contained a mean of 1,074 t C ha^{-1}; oceanic sites contained 990±96 t C ha^{-1}. Above-ground carbon pools were sizeable (mean 159 t C ha^{-1}, maximum 435 t C ha^{-1}), but below-ground storage in soils dominated, accounting for 71-98 per cent and 49-90 per cent of total storage in estuarine and oceanic sites, respectively (Daniel, 2011). Further, Ong (2002) has reported that the sediments in mangrove forests held 700 tons of carbon per meter depth per hectare.

The SOC of mangrove soil of (*R. apiculata* crop) at Tambol Yisan, Samut Songkram Province has been estimated as 50-70 t ha^{-1} for a 9-10 yr old plantation while for younger plantation it was reported as 50 t ha^{-1} (Kridiborworn, unpublished data). The present work has estimated 87.83 t ha^{-1}, 36.99 t ha^{-1} and 44.08 t ha^{-1} carbon sequestration for dense, moderate and sparse mangroves of Gujarat. Considering the four mangrove regions, the SOC was found 36.60 t ha^{-1}, 70.64 t ha^{-1}, 75.31 t ha^{-1} and 162.02 t ha^{-1} for Kachchh, Gulf of Kachchh, Saurashtra and South Gujarat. The Indian council of Forestry Research and Education has estimated the SOC for 14 types of Forests of India among them the soils of littoral swamp forests of have been reported to sequester the highest carbon *i.e.* 155.22 t ha^{-1} (Kishwan., 2009).

The SOC of mangrove forests of Khanom district of Thailand has been found to increase from 110 t ha^{-1} to 160 t ha^{-1} after hydraulic restoration (by opening shrimp pond bank) which was very low as compared to the natural mangrove forests *i.e.* 370 t ha^{-1} to 553 t ha^{-1} (Matsui, *et al.*, 2010). Higher temperatures and wet environments result in increased decomposition rates in wetlands soils (Chmura, *et al.*, 2003). Efficiency of carbon sequestration in sediments improves with the age of the mangrove forests, from 16 per cent for a 5-year old stand to 27 per cent for an 85 year old stand (Alongi, 2004; Lafolley and Grimsditch, 2009). However, mangrove SOC is suseptible to decomposition due to the abundance of aliphatic-rich humic acid which easily change to carboxilic humic substances by humification (Matsiu, 2007; Orlov, 1995; Yonebayashi, 1994). This intrinsic character of mangrove SOC may lead to accelerated carbon decomposion in the soil which is exposed to air (Matsui, *et al.*, 2010). Therefore, inappropriate land use in mangrove area risks including a significant carbon loss transforming mangrove areas to source of CO$_2$ emission. In the mangrove areas which receive proper and adequate inundation (>700 hrs per year), the SOC is protected from severe decomposition (Matsui, *et al.*, 2010). However, in the mangrove areas receiving <200 hr per year inundation, the SOC is very low (Matsui, *et al.*, 2010).

The carbon sequestration by Matang mangrove forests of Malaysia has been estimated as 1.5 t h^{-1} Yr^{-1}. The above carbon pool in 25 mangrove systems across the Indo-Pacific has been estimated as 159 t ha^{-1}. The carbon sequestered by undisturbed mangroves of Términos Lagoon, Campeche has been estimated as 396.1 t ha^{-1} (Cerón-Bretón, *et al.*, 2010). The carbon sequestered by the mangrove forests of Everglades National Park, South Florida, USA has been estimated as 5.6X 10^9 kg or 5.6 million tons (Marc, *et al.*, 2006). The present work has reported that carbon sequestered in the mangrove forests (plants and soil) of Gujarat is 8.116 million ton.

Kishwan (2009) has estimated carbon sequestration per unit area by (littoral and) swam forests of India as 106.9 t ha^{-1}. Present work has estimated the carbon sequestration per unit area by Kachchh, Gulf of Kachchh, Saurashtra and South Gujarat as 67.73 t ha^{-1}, 82.90 t ha^{-1}, 83.42 t ha^{-1} and 180.24 t ha^{-1} respectively. In terms of dense, moderate and sparse mangrove forests of the state, it is 123.30 t ha^{-1}, 55.35 t ha^{-1} and 53.95 t ha^{-1} respectively.

31.6 Acknowledgements

We acknowledge GEER Foundation, Gandhinagar for providing necessary support for implementing the project. We also acknowledge the Ministry of Forest and Environment, Gov. of Gujarat for providing financial assistance. We are thankful to the officials and staff of Gujarat Forest department for providing the local help and the permissions for field works. The thanks are also due to Dr. Harshad Salvi (Research Associate, GEER Foundation), Mr. Mukesh Mali (Senior Research Fellow,), Mr. Bipin Khokhariya (Senior Research Fellow), Mr. Bhargav Brahmbhatt (Junior Research Fellow), Ms. Reshma Bobeda (Senior Research Fellow, GEER Foundation), Ms. Jagruti Jaggiwala (Junior Research Fellow), for helping field works.

References

Aboal, J. R.; Arevalo, J. R. and Fernandez, A., 2005. Allometric relationships of different tree species and stand above ground biomass in the Gomera laurel forest (Canary Island). *Flora-Morphology, Distribution, Functional Ecology of Plants*, 200(3): 264-274.

Alongi, D. M. 2004. Sediment accumulation and organic material flux in a managed mangrove ecosystem: Estimates of land-ocean-atmosphere exchange in peninsular Malaysia. *Mar. Geol.*, Vol: 208, 383-402.

Antonio, B. Mendoza, Danilo, Alura P. 1999. Mangrove structure on the eastern coast of Samar Island, Philippines. In D. E. Stott, *Sustaining the Globe* (pp. 423-425). Perdue University and USDA-ARS National Soil Erosion Research Laboratory.

Araujo, T. M., Higuchi, N. and Carvalho, J. A., 1999. Comparison of formulae for biomass content determination in a tropical rain forest site in the state of Para, Brazil. *Forest Ecology and Management*, 117 (1-3): 43-52.

Brown, S. 1997. *Estimating biomass and biomass change of tropical forests: a primer* (*FAO forestry paper-134*). Rome: FAO, United Nations.

Cerón-Bretón, J.G., Cerón-Bretón, R.M., Rangel-Marrón, M. and Estrella-Cahuich, A., 2010. Evaluation of carbon sequestration potential in undisturbed mangrove forest in Términos Lagoon, Campeche. *Development, Energy, Environment, Economics*, 295-300.

Chave, J.; Andalo, C.; Brown, S.; Cairns, M. A.; Chambers, J. Q.; Eamus, D.; Folster, H.; Fromard, F.; Higuchi, N.; Kira, T.; Lescure, J. P.; Nelson, B. W.; Ogawa, H.; Puig, H.; Riera, B and Yamakura, T., 2005. Tree allometry and improved estimation of carbon stocks and balance in tropical forests. *Oecologia*, 145:87-99.

Chmura, G. L., Anisfeld, S. C., Cahoon, D. R. and Lynch, J. C., 2003. Global carbon sequestration in tidal, saline wetland soils. *Glob. Biogeochem. Cycles*, Vol: 17, 1111.

Christophe Proisy, Anthea Mitchell, Richard Lucas, François Fromard, Eric Mougin, (2003). Estimation of Mangrove Biomass using Multifrequency Radar Data-Application to Mangroves of French Guiana and Northern Australia. *Proceeding of the Mangrove 2003 conference*, (pp. 1-9). Salvador, Bahia, Brazil.

Clark, D. B. 2001. Net primary production in tropical forests: an evaluation and synthesis of existing field data. *Ecol. Appl.*, 11, 371-374.

Cloug, B. F.; Dixon, P. and Dalhaus, O., 1997. Allometric relationships for estimating biomass in multi-stemmed mangrove trees. *Australian Journal of Botany*, 45:1023-1031.

Comley, B. W. T. and McGuinness, K. A., 2005. Above-and below ground biomass and allometry of four common northern Australian mangroves. *Australian Journal of Botany*, 53: 421-436.

Daniel C. Donato, J. B. (2011). Mangroves among the most carbon-rich forests in the tropics. *Nature Geoscince*, 1-5.

Daniel P. L. Walthert, S., and Luscher, P. (2000). Contemporary carbon stocks of mineral forest soils in the Swiss Alps. *Biogeochemistry*, Vol.50: 111–136.

Duke, N. C. (2007). A world without mangroves? *Science*, 317, 41-42.

FAO - Food and Agriculture Organization (2007). FAO Forestry Paper 153.

Fujimoto, K. (1999). Belowground carbon storage of Micronesian mangrove forests. *Ecol. Res.*, Vol: 14, 409-413.

Golley, F., Odum, H. T. and Wilson, R. (1962). The structure and metabolism of a Puerto Rican red mangrove forest in May. *Ecology*, vol:43, 9-19.

Kishwan, J. (2009). Estimation of Forest Carbon Stocks in India: A Methodology based on National Forest Inventory. *CFRN-ICFRE International Workshop* (April 27th). Dehradun, India.

Komiyama, A.; Poungparn, S. and Kato, S. (2005). Common allometric equations for estimating the tree weight of mangroves. *Journal of Tropical Ecology*, 21:471-477.

Komiyana, A., Ong, J. E. and Poungparn, S. (2008). Allometry, biomass and productivity of mangrove forests: A review. *Aquatic Botany*, 89: 128-137.

Kraenzel, M.; Castillo, A., Moore, T. and Potvin, C. (2003). Carbon storage of harvest age teak (*Tectona grandis*) plantation, Panama. *Forest Ecology and Management*, 173(1-3):213-225.

Kristensen, E., Bouillon, S., Dittmar, T. and Marchand, C. (2008). Organic carbon dynamics in mangrove ecosystems. *Aquat. Bot.*, Vol: 89, 201-219.

Laclau, P. (2003). Biomass and carbon sequestration of ponderosa pine plantations and native cypress forests in northwest Patagonia. *Forest Ecology and Management*, 180(1-3):317-333.

Laffoley, D.d'A. and Grimsditch, G. (eds) (2009). *The management of natural coastal carbon sinks.* Switzerland: IUCN, Gland.

Lasco, R. D. (2004). The clean development mechanism and LULUCF projects in the Philippines. *International Symposium/Workshop on the Kyoto Mechanism and the Conservation of Tropical Forest Ecosystems,* pp. 53-57, Waseda University.

Losi, C. J.; Siccama, T. G.; Condit, R. and Morales, J. E. (2003). Analysis of alternative methods for estimating carbon stock in young tropical plantations. *Forest Ecology and Mangement*, 184(1-3):355-368.

Lu, D. Mausel, P.; Brondizio, E. and Moren, E. (2004). Relationships between forest stand parameters and landsat TM spectral responses in the Brazilian Amazon Basin. *Forest Ecology and Management*, 22:459-470.

Marc Simard, Keqi Zhang, Victor H. Rivera-Monroy, Michael S. Ross, Pablo L. Ruiz, Edward Castañeda-Moya, Robert R. Twilley and Ernesto Rodriguez. (2006). Mapping Height and Biomass of Mangrove Forests in Everglades National Park with SRTM Elevation Data. *Photogrammetric Engineering and Remote Sensing*, Vol. 72, No. 3, 299–311.

Matsiu, N. and Kosaki, T. (2007). Quantitative and qualitative evaluation of stored carbon of mangrove ecosystems in Chumphon, Thailand, *Mangrove Sci.*, Vol. 5, 13-19.

Matsui, M. Suekuni, J., Nogami, M.,Havanond, S. Salikul, P. (2010). Mangrove rehabilitation dynamics and soil organic carbon change as a result of full hydraulic restoration and regarding of a previously intensively managed shrimp pond. *Wetland Ecol. Manag.*, Vol. 18: 233-242.

Matsui, N. (1998). Estimated stocks of organic carbon in mangrove roots and sediments in Hinchinbrook Channel, Australia. *Mangr. Salt Marsh.*, Vol. 2: 199-204.

Michael, S. Ross, Pablo, L. Ruiz, Guy, J. Telesnicki, John, F. Meeder. (2001). Estimating above ground biomass and production in mangrove communities of Biscayne National Park, Florida (U.S.A.). *Wetland Ecology and Management*, 9:27-37.

Montagu, K. D.; Duttmer, K.; Barton, C. V. M. and Cowie, A. L. (2005). Developing general allometric relationships for regional estimates of carbon sequestration-an example using *Eucalyptus piplularis* from seven contrasting sites. *Forest Ecology and Management*, 204(1): 115-129.

Murdiyarso, D. M., Hergoualc'h, K. and Verchot, L. V. (2010). Opportunities for reducing greenhouse gas emissions in tropical peatlands. *Proc. Natl Acad. Sci. USA*, 107, 19655-19660.

Nayar. T. S. (2008). Plant crab association in mangrove ecosystem with a case study from Kerala. *Towards conservation and management of mangroves ecosystems in India* (pp. 67-80). Gandhinagar: GEER Foundation.

Ong, J. (2002). *The Hidden Costs of Mangrove Services: Use of Mangroves for Shrimp Aquaculture.* Bali, Indonesia: Aquaculture. International Science Roundtable for the Media.

Ong, J. E., Gong, W. K., Wong, C. H. (2004). Allometry and partitioning of the mangrove - *Rhizophora apiculata. Forest Ecology and Management*, 188: 395-408.

Orlov, D. S. 1995. *Humic substances of soil and general theory of humification.* Moscow: A. A. Balkema, Rotterdam, Brookfield.

Page, S. E., Rieley, J. O. and Banks, C. J., 2001. Global and regional importance of the tropical peatland carbon pool. *Glob. Change Biol.*, Vol.17, 798-818.

Page, S. E. 2002. The amount of carbon released from peat and forest fires in Indonesia during 1997. *Nature*, Vol. 420, 61-65.

Pandey C. N. and R. Pandey 2009. Study of floristic diversity and natural recruitment of mangroves in selected habitats of south Gujarat. Gujarat Ecological Education and Research (GEER) Foundation, Gandhinagar.

Pandey, C. N. and R. Pandey (2010). Study of pollination biology and reproductive ecology of major mangroves of Gujarat, Gujarat Ecological Education and Research (GEER) Foundation, Gandhinagar. Pp. 919

Pansu, M. and Gautheyrou, J. (2006). *Handbook of soil analysis-mineralogical, organic and inorganic methods.* Netherlands: Springer.

Saenger, P. (2002). *Mangrove ecology, silviculture and conservation.* The Netherlands, pp. 360: Kluwer Academic Press.

Samantha, S., Linwood, P. and Brian C. Murray. (2011). *State of the Science on Coastal Blue Carbon: A Summary for Policy Makers.* Nicholas Institute, For Environment Policy solution, Duke University.

Smith III, T. J. and Whelan, K. R. T. (2006). Development of allometric relations for three mangrove species in south Florida for use in the Greater Everglades Ecosystem restoration. *Wetland Ecology and Management,* 14: 409-419.

Soares, M. L. and Yara, Schaeffer-Novelli, 2005. Above-ground biomass of mangrove species- Analysis of models. *Estuarine, Coastal and Shelf Science,* 65 : 1-18.

Steven, Bouillon, Alberto V, Borges, Edward Casteneda-Moya, Karen Diele, Throsten Dittmar, Norman C. Duke, Erik Kristensen, Shing Y. Lee, Cyril Marchand, Jack J. Middleberg, Victor H. Riviera-Monroy and Thomos J. Smith, 2008. Mangrove production and carbon sinks: A revision of global budget estimates. *Global Biochemical Cycles,* 22, 1-12.

Tam, N. F. Y.; Wong, Y. S.; Lan, C. Y and Chen, G. Z. (1995). Community structure and standing crop biomass of a mangrove forest in Futian Nature Reserve, Shenzhen, China. *Hydrobiologia,* 295: 193-201.

Twilley, R. R., R. H. Chen and T. Hargis. (1992). Carbon sinks in mangrove forests and their implications to the carbon budget of tropical coastal ecosystems. *Water, Air Soil Pollu.,* 64: 265-288.

Yang Yuhai, Chen Yaning, Li Weihong and Chen Yapeng. (2010). Distribution of soil organic carbon under different vegetation zones in the Ili River Valley, Xinjiang. *J. Geogr. Sci.,* Vol. 20(5): 729-740.

Yee, S. M. (2010). *REDD and BLUE Carbon: Carbon Payments for Mangrove Conservation.* MAS Marine Biodiversity and Conservation Capstone Project.

Yonebayashi, K. (1994). Humic component distribution of humic acids as shown by adsoruption chromatography using XAD-8 resin. In N. M. Senesi, *Humic substances in the global environment and implications on human health* (pp. 181-186). Amsterdam: Elsevier Science.

Zhang, J.; Rivard, B.; Sanchez-Azofeifa, A. and Castro-Easu, K. (2006). Intra-and inter-class spectral variability of tropical tree species at La Selva, Costa Rica: Implications for species identification using HYDICE imagery. *Remote Sensing of Environment,* 105(2):129-141.

2015, **Impact of Global Warming and Climate Change on**
        Human and Plant Health
Editors: **Dr. Arun Arya and Prof. V.S. Patel**
Published by: **DAYA PUBLISHING HOUSE, NEW DELHI**

Pages ***341–345***

Chapter 32

Biofertilizers: A Status Update of Past, Present and Future Applications and its Role in Organic Farming and Sustainable Agriculture

Sheuli Dasgupta*

Department of Microbiology,
Gurudas College, Kolkata – 700 054, W.B., India

ABSTRACT

Biofertilizers are defined as preparations containing living cells or latent cells of efficient strains of microorganisms that help crops actively uptake nutrients by their interactions with the rhizosphere when applied through seed or soil. Micronutrients are needed by plants in agriculture or forest areas. The use of chemical fertilizers can be reduced by application of biofertilizers. The demand for organic food is steadily increasing both in developing and developed world. There is a large demand of such food crops produced by 130 countries. Use of biofertilizers and agro waste will conserve the fertility of soil for a longer time. The organic farming by involvement of traditional practices and preventive pest management will empower the farmers and village communities.

Keywords: *Biofertilizers, Past, present and future applications, Organic farming, Sustainable agriculture.*

———

* E-mail: sheulidasgupta@yahoo.co.in

32.1 Introduction

Environmental and health problems related to agriculture have been well documented, but it is very recently the calculated.the cost of agriculture in U.K. is 1.1-3.9 billion pounds per annum (Pretty,2000). The external costs of farming are not internationalized in the price of food, tax payers will have to pay the bill. This alarming rate of use of chemicals and pesticides has polluted the crops as well as the soil. Global consumers are looking towards green food organic food. Production of such food demands use of organic fertilizer, compost or biofertilizers.

Use of such fertilizers accelerate certain microbial processes in the soil which augment the extent of availability of nutrients in a form easily assimilated by plants. Biofertilizers are such as *Rhizobium, Azospirillum* and Phosphobacteria provide nitrogen and phosphorous nutrients to crop plants through nitrogen fixation and phosphorous solubilization processes. These biofertilizers could be effectively utilized for rice, pulses, millets, cotton, sugarcane, vegetable and many horticulture crops. Biofertilizers are one of the prime inputs in organic farming and other sustainable agricultural practices that not only enhance the crop growth and yield; but also improve the soil health and sustain soil fertility. At present, biofertilizers are supplied to the farmers as carrier based inoculants.

Very often microorganisms are not as efficient in natural surroundings as one would expect them to be. Therefore artificially multiplied cultures of efficient selected microorganisms play a vital role in accelerating the microbial processes in soil. Use of biofertilizers is one of the important components of integrated nutrient management, as they are cost effective and renewable source of plant nutrients to supplement the chemical fertilizers for sustainable agricultural practices. Several microorganisms and their association with crop plants are being exploited in the production of biofertilizers. *Rhizobium* is a soil bacterium, which can colonize the legume roots and fix atmospheric nitrogen symbiotically. The morphology and physiology of *Rhizobium* varies from free-living condition to the bacteroid in the nodules. They are the most efficient biofertilizer as per the quantity of nitrogen fixed is concerned. They represent seven genera and are highly specific to form nodule in legumes, referred to as cross inoculation group. Initially, due to absence of efficient Bradyrhizobial strains in soil, soybean inoculation resulted in bumper crop production. However, due to unrestricted use of inoculation in the past by US farmers has resulted in the build up of a plethora of inefficient strains in the soil whose replacement by efficient strains of Bradyrhizobia has now turned into a challenging problem. The bacterium produces abundant slime that results in soil aggregation.

32.2 Plant Growth Promoting Rhizobacteria

The group of bacteria that colonize roots or rhizosphere soil and beneficial to crops are referred to as Plant Growth Promoting Rhizobacteria (PGPR). The PGPR inoculants currently commercialized seem to promote growth through at least one mechanism; suppression of plant disease (termed Bioprotectants), improved nutrient acquisition (termed Biofertilizers), or phytohormone production (termed Biostimulants). Several species of *Pseudomonas* and *Bacillus* can produce as yet not well characterized phytohormones or growth regulators that cause crops to have greater amounts of fine roots. This helps in increasing the absorptive surface of plant roots for the uptake of water and nutrients. These PGPRs are referred to as Biostimulants and the phytohormones they produce include indole-acetic acid, cytokinins, gibberellins and ethylene inhibitors. Recent advances in molecular techniques are also encouraging in that several tools are becoming available to determine the mechanism by which crop performance is improved using PGPR. This is also helping to track the survival and activities of the growth promoting rhizoidal organisms in both soil and roots.

Despite promising results, biofertilizers has not received widespread application in agriculture mainly because of the variable response of plant species or genotypes/germplasms to inoculation depending on the bacterial strain used. Differential rhizosphere effect of crops in harboring a target strain or even the modulation of the bacterial nitrogen fixing and phosphate solubilizing capacity by specific root exudates may account for the observed differences. On the other hand, good competitive ability and high saprophytic competence are the major factors determining the success of a bacterial strain as inoculants. Studies to know the synergistic activities and persistence of specific microbial populations in complex environments, such as the rhizosphere, should be addressed in order to obtain efficient inoculants. Though the biofertilizer technology is a low cost, eco-friendly technology; several constraints limit the application or implementation of the technology. These constraints may be environmental, technological, infrastructural, financial, human resources, unawareness, quality, marketing, and related factors. These different constraints in one way or other affect the development of new technology, marketing or usage. However, the current focus on sustainable agriculture will certainly encourage the use of biofertilizers in crop production systems in near future.

Study to know the synergistic activities and persistence of specific microbial polpulations in complex environments such as the rhizosphere.

32.3 Microbes Used to Produce good fertilizer

Microbes play a crucial role in functioning of all ecosystems. Microbes can be an important food resource, are responsible for decomposition of organic matter, and have unique capabilities for transforming nutrients (N in particular) from one form to another. The term, *microbe*, refers to a size class of organisms, rather than to a taxonomic group. More genotypic and phenotypic variability exists within microbes than among all other organisms combined. Microbes include Bacteria, Archaea, and Eukarya (Pace, 2006), viruses, autotrophs, heterotrophs, parasites, and predators. Some of these organisms have macroscopic features, such as mushrooms and algal colonies, whereas others are individual cells only a fraction of a micrometer in size. Ecosystems exist where microbes are the only form of life (deep within geological formations), but even in more diverse ecosystems, microbial biomass and metabolic functioning can dominate mass and process budgets. For the purpose of reviewing the past 25 y of progress in microbial ecology.

Microorganisms collectively have tremendous capabilities for degrading organic C compounds, and they can use a wide array of electron acceptors. As autotrophs, they are capable of fixing C, and they can use sunlight or many different electron donors. One challenge is understanding how much of this observed functional variability is caused by gene expression within a taxonomic group and how much is a consequence of representation of different groups within assemblages. Microbes are unique in that detecting and measuring what they do (*e.g.*, decay organic matter, immobilize NH_4^+) is easier than determining their standing stock or identity. Techniques for quantifying the net result of their metabolism (*e.g.*, light–dark bottles, Winogradsky columns) have been in common use for well over a century, but only now are we developing a good sense of how many taxa are present in any ecosystem or whether regular patterns in taxonomic composition exist over time or across systems.

For many years, the field of microbial ecology was hampered by a lack of techniques to estimate the biomass of different microbial groups, but this limitation has been largely resolved by application of direct microscopy for estimating bacterial biomass (Findlay and Arsuffi 1989) and ergosterol assays to quantify fungi (Gessner and Chauvet 1993). Molecular techniques can provide information on microbial identity at levels of resolution ranging from deoxyribonucleic acid (DNA) sequencing at the fine scale to various fingerprinting options to measure community similarity / dissimilarity at the broad scale (Logue *et al.*, 2008). The next frontiers are to connect presence and process by exploring the abundance and regulation of functional genes that are part of the genetic identity of microorganisms and to identify the actual mechanisms by which they acquire energy, transform elements, and alter their environment (Zak *et al.*, 2006).

The microbial ecologists studying soil systems share many of the problems of scientists working in streams. In both systems, most bacterial cells and fungal hyphae are tightly associated with particles or surfaces. This association causes difficulties when isolating cells or extracting compounds quantitatively. Moreover, significant interference is encountered in both systems from noncellular materials, such as humics. Second, stream ecologists have a strong interest in process-derived questions, such as the balance between transport and transformation. This focus has led to a preponderance of questions that deal with the net effect of organisms rather than with links between organisms and processes. Perhaps stream ecologists have simply by-passed questions of microbial community structure and interactions.

Patterns in microbial composition associated with geographic distance and variation in function clearly do exist (see below), and the wealth of information on stream microbial function might encourage future research that bridges the gap between the processes that are occurring and which microbes are present (Knapp *et al.*, 2009). In a fashion, the past emphasis on function could have made streams, with their wide diversity of habitats and rates, perfect systems in which to consider whether functional variability is associated with differences in microbial community structure. Thus, we might expect more work on these important questions to be done in streams in the near future.

The quantity of bacterial and fungal biomass as C in the detritus–microbe complex is typically a only a small percentage of total leaf mass and reaches a maximum within a few weeks (Methvin and Suberkropp, 2003). Therefore, absolute mass of substrate is so much greater than the mass of microbial C that leaf material might make the largest contribution to consumer C demand, despite the fact that the assimilation efficiency for microbial C is much greater (perhaps 5–10× higher) than assimilation efficiency for the nonliving leaf substrate. Selective feeding by larger consumers occurs at relatively large scales (*i.e.*, one leaf vs another) and little opportunity exists to avoid consumption of the nonliving substrate. Therefore, assimilated C generally will be derived from both sources. However, smaller consumers might have finer-scale selective ability, such that they can feed on microbial biomass itself or on enriched patches on an individual leaf. Hall and Meyer (1998) found that many smaller taxa could derive a significant portion of their C demand from bacterial biomass.

References

Hall, R.O. Jr. and Meyer, J.L. (1998). The trophic significance of bacteria in a detritus-based stream web. *Ecology,* 79: 1995-2012.

Findlay, S.E.G. and Arsuffi, T.L. (1989). Microbial growth and detritus transformational during decomposition of leaf litter in a stream. *Freshwater Biology,* 21: 261-269.

Kavino, M., Harish, S., Kumar, N. and Samiyappan, R. (2010). Biological hardening of micropropagated banana (*Musa* sp.) plantlets in rhizosphere endophytic bacteria to manage banana bunchy top virus (BBTV). In: Plant Growth Promotion by Rhizobacteria (Eds.) Reddy *et al.* Proc. of First Asian PGPR Congress, pp. 279-287.

Methvin, B.R. and Suberkropp, K. (2003). Annual production of leaf-decaying fungi in two streams. *J. of North American Benthological Society,* 22: 554-564.

Zak, D.R., Pregitzer, K.S., Curtis, P.S., Teeri, J.A. Fogel, R. an Randlett, D.L. (1993). Elevated atmospheric CO2 and feedback between carbon and nitrogen cycles. *Plant Soil,* 151: 105-117.

2015, Impact of Global Warming and Climate Change on
    Human and Plant Health
Editors: **Dr. Arun Arya and Prof. V.S. Patel**
Published by: **DAYA PUBLISHING HOUSE, NEW DELHI**

Pages 346–351

Chapter 33

Plantation to Reduce the Pollution in the Leading Eco-City: Vadodara

Arun Arya*

Department of Botany,
The M.S. University of Baroda, Vadodara, Gujarat, India

ABSTRACT

Baroda has many magnificent buildings like Nyaya mandir, Kala bhawan and Baroda Museum and Picture gallery etc.. The museum has Greco Roman structural panels made in Indo Sarascenic style. Building of Laxmi Vilas Palace and Baroda college (144 feet high building) constructed in 1881 has second highest dome in Asia. This building was designed by architect Robert Chisholm. The city has a large number of gardens. Sayaji baug earlier called as Kamatti baug is a large garden in the heart of city.

A large number of trees can be seen in gardens and on both sides of roads. These trees help in reducing the temperature in hot summer and dust pollution throughout the year. Dust particles are major source of pollution in all the cities of country. The concentration of these particles may vary from season to season or areas where higher construction activities are taking place, vehicles playing on roads also secrete such suspended particles. An assessment of dust carrying capacity of nine common road side plants in Vadodara was done during 2010-2011. Leaves of Vad (*Ficus benghalensis*) and Mango (*Mangifera indica*) showed more accumulation of dust or SPM as compared to plants like Ashoka (*Polyalthia longifolia*) and Saptparni (*Alstonia scholaris*).The plants were able to accumulate 1.4 mg/cm^2

---

* E-mail: aryaarunarya@rediffmail.com

to 3.3 mg/cm² dust and also helped in reduction of noise. Plants with higher chlorophyll and ascorbic acid contents are thought to have more pollution scavenging potential. Efforts are needed to introduce more plants on busy roads of Raopura, Dandia bajar, Alkapuri etc.

Keywords*: Plantation, Pollution, Eco-city, Vadodara, Busy roads, Raopura, Alkapuri, Ficus benghalensis*

33.1 Introduction

The United Nations (1991) predicted that by the year 2000, more than 50 per cent of the World's population would live in urban areas and that this would increase to 65 per cent by 2025. Urbanization would lead to development of many mega cities and increasing urban sprawl. Such changes are of special interest to social scientists and urban planners. Urbanization may provide comforts to citizens but adversely affects environment and may lead to loss of biological diversity. Recent problems of increase in temperature and global warming calls for collaborative efforts leading to more plantation and reducing the city pollution due to vehicles.

From nomadic life, human beings settled into groups to safe guard themselves from beasts and natural calamities like fire and floods. He practiced agriculture and slowly and slowly from small villages, towns, settlements, he built houses, flats, high rise buildings and thus formed cities. Industrialization lead to development of metro cities like Delhi, Kolkata and Mumbai in India, New York, Tokyo, Chicago, Barcelona and Canada's most lovable city Vancouver. This city has an ambitious 100 year plan for clean and green living. The city makes us of hydroelectric energy along with solar, wind, wave and tidal energy. City has solar powered trash compactors which reduces the number of garbage trucks on roads.

It has long been suspected that increase in respiratory allergy in recent years is not only due to changes in diagnostic practices but air pollution also play an important role in the manifestation of airway diseases. According to Harnam (2001) from Malaysia, allergy cases have increased from less than 10 per cent up to 40 per cent over the last three decades, the incidence is a frightening scenario, whereby there will hardly be a family without someone suffering from this plague of the 21st century.

Unlike foreign countries traffic speed is slow in most of Indian cities and vehicles further slow down near signals. It is suspected that fine particles in air function as nuclei for deposition of pollutants and intensity of pollution hazard. This is also a major health issue for children and senior citizens. It is interesting to note that SPM is high at 255 mg/m³, as against the permissible limit of 140 mg in almost all cities of the country. Use of green belts and plantation where ever possible can reduce the pollution levels.

33.2 Increase in City Population

Perhaps the most fundamental thing that man has so far failed to do is to control his own population. Despite having necessary means and knowledge of how to undertake it humanely, so the remedy itself is left to nature's way of famine and/or pestilence, or to man's way of increased violence.

The recent DNA studies have indicated that about 70,000 years ago the human population reduced to about 10,000 people and then increased to present level. Indian author Manu also described that some flood like calamites have occurred in past. Researchers have shown that changes in population growth, age structure, and special distribution interact closely with the environment and with development; rapid population grown causes lead degradation and depletion of resources. Global warming is also one of the out come of massive industrialization and burning of fossil fuel in past, empowerment through education will make him healthy and wealthy.

In 1961 the population of Vadodara was 2,98,398 which increased to 10,31,346 in 1991. Currently (in 2011) the population of Vadodara is 41,66,703. This increase in population have increased the more utilization of natural and energy utilization, and more numbers of vehicles plying on roads increasing vehicular pollution. There is a need to have more responsible human behavior. The proper use of water and energy resources is desired. We should check pollution at all levels.

33.3 Damage to Heritage Buildings

Baroda has many magnificent buildings like Nyaya mandir, Kala bhawan and Baroda Museum and Picture gallery. The museum has Greco Roman structural panels made in Indo Sarascenic style. Building of Laxmi Vilas Palace and Baroda college (144 feet high building) constructed in 1880 has second highest dome in Asia. This building was designed by architect Robert Chisholm at the cost of Rs. Eight lakhs. The buildings are looking ugly due to growth of bacteria and cyanobacteria. The dust and dirt also adding to loss of aesthetic appeal. A survey was conducted of some old building like Nyay mandir, Hazira and Baroda College showed the presence of herbs and weeds on walls and roof tops. The bigger plants like Pipal and Vad (*Ficus* spp.) were also observed at certain places.

The four historical gates Laheripura gate, Panigate, Gandi gate, Champanar gate are situated in eastern, western, northern, and southern direction of Mandvi gate, respectively. These four gates are closely related with the history and culture of the city. Mehmud Begda the sultan of Gujarat died in 15th century A.D. and the Sultanate passed to his son Khalikhan who assumed the name Muzaffar and constructed a fort to the East of ancient vadapadraka which later on evolved into Vadodara. The fort was known are Kila-e-Daulatabad and present day Laheripura gate in part of the western ramparts of the fort. Under its august façade the British and the Arab soldiers of the Gaekwad fought a bloody battle in 1802. Coppersmiths residing in that area where known as Laheris. The word Laheripura is a variant of the original Lahari. The stone areas of these gates were subjected to lime coats and OBD (Oil Bond Distemper) coats in past. These coats are not only concealing the architectural details but also disturbing aesthetic view of the gates, besides causing damage to the stone surface. The theft of beams and iron support was cause of concern in last few days. It may further damage the gate.

The microorganisms present on external stone walls of building are fungi, bacteria, algae (including cyanobacteria), lichens and protozoa. Microorganisms are able to obtain different elements for their metabolism, *e.g.,* calcium, aluminum, silicon, iron and potassium, by biosolubilization from the stone matrix. Such biosolubilization involves the production of organic and inorganic acids by the metabolic activity of microbes. The deposit covered area seems to be very old due to the formation of secondary dull green, pale, white lichens, present all over the stone surface. Due to these deposits the aesthetic beauty of the monument is seriously affected.

33.4 Is Vadodara a Clean and Green City?

Vadodara is located at 22.07′5.9′′°N 73.15′8′′°E in western India at an elevation of 39 m. It is the 18th largest city of the country with an area of 148.95 km² and a population of 4.1 million according to 2010-11 censuses. During medieval period the city of Baroda was aptly described as a "Tilaka on the brow of Lata" by Jain writer named Viragani (1104 A.D.) According to traditions, Baroda was known as Chandanvati, Vatapadra, Vadapadra and Virapur. The evidence from earliest settled life at Baroda appears from about 2200 years ago. In the 17th century A.D., the British and Dutch rulers thought of establishing their cloth factories in Baroda. The textile industry occupied the north western side of the Vishwamitri river, which led to the development of Cantonment area of the present day Fatehgunj. The Green cosmopolitan city Vadodara is popularly known as Banyan City. The city represents perfect amalgamation of the past and the present. Friendly people can breathe fresh air and live in naturally air-conditioned pleasant societies. Vadodara has about 7,47,193 trees spread over in an area of 16,261 hectares. There are 175 varieties of trees.

According to CPCB data (2009-10), the RSPM (91.27) and SPM (293.65) were more in GIDC area of Makarpura as compared to Dandia bazaar (SPM- 81.34) and Harinagar (SPM- 138). In a study, Dhar (2011) found that accumulation of dust was more in trees with compound leaves such as margosa (4mg/cm²), *Samanea* (3mg/cm²) as compared to *Nerium* (1mg/cm²) and rain tree (2mg/cm²). Vad and *Samanea* leaves showed presence of trichomes on lower surface of leaves which hold the large amount of dust particles. Trees act as dust filters, sound barriers and provide suitable shade and temperature in most part of the city.

33.5 Plants as Pollution Scavengers

Ambient air constitutes solid particles of various shapes and sizes commonly referred as SPM. Dust is major component of SPM which is continuously deposited on various surfaces. Researchers have shown that leaves can act as biological filters in urban setup. The higher the concentration of SPM in area, the higher will be concentration of different leaf surfaces. The accumulation of SPM may be influenced by shape and arrangement (Phyllotaxy) of leaves.

It is estimated that almost 60 per cent of air pollution in Mumbai city is caused due to automobiles. On a national scale the figure stands somewhere near 40 per cent it is increasing rapidly (NEERI, 2005). Road dust formed due to wear and tear of road surface and tires, combined with emission and frequent workings of road surfaces by different government agencies, remains airborne for long duration due to heavy traffic. Almost eight months of dry season in the year further aggravates the dust problem. It is the dust pollution problem on and along roads that has been grossly neglected both by the authorities as well as researchers in this field.

The concentration of these particles may vary from season to season or areas where higher construction activities are taking place, vehicles playing on roads also secrete such dust particles. An assessment of dust carrying capacity of nine common road side plants in Vadodara was done during 2010-2011. Leaves of Vad (*Ficus benghalensis*) and Mango (*Mangifera indica*) showed more accumulation of dust or SPM as compared to plants like Ashoka (*Polyalthia longifolia*) and Saptparni (*Alstonia scholaris*).The plants were able to accumulate 1.4 mg/cm² to 3.3 mg/cm² dust and also helped in reduction of noise. Plants with higher chlorophyll and ascorbic acid contents are thought to have more pollution scavenging potential. Efforts are needed to introduce more plants on busy roads of Raopura, Dandia bajar, Alkapuri etc. There is need to have proper balance of ornamental plants, shade providing and certain fruit trees.

33.5.1 Which Trees Should be Preferred for Road Side Plantation?

Urban trees are the powerful symbols (architectural objects), which can produce poetry or even inspire depending on the way they are handled.

1. The arrangement, spacing, location and type of species can be determined on the basis of availability of soil, water and sufficient sunlight.
2. Trees should be preferred over shrubs and herbs, many pollution scavenging plants are identified then can be plated.
3. The plantation practice should be changed from one tree to at least 3-4 trees in a single strip of load which will reduce water loss.
4. Monoculture and introduction of exotic trees should be minimized. Fruit trees and flowering trees should be preferred as they may provide shelter to birds and other creatures.
5. Cleaning of road side trees should be stated by VMSS as many places turn very ugly due to dust particles.
6. Plants which can tolerate ammonium nitrate are *Bambusa*, *Pithecelobium dulci* (Goras amli), mango, *Eucalyptus* and *Zizyphus* (bor). Plants with chloride resistance are Peepal, Kachnar, Shisham and margosa tree. Plants which can survive in more polluted atmosphere due to sulphur are *Bauhinia* (Kachnar), bamboo, shisham, *Ficus* and silver oak etc. In polluted areas of city such plants can be preferred.

33.6 Beauty of Parks and Gardens

Many of the gardens still look beautiful in spite of very poor designing is due to the fact that the raw materials of gardens – all living plants are intrinsically so lovely that they cannot look ugly, no matter how badly they are placed. Sayajibaug and Ajwa gardens are most beautiful gardens of the city. Beauty of Sayaji baug is due to flowers of *Bobax ceiba, Bauhinia variagata, Delonix regia, Spathodea campanulata, Taebebuia argentea, Couropita guianensis* etc. There is a need to develop biodiversity park or butterfly park. In comparison to a zoo, the birds and animals can be kept in an open area in a restricted area. Government is constructing a butterfly park away from city in place called Saputara. On the lines of Jurong Bird park of Singapore, we can think of developing a biodiversity park in the city. The development of such park will provide aesthetic beauty, knowledge to visitors about birds/animals and people will be motivated to protect them.

33.7 Which Trees should be Preferred for Road Side Plantation?

Urban trees are the powerful symbols (architectural objects), which can produce poetry or even inspire depending on the way they are handled.

1. The arrangement, spacing, location and type of species can be determined on the basis of availability of soil, water and sufficient sunlight.

2. Trees should be preferred over shrubs and herbs. The plantation practice should be changed from one tree to at least 3-4 trees in a single strip of load which will reduce water loss.

3. Monoculture and introduction of exotic trees should be minimized. Fruit trees and flowering trees should be preferred as they may provide shelter to birds and other creatures.

4. Cleaning of road side trees should be done time to time by VMSS as many places turn very ugly due to dust particles.

5. Plants which can tolerate ammonium nitrate are *Bambusa, Pithecelobium* (Goras amli), mango, *Eucalyptus* and *Zizyphus* (bor). Plants with chloride resistance are Peepal, Kachnar, Shisham and margosa tree. Plants which can survive in more polluted atmosphere due to sulphur include *Bauhinia* (Kachnar), bamboo, shisham, *Ficus* and silver oak etc. In polluted areas of city such plants should be preferred.

Acknowledgements

Thanks are due to Head, Department of Botany for providing lab facilities and to UGC for providing Financial support in form of DRS project.

References

Bhatt B. 2005 Noise pollution: Issues and implications. (Eds). Arya A. *et al.* In: Urban pollution issues and solutions. pp.195-202.

Bhanarkar, A.D., Rao, B.P.S., Gajghate, D.G. and Nema, P. 2005. Inventory of SO2 PM and toxic metals emissions from industrial sources in Greater Mumbai. *Atmospheric Environment*, 39(21): 3851-3864.

Harnam, D.S., 2001. Allergy: The immune system, and antiaging. In: Antiaging medical therapeutics Vol. 5. Pub by Am. Acad. of anti aging medicine, pp. 247-253.

2015, Impact of Global Warming and Climate Change on
    Human and Plant Health
Editors: Dr. Arun Arya and Prof. V.S. Patel
Published by: DAYA PUBLISHING HOUSE, NEW DELHI

Pages 352–363

Chapter 34

Silvopastoral Systems and their Contribution to Carbon Sequestration and Livestock Methane Emissions Reduction under Tropical Conditions

S. F. J. Solorio[1*], S.B. Solorio[1], S.K. Basu[2], L.F. Casanova[1],
V.J.C. Ku[1], P.C. Agilar[1], A.L. Ramírez[1], and B.A. Ayala[1]

[1]*Campus de Ciencias Biológicas y Agropecuarias.
Universidad Autónoma de Yucatán. Carretera Mérida-Xmatkuil Km.
15.5. C.P. 97100, Mérida, Yucatán, México*
[2]*Department of Biological Sciences, University of Lethbridge,
T1K 3M4 Lethbridge, Alberta, Canada*

ABSTRACT

Silvopastoral system is an efficient and integrated land use management system that has emerged as a valuable strategy to develop livestock systems. There is also, evidence that silvopastoral systems have an important role in providing environmental services; it involves interactions of woody perennial species with grasses or other crops and livestock. The objective of this review is to discuss the role of silvopastoral systems in providing environmental services, including more diverse and sustainable livestock (milk and meat production), increased carbon stocks, biodiversity conservation, improved soil fertility and

* *Corresponding author.* E-mail: ssolorio@uady.mx

atmospheric nitrogen fixation. The adoption of silvopastoral systems contributes to reduced carbon dioxide emissions, diminishes the pressure on vulnerable ecosystems and substantially improves forage quality and livestock production.

Keywords: *Carbon sequestration, Forage, Silvopastoral systems, Livestock, Methane, Monoculture, Nitrogen fixation, Soil fertility.*

List of Abbreviations

AFS: Agro forestry; GHG: Greeenhouse Gases; ISPS: Intensive Silvopastoral Systems

34.1 Introduction

Silvopastoral systems combine fodder plants such grasses and leguminous herbs with shrubs and trees for animal production and environmental services. The silvopastoral systems provide a diverse ecosystem services, they favor biodiversity by creating complex habitat that can support a wide variety of species of plants and animals (Castro, 2009; Moreno and Pulido, 2009). Under humid, tropical conditions, the silvopastoral systems can sequester more carbon and fix atmospheric nitrogen than traditional monocrop pasture systems (Nair *et al.*, 2009). The combination of grasses and trees contribute comprehensively to retain soil and water, there by supporting soil and watershed protection (Ibrahim *et al.*, 2006). Silvopastoral systems are an excellent way of reforestation and recuperation of degraded lands.

Evidence supporting these benefits have been gathered only recently (Murgueitio *et al.*, 2011). Although scientific reports supporting these benefits have increased noticeably within the last few decade, they have generally focused on single agroforestry ecosystem services; for instance, impacts on biodiversity conservation in tropical landscapes (Schroth *et al.*, 2004), soil fertility (Schroth and Sinclair, 2003) or potential carbon sequestration (Montagnini, 2006). Recently in Mexico, there has been an increasing interest towards intensive silvopastoral systems or ISPS. This is an important approach in recovering degraded pastures or dry tropical areas, increasing the availability and quality of fodder generated and holds great opportunities for economic value addition resulting from sustainable and healthy animal production (milk, cheese and meat) further contributing to environmental protection and sustainable ecosystem generation.

In this context, there have been significant advances in the understanding of the methodologies used for the silvopastoral systems benefitting integrated systems with joint ruminants and tree crop production, especially in association with citrus members, coconuts palm and mangoes among others. The main areas in which research have been currently undertaken include:

☆ Characterization of environmental and soil conditions within plantations.

☆ Assessment of forage availability and quality, as well as seasonality of production.

☆ Carbon sequestration and atmospheric nitrogen fixation.

☆ Measurement of animal performance for milk and meat production and quality.

☆ Measurement for animal behavior.

☆ Analyses of the economic benefits of the silvopastoral systems.

34.2 Intensive Silvopastoral Systems (ISPS)

ISPS refers to the integration of shrubs mainly leguminous plants at high densities in association with grasses and trees species for animal production. Fodder production and accessibility is improved by using high density *Leucaena* shrubs at narrow spacing (Figure 34.1). Rows are established about 1.6 m apart and the row spacing of shrubs varies from 20-30 cm. Ideally, rows are oriented along the contours in east-west direction. Once the *Leucaena* are well established, grass should be allowed to grow in the area between the rows.

Figure 34.1: ISPS Established with A. *Leucaena leucocephala* (Lam.) de Wit and B. *Panicum maximum* Jacq.

Figure 34.2: ISPS Examples. A. High density *L. leucocephala* and guinea grass (*P. maximum*) farm, Ejido la Concha.); B. Livestock for milk production, Huarinches farm, Tepalcatepec; C. Steer for meat production, Uricho farm, and D. *P. maximum* farm, La Concha municipality.

Contd...

Figure 34.2–*Contd...*

There must be a significant positive interactions between shrubs-trees and grasses in the system ecologically and economically. The integration of shrubs and tree crops or timber trees with pastures can increase soil fertility, reduce soil erosion, favor biodiversity and increase carbon capture and fixation of atmospheric nitrogen. Among ruminants, cattle are well suited to integration with tree crops such coconuts, citrus species and mango trees (Figure 34.2). Murgueitio *et al.* (2011) described intensive ISPS as an advanced form of agroforestry (AFS) for animal production that integrates fodder shrubs planted at high densities (>10,000 plants/ha), intercropped with improved, highly productive pastures and timber trees all combined in a systems that can be directly grazed by livestock.

34.3 Environmental Services from Agroforestry Systems

Carbon Sequestration and Methane Emissions

Carbon sequestration is the capture and storage of atmospheric carbon into carbon sinks (*e.g.* oceans, vegetation, or soils) through physical and biological processes (Ibrahim *et al.*, 2005). Incorporating trees and shrubs into ISPS will increase carbon sequestration, compared with other systems like monoculture pastures. Besides storing important amounts of carbon in above ground biomass, they can also store greater amounts of carbon in below ground biomass (Casanova *et al.*, 2010). However, this is an area that has not been addressed in Mexico, the expanding areas under tree crops will provide good opportunities for carbon sequestration trough more widespread use of leguminous shrubs in association with grasses including the improved forage management practices which will result in decreased carbon atmospheric emissions and global warming.

Casanova *et al.* (2011) has calculated that in silvopastoral systems (mixed fodder bank systems) the carbon sequestered per hectare was from 14.9 to 21.8 t C/ha/yr using a combination of leguminous with non-leguminous shrubs, root biomass is estimated to be 20-30 per cent of the aboveground shrubs carbon stock. This is an area has not been fully addressed in Mexico concern carbon sequestration, the expanding land areas under silvopastoral systems with tree crops provide good opportunities for carbon sequestration through more widespread use of grasses and legumes shrubs, with resultant decreased carbon atmospheric emissions and global warming.

On the other hand the wide, deep root systems of *L. leucocephala* (Figure 34.3) in the ISPS increases the available area for nutrient capture and help maintain nutrient stock by reducing leaching losses or by taking up nutrient from deeper soil layer

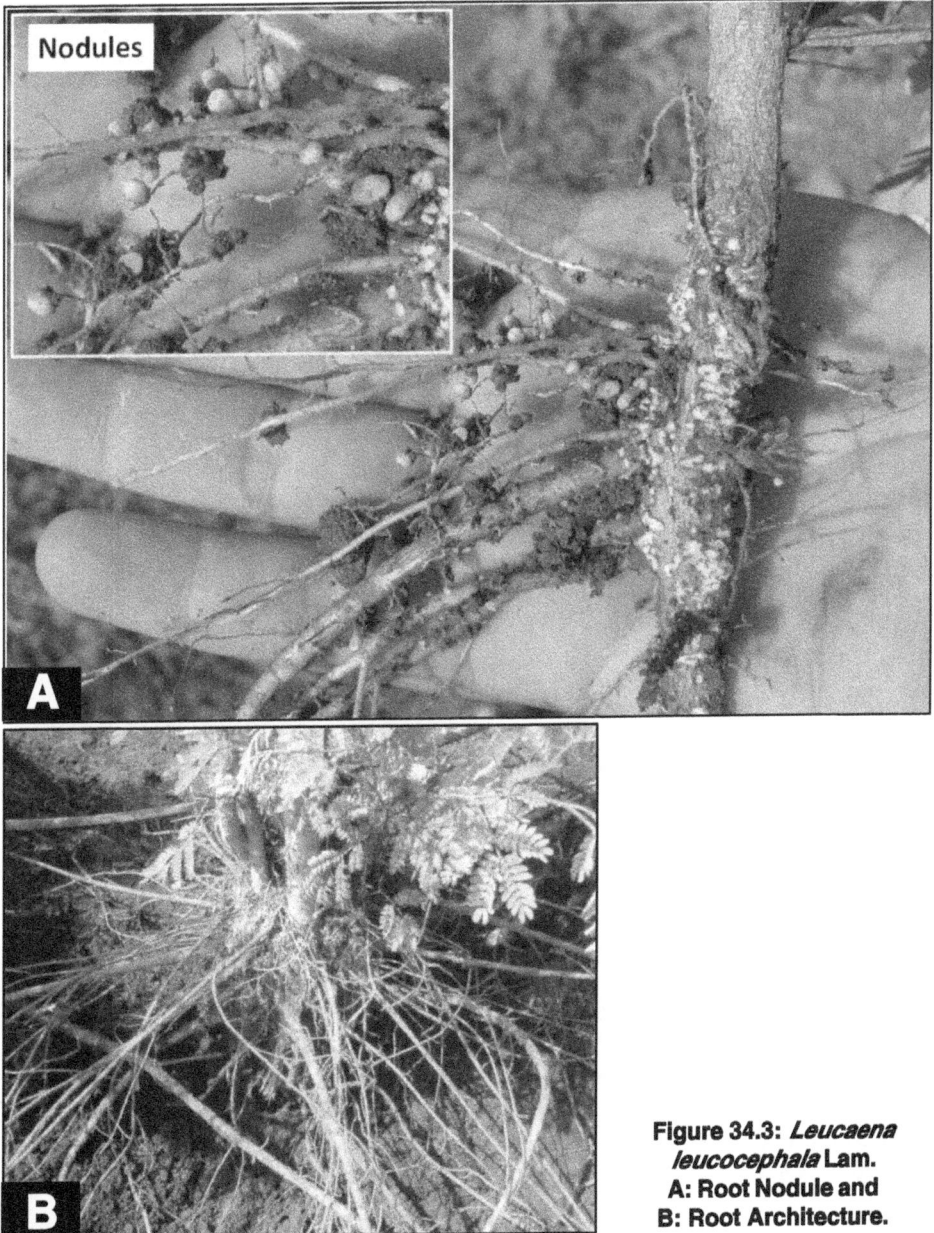

Figure 34.3: *Leucaena leucocephala* Lam. A: Root Nodule and B: Root Architecture.

including the C storage, (Casanova, 2012). In addition, organic matter is incorporated gradually into the soil, this help to improve soil stability, mineralization and availability of soil nutrients, (Petit, 2012).

Associated with above is the issue of Greeenhouse Gases (GHG) emissions mainly methane (CH_4) and its effects on climate change. Improved monocrop pastures with leguminous shrubs to feed grazing ruminants will have beneficial effect, the CH_4 yield tends to decrease as feed quality increases (Moss *et al.,* 2000). In response to possible effects on climate change, mitigation efforts have therefore concentrated on ways of reducing the GHG emissions in which some strategies to include enhance feed quality, supplemental with foliage and pods of tropical legumes may contributed to improvement of ruminant productive performance. The secondary metabolites (such as tannins and saponins) present in the pods of *Acacia pennatula* (Schlecht. and Cham.) Benth and *Enterolobium cyclocarpum* (Jacq.) Griseb. may contribute towards reducing CH_4 production in the rumen (Briceño-Poot *et al.,* 2012).

To reiterate, more widespread use of high quality grass-legume (such as the association of guinea grass with *L. leucocephala*) and improved forage management practices (grazing rotation, adequate stocking rate); including the introduction of best, available animal breeds (genotypes) for grazing under tropical conditions will be necessary. Recently, it has been reported that it is possible to reduce CH_4 production by 27 per cent in the rumen of sheep fed with saponin from tea leaves (Mao *et al.,* 2010). As millions of cattle graze low nutritive quality tropical pastures in Central and South America, contributing to global GHG emissions, there is scope to reduce enteric CH_4 production in grazing ruminants under ISPS based in the association of leguminous shrubs with tropical grasses. Thus, mitigation of CH_4 emissions in ruminants could reduce their contribution to GHG emission while improving feed efficiency for ruminants. However, better understanding is necessary for the relative GHG emissions from improved grass-legume pastures. Hence, strategies including the use of leguminous shrubs with the capacity to fix atmospheric nitrogen in order to reduce fertilizer-related emissions and for reducing GHG emissions from land use through better carbon sequestering, ISPS offer promising sustainable alternatives both for the local economy as well as environment.

34.4 Soil Fertility

It is indisputable that decreasing vegetation cover has caused a reduction in nutrient cycling and soil fertility (Iridiondo *et al.,* 1998). Livestock production is frequently referred to as a major driver of tropical deforestation. Mexico is a representative example within the regional context. The estimates for the rate of deforestation in Mexico ranged from 400,000-1,500,000 ha/year and there is much more deforestation in the tropical region than in the temperate areas. The dominant impact of forest conversion has been made by the rapid expansion of livestock production and the increased demand for pasture land, particularly in tropical areas (Barbier and Burgess, 1996). Forest ecosystems are closed and efficient systems (Petit *et al.,* 2009). They have high rates of return and low rates of losses, and are thus self-sustaining. On the other hand, many conventional agroecosystems (*e.g.,* monocrops) are open or permeable, with relatively low rates of return and high rates of losses (Nair, 1993).

Agroforestry (AFS) is positioned between these two extremes, with more efficient nutrient cycling than conventional agricultural systems and similar productivity to forest ecosystems. Nair (1993) claimed that the difference between AFS and other agricultural land use practices lies in the transference or recovery of nutrients into the system from one component to another and the possibility of managing the system or its components to increase nutrient recycling rates without affecting total productivity. The incorporation of shrubs and trees in the silvopastoral systems increases soil fertility, and improves soil structure. Trees have deep root systems which serve as an underground net through which nutrients can be captured from deep within the soil profile. These nutrients are returned to the soil via leaf litter, increasing the nutrient recycling efficiency of the system (van Noordwijk *et al.*, 1996; Allen *et al.*, 2004).

The incorporation of *L. leucocephala* in pastures led to an increase in the content of soil nutrient such as nitrogen (N), phosphorus (P) and carbon (C). These results have been explained by the amount of good quality litter decomposition by tropical shrubs in silvopastoral systems. *Moringa* shrubs (M) residue decomposed significantly faster than the other shrubs followed closely by the combination of *Leucaena + Moringa*, (Figure 34.4) *Leucaena* leaves released their organic matters at intermediate rates. These differences in organic matters released have grave importance with respect to synchronization with nutrients demand by either crops or grass in silvopastoral systems.

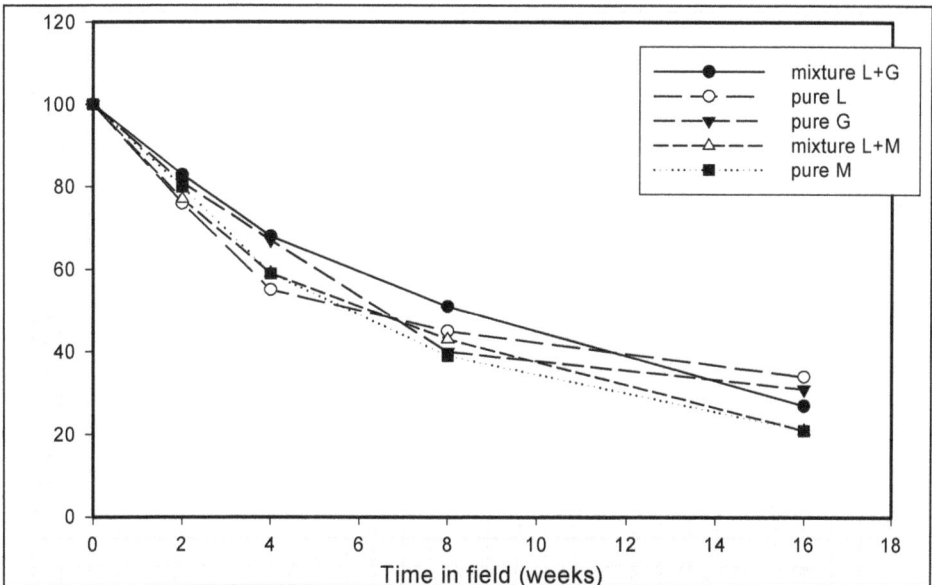

Figure 34.4: Litter Decomposition from different Tropical Shrubs Species under Silvopastoral Systems Arrangement. *Guazuma ulmifolia* Lam. (G), *L. leucocephala* (L), *Moringa oleifera* Lam. (M) and the combinations of *Leucaena* with *Guazuma* (L + G) and *Moringa* with *Leucaena* (L + M).

Additionally, most shrubs and trees used in silvopastoral systems are legumes that have the capacity to fix nitrogen through the association of bacteria living in the roots nodules. These bacteria can change inert N_2 to biologically useful NH_3, which is then converted to protein in the plant. Under ISPS, *L. leucocephala* provide the main input of nitrogen for pastures. Furthermore, there is a reduction in the use of nitrogen fertilizers in pastures and have the extra benefit of improving feed quality for grazing animal. Bacab *et al.* (2012) found high productivity of forages without using fertilizer with the introduction of high density ISPS using *L. leucoephala* in combination with guinea grass (Table 34.1).

Table 34.1: Fodder Production under ISPS and Monocrop Based Pasture

Indicator	ISPS (L. leucocephala + P. maximum)	P. maximum
Fodder production (t DM/ha/) dry period	14.5	5.0
Fodder production (t DM/ha/) wet period	24.2	7.0
Crude protein production (kg/ha)	2100-3500	360 -600

Source: Modified from Bacab *et al.* (2012).

34.5 ISPS for Cattle Production: Fodder Availability and Quality

Ruminants raised in the tropics of Mexico largely depend on seasonal grass feed resources which are relatively low in quality in terms of low crude protein. The introduction of *L. leucocephala* in pastures improves the forage quality of associated grasses compared to grasses under monocrop system. In one study carried out by Bacab *et al.* (2011) in the Tepalcatepec Valley, Michoacán, Mexico, an ISPS of *L. leucocephala* in association with *P. maximum* cv Tanzania showed a forage production of 8.2 t DM/ha/year without fertilization compared to 3.2 t DM/ha/year under traditional pasture system. These results demonstrate the great potential of ISPS to produce high quality biomass and iradicate problems of fodder scarcity. Bacab-Perez and Solorio-Sánchez, (2011) evaluated the animal performance (milk production and daily weight gain) for cattle grazing under ISPS. In all these cases, the stocking rate and milk productivity was reported increased. The increase in acerage of land use with high density ISPS *L. leucocephala* with *P. maximum* has led to an increase in the stocking rate and milk production substantially (Table 34.2). Silvopastoral systems can remain productive for longer periods than conventional pastures, thus reducing the pressure to clear more forests for agricultural purposes (Steinfield *et al.,* 2006).

In case of ISPS, cattle grazing under the shade of trees suffer less heat stress than in open pastures (Table 34.3). The animals graze more efficiently and have lower respiratory rates, thereby producing more milk and meat. In addition, the combination of trees, shrubs and grasses help to retain and use water more efficiently and help in proper soil and nutrient cycling. All the above conditions provide suitable habitat for insects and other litter decomposers that can quickly recycle the nutrients and for beneficial insects (predator and parasitoids) that control harmful insect in a biological control system without the application of expensive and non-environment friendly chemical pesticides (Murgueitio *et al.*, 2011).

Table 34.2: Animal Parameters under ISPS and Traditional Monoculture System

Parameters	SSPi (LI* + Pm**)	Traditional System (Pm monoculture)
Milk production (L/cow/day)	8.0	3.5
Weight gain (g/día)	900	500
Stocking (AU/ha/year)	4.0	2.0

* Ll: *L. leucocephala* cv Cunningham; ** Pm: *P. maximum* cv. Tanzania.

Source: Bacab-Perez and Solorio-Sánchez (2011)

Table 34.3: Environmental Indicators under Traditional Monoculture System and ISPS in the Apatzingan Valley, Michoacán

Indicators	Traditional System (Grass monoculture)	ISPS
Environmental temperature (°C)	34-38	30-34
Nutrient recycling (kg/ha N-P-K).	Less 15, 6, 17	More 22, 4, 2
Efficiency of water (per cent)	30	80-90
Organic Matter (kg/ha)	320	1000
N fixation (kg/ha/year)	0	300-500
Carbon storage (t/ha/year)	120	220

34.6 Conclusions

Silvopastoral systems are very important and implementation of silvopastoral systems on cattle farms has resulted in significant improvements in livestock productivity and environmental sustainability. Well managed silvopastoral systems increase biological diversity, soil and biomass carbon sequestration; as well as increase the capacity to fix atmospheric nitrogen more efficiently. The use of leguminous shrubs in combinations with grasses can replace the use of nitrogen fertilizer for sustaining pasture yields and reducing the impact on the local ecosystem and environment. A combination of clear polices and increased technology diffusion of silvopastoral systems can further accelerate the adoption of this innovative and sustainable approach among farmers in developing and under developed countries. Future research and development efforts are still expanding with enhanced emphasis on environmental protection and sustainability. Grazing management and socio-economic analysis on the adoption of this approach on local rural economy level will be important to properly asses the success of this approach under low input agricultural systems. Further research and developments will also be necessary under different dry tropical agro-ecosystems as well as in the proper and scientific integration of livestock management with tree crop productions for making ISPS a viable solution for dry and tropical agriculture.

Acknowledgments

We are grateful to the National Council of Science and Technology (Fordecyt-CONACyT) for their generous support, to Fundacion Produce Michoacan funded our research.

References

Bacab, H. M.; Solorio, F. J. y Solorio, S. B. 2012. Efecto de la altura de poda en Leucaena leucocephala y su influencia en el rebrote y rendimiento de Panicum maximum. *Avances en Investigación Agropecuaria* 16(1): 65-77.

Bacab-Perez, H. M., and Solorio-Sanchez, F. J. 2011. Forage offer and intake and milk production in dual purpose cattle managed under silvopastoral systems in Tepalcatepec, Michoacán. *Tropical and Subtropical Agroecosystems*, (13): 271-278.

Briceño-Poot, E. G.; Ruiz-González, A.; Chay-Canul, A. J.; Ayala-Burgos, A. J.; Aguilar-Pérez, C. F.; Solorio-Sánchez, F. J.; Ku-Vera, J. C. 2012. Voluntary intake, apparent digestibility and prediction of methane production by rumen stoichiometry in sheep fed pods of tropical legumes. *Animal Feed Science and Technology* 176: 117-122.

Casanova LF, Caamal MJ, Petit AJ, Solorio SF, Castillo CJ 2010. Acumulación de carbono en la biomasa de *Leucaena leucocephala* y *Guazuma ulmifolia* asociadas y en monocultivo. *Revista Forestal Venezolana* 54(1) 45-50.

Casanova-Lugo, F.; Caamal-Maldonado, J.; Petit-Aldana, J.; Solorio-Sánchez, F.; Castillo-Caamal, J. 2010. Acumulación de carbono en la biomasa de Leucaena leucocephala y Guazuma ulmifolia asociadas y en monocultivo. *Revista Forestal Venezolana* 54(1): 45-50.

Ibrahim M, Villanueva C, Casasola F, Rojas J 2006. Sistemas silvopastoriles como una herramienta para el mejoramiento de la productividad y restauración de la integridad ecológica de paisajes ganaderos. *Pastos y Forrajes* 29(4): 383-40.

Ibrahim M, Villanueva C, Mora J 2005. Traditional and Improved Silvopastoral Systems and their Importance in Sustainability of Livestock Farms. In: Silvopastoralism and sustainable land management. Mosquera-Losada MR, McAdam J. Rigueiro-Rodriguez A (eds.). CABI publishing.

Iriondo E, Álvarez E, Chinea A, Barroto D 1998. Experiencias campesinas sobre la utilización de árboles y arbustos en huertos caseros. III Taller Internacional Silvopastoril "Los Árboles y Arbustos en la Ganadería". EEPF "Indio Huatuey", Matanzas, Cuba.

Moss, A.R. Jouany, J.P. and Newbold, J. 2000. Methane production by ruminants: its contribution to global warming. *Ann. Zootech.* 49:231-253.

Murgueitio, E.; Calle, Z.; Uribe, F.; Calle, A.; Solorio, B. 2011. Native trees and shrubs for the productive rehabilitation of tropical cattle ranching lands. *Forest Ecology and Management* 261: 1654-1663.

Nair PKR, Kumar BM, Nair VD 2009. Agroforestry as a strategy for carbon sequestration. *J Plant Nutr Soil Sci* 172: 10-23.

Petit AJ, Casanova LF, Solorio SF 2009. Asociación de especies arbóreas forrajeras para mejorar la productividad y el reciclaje de nutrimentos. *Agric Téc Méx* 35(1): 113-122.

Petit-Aldana, J.; Uribe-Valle, G.; Casanova-Lugo, F.; Solorio-Sánchez, F. J.; Ramírez-Avilés, L. 2012. Descomposición y liberación de nitrógeno y materia orgánica en hojas de Leucaena leucocephala (Lam.) de Wit, Guazuma ulmifolia Lam. y Moringa oleifera Lam. en un banco mixto de forraje. *Revista Chapingo Serie Ciencias Forestales y del Ambiente* 18(1): 5-25.

Schroth G, da Fonseca GA, Harvey CA, Gascon C, Vasconcelos H, Izac AN 2004. Agroforestry and biodiversity conservation in tropical landscapes. Island Press, Washington, DC.

Schroth G, Sinclair F 2003. Trees crops and soil fertility: concepts and research methods. CABI, Wallingford, UK.

Segura-Rosel, A.; Casanova-Lugo, F.; Solorio-Sánchez, F. J.; Chay-Canul, A. J. 2012. Asociación de especies leñosas en bancos de forraje: influencia sobre el aporte de hojarasca, descomposición y liberación de nitrógeno. *Tropical and Subtropical Agroecosyst. J.*

2015, Impact of Global Warming and Climate Change on
        Human and Plant Health
Editors: Dr. Arun Arya and Prof. V.S. Patel
Published by: DAYA PUBLISHING HOUSE, NEW DELHI

Pages 364–374

Chapter 35

Effect of Air Pollution on Physiological and Biochemical Activities of Medicinally Important Neem Tree (*Azadirachta indica* A. Juss.)

Deepika K. Chandawat* and H.A. Solanki**

*Department of Botany, Gujarat University,
Ahmedabad – 380 009, Gujarat, India*

ABSTRACT

Even though, automobiles are the lifeline for the city dwellers. They serve as one of the major source of pollution in Indian metro cities and big towns. Vehicular pollution generally accounts for 60-70 per cent of total pollution loads of any city. To improve our environment and quality of life trees are on the job every day working for all of us by absorbing carbon dioxide and releasing oxygen in the presence of light through the phenomenon of photosynthesis.

Ahmedabad, a mega city of Gujarat, is continuously losing its grace and beauty under the intense densification of activities. The air is being continuously polluted in urban areas due to heavy traffic, industry, domestic fuel combustion; coal based thermal power plants and various agricultural activities from the adjoining areas.

E-mail: *deepikachandawat@gmail.com; **husolanki@yahoo.com

Considering the above facts the present investigation was done to find the effect of air pollution on medicinally important neem (*Azadirachta indica* A. Juss.) tree present at the various cross-roads of Ahmedabad city. Dust fall, pigment content *i.e.* total chlorophyll, chlorophyll a and chlorophyll b and Starch content were tested seasonally. Coefficient of correlation of dust fall with pigment content and starch were also analyzed. Tree at polluted sites shows higher dust on leaf surface, lower pigment and starch content in comparison to control site.

Keywords: *Cross-roads, Azadirachta indica, Dustfall, Chlorophyll pigments and starch.*

35.1 Introduction

India and other developing countries have experienced a progressive degradation in air quality due to industrialization, urbanization, lack of awareness, number of motor vehicles, use of fuels with poor environmental performance, badly maintained poor roads and ineffective environmental regulations (Joshi and Chauhan, 2008). In urban areas – both developing and developed countries, it is predominately mobile or vehicular pollution that contributes to air quality problem. The worst thing about vehicular pollution is that it cannot be avoided as the vehicular emissions are emitted at the near-ground level where we breathe. Motor vehicles are responsible for 60 to 70 per cent of the pollution found in an urban environment (Singh *et al.*, 1995).

The city of Ahmedabad has seen a rapid growth in two wheeler population in the last two decades, which has also resulted in rising pollution levels in the city. The root cause of air pollution in Ahmedabad is the two-stroke two wheelers and auto rickshaws, which contribute to the pollution load. The city has been identified as one of worst with regard to Air pollution by Honourable Supreme Court Committee. The permissible SPM and RSPM levels as per National Ambient Air Quality Standards (NAAQS) have been consistently high for the past decade in industrial and residential areas.

Among ancient civilizations, India has been known to be rich repository of medicinal plants. In Indian culture neem has been referred as an "air purifier" so it may be an avenue tree of choice in thickly populated areas, by its capacity to survive in adverse conditions, and absorb some of the environmental pollutants, and act as an "air freshener" by releasing oxygen and mild odor principle. Also it has medicinal values. Various parts of this plant are used to cure diseases such as headache, fever, chickenpox, intestinal problems, malaria, skin diseases and to purify blood. As this plant have versatile values so is variously known as "Sacred Tree," "Heal All," "Nature's Drugstore," "Village Pharmacy" and "Panacea for all diseases" in India [2].

Considering the medicinal value and role in abatement of air pollution, this study was carried out to assess the impact of air pollution on some biochemical parameters of neem tree (*Azadirachta indica* A. Juss.) growing at various cross-roads of Ahmedabad city. The chosen determinants were Dustfall, leaf pigment (total chlorophyll, chlorophyll *a*, and chlorophyll *b*) content and starch content.

35.2 Materials and Methods

Study Sites and Sampling Procedure

The present study was carried out at seven different polluted cross-roads of Ahmedabad city. At the height of two to three meters, fully expanded mature leaves were collected in the polythene bags and transported to the laboratory. Leaves of neem tree were also collected from Gujarat University, Campus which is used as control area. The leaf samples were collected on seasonal basis and this frequency was strictly maintained throughout the year (2009-2010). Table 35.1 gives the description of sites selected for the study.

Table 35.1: Description of Sites Selected

	Name of the Site	Location in the City	Characteristics of Site
Site-1	Power house	Northern region	Coal based thermal power emission region, light and heavy vehicles, vehicle density is less.
Site-2	Paldi	Western region	Heavy and light vehicles, frequent traffic jams, vehicle density is more.
Site-3	Lal-Darwaja	Central region	Market area, frequent congestion, traffic jams, light vehicles density is more.
Site-4	S.T bus stand	Central region	Heavy and light vehicles, frequent traffic jams, vehicle density is more.
Site-5	Naroda	North eastern region	Industrial area, Heavy and light vehicles, density is more.
Site-6	Railway station	Central region	Railway tract, light vehicles, vehicle density is more.
Site-7	Residential area	Northern region	Light vehicles, no traffic jams, vehicle density is less, open area.
Site-8	Control	Western region	Light vehicles, low- polluted area, open area.

Dust fall: Ten fully matured leaves were collected from all the plants at all the sites. The dust deposition on leaf surface was calculated by dry technique (Das and Pattanayak, 1977).

In dry technique, first the intact leaf was weighted (in mg) then dust particulates from leaf surfaces were gently collected with the help of camel brushes and the weight of leaf was measured again. The amount of dust deposition in mg/cm^{-2} was calculated.

$$\text{Dust content (mg/cm}^2) = \frac{\text{Wt. of intact leaf} - \text{initial wt. of leaf}}{\text{Total surface area of leaf (cm}^2)}$$

Chlorophyll pigments were estimated following the method of Arnon (1949). Pigment contents were calculated by following method.

Chlorophyll a = [(12.7 X OD at 663) – (2.69 X OD at 645)] X dilution factor

Chlorophyll b = [(22.9 X OD at 645) – (4.68 X OD at 663)] X dilution factor

Total chlorophyll = [(20.2 X OD at 645) – (8.02 X OD at 663)] X dilution factor.

Starch content was estimated following the method of Chinoy (1939).

35.3 Results and Discussion

Dust- Fall

Figure 35.1 shows dust fall on the leaves of *A. indica*, growing in polluted area was significantly high compared to those growing in the controlled area during summer and winter except at site-7 (Residential area) which did not show any significant result. In rainy season except at site-4 (ST bus stand) and site-5 (Naroda) no significant result was observed at any site Figure 35.1. Maximum dust load on leaf surface was observed in summer at site-5 (Naroda, 0.54mg/cm^2) and minimum in monsoon at site-1 (Power house, 0.03mg/cm^2). At all the sites maximum dust load was observed in summer season except at Paldi which shows maximum dust deposition in winter.

Analysis of the present investigation shows that in all the three seasons dust fall on the leaves of all the plants under study, was observed very high in polluted areas, which was due to more pollutants releasing through industries, congested market area, and traffic activities while in controlled area and low polluted region *i.e.* Residential area dust particles settled down on leaves generally come from the surrounding soils due to high wind speed. Same result of high dust deposition on the leaf surface in urban and industrial area have been reported by Rao and Pal (1979) and Shetye and Chaphekar (1980). They concluded that high dust deposition on leaf surface at road side with heavy vehicular traffic may be due to spray of unburnt oil residue of diesel or petrol on the leaf surface.

Bhatnagar *et al.* (1985) reported very high dust fall on the leaves of all the nine plants under study growing in industrial in comparison to those growing in non-industrial area. Varshney and Mitra (1993) concluded that the row of roadside hedges trapped nearly 40 per cent of particulate matter, most of which arises from the traffic

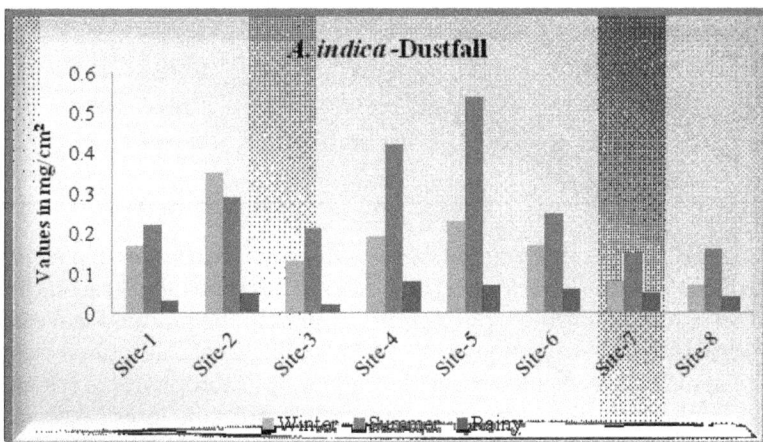

Figure 35.1: Total Dust Accumulated in the Leaves of *A. indica*.

movement. Finally we can conclude that high dust fall on the leaves of plants at polluted sites might be due to dense traffic movement, industries, power plant and congested infrastructure and market area.

Seasonal distribution shows maximum dust is in summer/winter and minimum in rainy. Higher concentration of dust in summer, high wind speed may be the reason for the relatively higher dust accumulation in the summer than in winter. In summer entire leaf was covered with the vehicular and industrial dust. The high dust accumulation in the winter season may be due to wet surfaces of leaves which help in capturing dust, with a gentle breeze and foggy condition preventing particulate dispersion. In the rainy season the least dust accumulation is reported because of washing of leaves and settling of particulates due to rain. Similar result of higher dust accumulation in summer and least in winter was obtained by Saha and Padhy, (2011) *Shorea robusta* and *Madhuca indica* foliage in Lalpahari forest.

35.4 Chlorophyll Pigments

The composition of chlorophyll pigments in the leaves of *A. indica* growing in all the areas has been shown in Figures 35.2a-c. All the three pigments were maximum in rainy followed by summer and least in winter season. As compared to the site-8 (control area), at all the sites pigment contents (total chlorophyll, chlorophyll *a*, and chlorophyll *b*), decreased significantly. Maximum percent reduction was found at site-5 while minimum at site-7 (Table 35.2).

Table 35.2: Per cent Reduction of Chlorophyll Pigment and Starch

	Total Chlorophyll		Chlorophyll a		Chlorophyll b		Starch	
	Mean Value	Per cent Reduction	Mean Value	Per cent Reduction	Mean Value	Per cent Reduction	Mean Value	Per cent Reduction
Site-8 (Control)	2.12		1.79		1.40		67.26	
Site-1	1.60	24.52	1.49	16.75	1.00	28.57	62.74	6.69
Site-2	1.37	35.37	0.91	49.16	0.50	64.28	61.35	8.77
Site-3	1.74	17.92	1.43	20.11	0.89	36.42	60.07	10.71
Site-4	1.28	39.62	0.93	48.04	0.54	61.42	58.47	13.09
Site-5	0.87	58.72	0.69	61.45	0.35	75	48.16	28.42
Site-6	0.99	53.16	0.71	60.33	0.41	70.71	55.36	17.7
Site-7	1.95	8.01	1.50	16.2	1.12	20	63.46	5.65

Bhatnagar *et al.* (1985) concluded that less chlorophyll in leaves of plants growing in polluted area was due to toxic effect of industrial dust and other gaseous pollutants on leaf. The reduction in chlorophyll concentration in the polluted leaves could be due to chloroplast damage (Pandey *et al.,* 1991), inhibition of chlorophyll biosynthesis (Esmat 1993) or enhanced chlorophyll degradation. Chlorophyll *a* is assumed to be degraded to phaeophytin, whereas chlorophyll *b* molecule loses its phytol group (Rao and Le Blanc 1966).

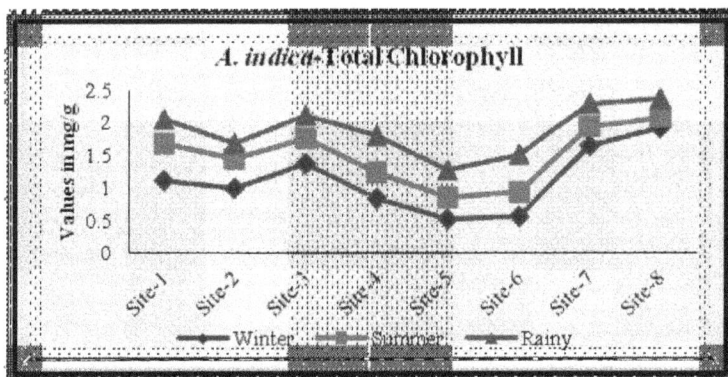

Figure 35.2a: Total Chlorophyll in the Leaves of *A. indica.*

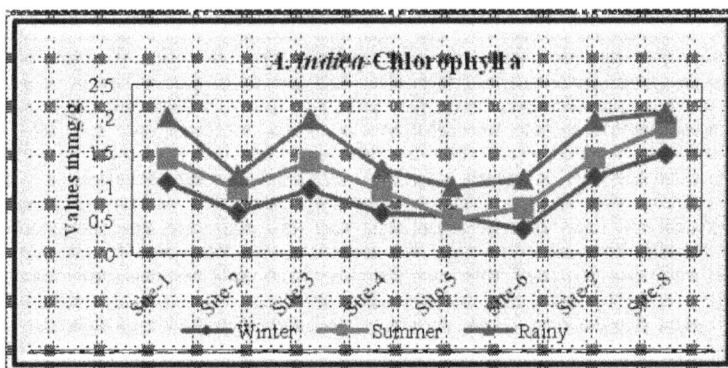

Figure 35.2b: Chlorophyll a in the Leaves of *A. indica.*

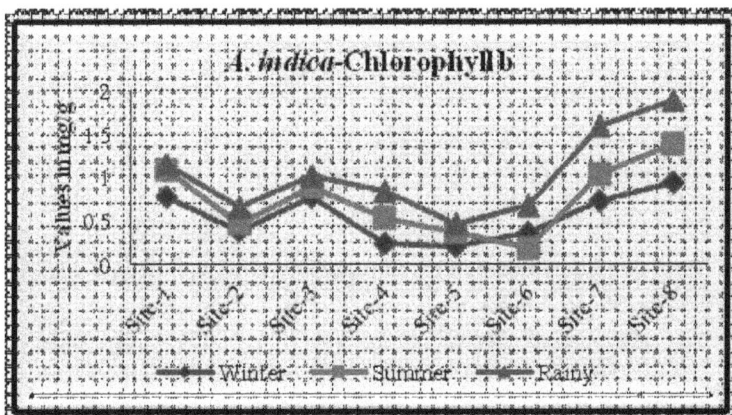

Figure 35.2c: Presence of Chlorophyll b in the Leaves of *A. indica.*

Less chlorophyll contents in the leaves of all the plants growing at the cross-roads of polluted region may be due to long term exposure of these plants to pollutants like SO_2, NO_2, and SPM. The synergistic effects of these pollutants caused foliar injury *i.e.* chlorosis and necrosis, which degrade the chlorophyll pigments. The same view has been reported by Rao and Le Black (1966), Puckett *et al.* (1973) and Malhotra (1976). The shading effects due to deposition of suspended particulate matter on the leaf surface might be responsible for this decrease in the concentration of chlorophyll in polluted area. It might clog the stomata thus interfering with the gaseous exchange, which leads to increase in leaf temperature which may consequently retard chlorophyll synthesis. Dusted or encrusted leaf surface is responsible for reduced photosynthesis and thereby causing reduction in chlorophyll content (Joshi and Swami, 2009).

Chlorophyll content was higher in monsoon season which might be due to the washout of dust particles from the leaf surface which will increase photosynthetic activity. Low chlorophyll content in winter season might be due to the high pollution level, temperature stress, low sunlight intensity, senescence period of plant and short photoperiod. Concentration of dust load on the leaves was high in polluted area in comparison to that of less polluted area. Numerous chemical reactions occur on the leaf surface in the presence of moisture which alters the composition of photosynthetic pigments. Toxic particulates are deposited on the surface of plant foliage and must either dissolve the cuticle or move through open stomata in a solution to produce acute injury. Taylor (1973), Singh and Rao (1978 a, b, 1981), Prasad *et al.* (1979) and Prasad and Rao (1981) reported a decrease of chlorophyll contents due to dust pollution. Similar study of seasonal variation of pigment content in plant species exposed to urban particulates pollutants was studied by Prajapati and Tripathi (2006), at Varanasi. They also found higher content of chlorophyll in monsoon season.

35.5 Starch

The starch content in the leaves of *A. indica* growing at all the sites has been shown in the above Figure 35.3. As compared to the control area and low polluted area, at all the sites starch, decreased significantly, in different seasons. Starch content

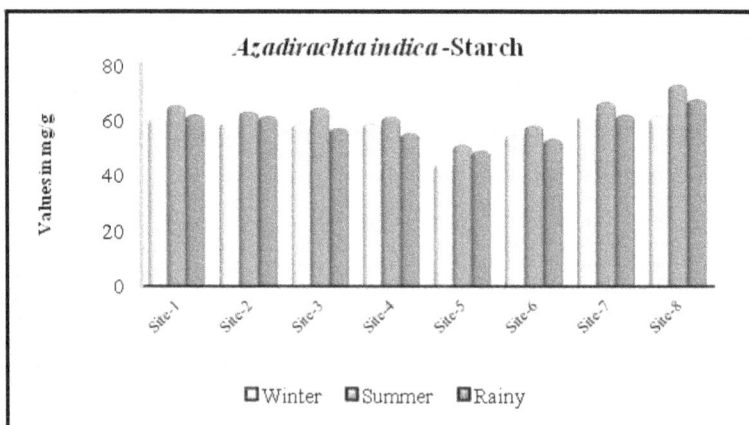

Figure 35.3: Starch in the Leaves of *A. Indica.*
Site 1: Powerhouse; Site 2: Paldi; Site 3: Laldarwaja; Site 4: ST busstand;
Site 5: Naroda; Site 6: Railway station; Site 7: Residential area; Site 8: Control.

was highest at all the sites in summer season followed by winter/rainy season. At site-7 (Residential area) no change was observed in winter and rainy. Maximum percent reduction was found at site-5 and minimum at site-7 (Table 35.2).

Table 35.3: Coefficient of Correlation

Winter	DF	TC	CA	CB	SH
DF	1	−0.639	−0.663	−0.682	−0.342
TC	−0.639	1	0.941**	0.886**	0.721*
CA	−0.663	0.941**	1	0.910**	0.595
CB	−0.682	0.886**	0.910**	1	0.663
SH	−0.342	0.721*	0.595	0.663	1

Table 35.4: Coefficient of Correlation

Summer	DF	TC	CA	CB	SH
DF	1	−0.804*	−0.793*	−0.661	−0.841**
TC	−0.804*	1	0.969**	0.928**	0.937**
CA	−0.793*	0.969**	1	0.965**	0.956**
CB	−0.661	0.928**	0.965**	1	0.859**
SH	−0.841**	0.937**	0.956**	0.859**	1

Table 35.5: Coefficient of Correlation

Rainy	DF	TC	CA	CB	SH
DF	1	−0.617	−0.779*	−0.461	−0.560
TC	−0.617	1	0.937**	0.933**	0.842**
CA	−0.779*	0.937**	1	0.859**	0.737*
CB	−0.461	0.933**	0.859**	1	0.831*
SH	−0.560	0.842**	0.737*	0.831*	1

*: Correlation is significant at the 0.05 level (2-tailed).

**: Correlation is significant at the 0.01 level (2-tailed).

DF: Dustfall; TC: Total chlorophyll; CA: Chlorophyll a; CB: Chlorophyll b; SH: Starch.

Reduction in starch content at polluted sites can be attributed to increased respiration and decreased CO_2 fixation because of chlorophyll deterioration. It has been mentioned that pollutants like SO_2, NO_2 and H_2S under hardening conditions can cause more depletion of soluble sugars in the leaves of plants grown in polluted area. The reaction of sulfite with aldehydes and ketoses of carbohydrates can also cause reduction in carbohydrate content (Tripathi and Gautam, 2007).

In the present investigation higher amount of starch was found in the summer season and lower in the winter this might be because of temperature effect. Since, at higher temperature respiration increased, starch is converted to glucose. Therefore as an adaptation may be higher starch is synthesized. Similar conclusion was made by Worley (1937) that the relative amount of starch decreases with a decrease in temperature and increases again with a return to higher temperatures.

35.6 Statistical Analysis

The Pearson correlation coefficient values of dust fall with total chlorophyll, chlorophyll *a*, chlorophyll *b*, and starch content of *A. indica* L. leaves in different seasons are presented in Tables 35.3–35.5. Significant negative correlation of dust fall was found with total chlorophyll, chlorophyll a, and starch during summer season and with chlorophyll a in rainy season.

35.7 Conclusion

It is evident from the above discussion that trees of *Azadirachta indica* present at various cross-roads suffers from air pollutants such as RSPM, SPM, SO_2 and NO_2, CO_2, HC etc released from various sources such as automobile exhaust, industries power plant, Railways etc. Significant reduction was obtained in pigment content and starch content also trees of polluted sites shows higher dust accumulation. Still they are flourishing well hence can be used as bio-marker and litigators of urban environment.

Acknowledgements

We take this opportunity to express my sincere thanks to our head and soul of the Botany Department, Prof. Dr. Y.T. Jasrai. He encourages us with all the supports including that of infrastructural facilities. We thank him from the depth of our heart for all that he has done to guide me.

References

Arnon, D.I. 1949. Coenzyme in isolated chloroplast. Polyphenol oxidase in *Beta vulgaris. Plant Physiology,* **24**: 1 - 15.

Bhatnagar, A.R., Vora, A.B. and Patel, T. S. 1985. Measurement of dust fall on leaves in Ahmedabad and its effect on chlorophyll. *Indian. J. Air Poll. Cont.* 6(2): 77 – 90.

Chinoy, J.J. 1939. A new method for the determination of starch to soluble starch, natural starches and flour. Part 1. Colorimetric determination of soluble starch. *Microchemie.* 26: p.132.

Das, T.N. and Pattanayak, P. 1977. The nature and pattern of deposition of air borne particles on leaf surface of plants. *Proc. Seminar on Afforestation*, Inst. of P.H.P. pp 56 - 62.

Esmat, A.S. 1993. Damage to plants due to industrial pollution and their use as bio indicators in Egypt. *Environmental Pollution,* 81: 251 – 255.

Joshi, P.C., and Swami, A. 2009. Air pollution induced changes in the photosynthetic pigments of selected plant species. *J. Environ. Biol.*, 30: 295-298.

Joshi and Chauhan, A. 2008. Performance of locally grown rice plants (*Oryza sativa* L.) exposed to air pollutants in a rapidly growing industrial area of district Haridwar, Uttarakhand, India. *Life Science Journal,* 5(3): 41 - 45.

Malhotra, S.S., 1976. Effect of sulphur dioxide on biochemical activity and ultrastructural organization of pine needle chloroplasts. *ibid,* **76**: 239 – 246.

Pandey, D.D., Sinha, C.S., and Tiwari, M.G. 1991. Impact of coal dust pollution on biomass, chlorophyll and grain characteristics of rice. *Journal of Biology,* **3**: 51 – 55.

Prasad, B.J., Rao, D.N., Singh, N. and Pandey, S. N 1979. A survey of Polygenic effects on vegetation around a petroleum refinery. *In: Proc. Sym. Environ. Biol,* 212 - 216.

Singh N. and Rao, D.N. 1981. Effects of SO_2 exposure on carbohydrate contents, phytomass and caloric values of wheat plants. *Water, Air and Soil pollution.* **16**: 287 - 291.

Prajapati, S.K., and Tripathi, B.D. 2006. Seasonal Variation of Leaf Dust Accumulation and Pigment Content in Plant Species Exposed to Urban Particulates Pollution. *Journal of Environmental Quality.* 37(3): 865 - 870.

Puckett, K.J., Nieboer, E., Flora, W.P., and Richardson, D.H.S. 1973. Sulphur dioxide: Its effect on photosynthetic 14C fixation in lichens and suggested mechanism of phytotoxicity. *The New Phytologist,* 72: 141

Rao, D.N. and Leblanc., F. 1966. Effects of SO2 on the lichens alga with special reference to chlorophyll. *Bryologist,* 70: 69 - 75.

Rao, D.N. and Pal, D. 1979. The effects of fluoride pollution on cattle. In: *Symposium volume on environmental pollution and toxicology,* Today and Tomorrows Publication, New Delhi, pp. 281 - 290.

Saha, D.C. and Padhy, P.K. 2011. Effects of stone crushing industry on *Shorea robusta* and *Madhuca indica* foliage in Lalpahari forest. *Atmospheric Pollution Research,* doi: 10.5094/APR.2011.053.

Shetye, R.P. and Chaphekar, S.B. 1980. Some estimation on dust fall in the city of Bombay, using Plants. *Proc. Seminar on progress in Ecology,* Today and Tomorrows Publication, New Delhi. **4**: 61 - 70.

Singh, H.B. 2004. Urban transport and vehicular pollution in Ahmedabad, Gujarat. *T P I JOURNAL,* **1(1)**: 45 - 49.

Singh, N., Yunus, M., Srivastava, K., Singh, S.N., Pandey, V., Misra, J. and Ahmad, K.J. 1995. Monitoring of auto exhaust pollution by road side plants. *Environmental Monitoring and Assessment.* 34: 13 - 25.

Singh, S.N. and Rao, D.N. 1978a. Possibilities of using chlorophyll and potassium contents in plants to detect cement dust pollution. *J. IPNE,* India. **1**: 10 - 13.

Singh, S.N. and Rao, D.N. 1978b. Effects of cement dust pollution on soil properties and wheat plants. *Ind. J. Environ. Heth.* 20: 258 - 267.

Singh, S.N. and Rao, D.N. 1981. Certain responses of wheat plants to cement dust pollution. *Environ. Pollut.* 24: 75 - 81.

Taylor, O.D. 1973. Acute responses of plants to aerial pollutants. *In: Air pollution damage to vegetation.* (Eds) J. A. Naegele, pp. 9 - 20.

Tripathi, A.K. and Gautam, M. 2007. Biochemical parameters of plants as indicators of air pollution. *J. Environ. Biol,* 28: 127 - 132.

Varshnay, C.K. and Mitra, I. 1993. Importance of hedges in improving urban air quality. *Landscape and Urban Plann,* 25: 75 - 83.

Worley, C.L. 1937. Carbohydrate changes within the needles of *Pinus ponderosa* and *Pseudotsuga taxifolia. Plant Physiol,* 12: 755 - 770.

www.AMC_CDP.com

2015, Impact of Global Warming and Climate Change on
Human and Plant Health
Editors: **Dr. Arun Arya and Prof. V.S. Patel**
Published by: **DAYA PUBLISHING HOUSE, NEW DELHI**

Pages 375–384

Chapter 36

Bioplastic: Demand of Globally Warmed Earth

Abhipsa R. Makwana*

*Civil Engineering Department, Faculty of Technology and Engineering
The M. S. University of Baroda, Vadodara - 390 001, Gujarat, India*

ABSTRACT

Fast urbanization and deforestation plus developmental activities generating construction waste, nuclear waste medical waste which gave rise to large quantity of plastic as a solid waste. In the production of polyolefin plastic (plastic that float-polyethylene and polystyrene), resin are polymerized and then highly stabilized into very large bio-inert carbon chains. While over a long period of time they will naturally oxidize and degrade, the time frame is simply not convenient for our societal needs. Plus our Earth is facing problem of environmental degradation along with global warming and plastic is also one of the major contributing factor for that.

This problem leads us towards thinking and development of some other type of plastic, which can get degraded by natural biological factors when disposed in the atmosphere called "bio-plastic". These bio-plastic is are not harmful to environment. This write up is a glimpse of emerging trends towards use of bio-plastics which are helpful for sustainable environment.

Keywords: *Petroplastic, Bio-plastic, Oxobiodegradable, Hydro-biodegradable.*

---

* E-mail: abhipsamakwana@gmail.com

36.1 Introduction

The term "*plastics*" covers a range of synthetic or semi-synthetic polymerization products. They are composed of organic condensation or addition polymers and may (often) contain other substances to improve performance or economics. There are few natural polymers generally considered to be "*plastics*". Plastics can be formed into objects or films or fibers. Their name is derived from the fact that many are malleable, having the property of plasticity. Plastics are designed with immense variation in properties such as heat tolerance, hardness, resiliency and many others. Combined with this adaptability, the general uniformity of composition and lightweight of plastics ensures their use in almost all industrial segments.

36.2 Characteristics of Ordinary Plastics

Petroleum was created hundreds of millions years ago when organic matter such as plankton was compressed and trapped in huge carbon reservoir as petroleum. It is a naturally occurring mixture composed primarily of hydrocarbons in a gaseous, liquid or solid state. "Organic" carbon is different than carbon metal in that its origins are primarily organic plant tissue and like any organic material in our environment, it is the food source for the tens of thousands of different micro organisms (bacteria and fungi) that survive and thrive in our air, soil and water.

Every combination of element has a gram molecular weight (GMW). A water molecule has GMW of 18. Carbon dioxide has a GMW of 44. Ethane has GMW of 14, which is highly polymerized. A plastic molecule with a GMW of 100,000 to 1,000,000 or more is formed through a process of combining smaller molecules to form larger molecules. The polymerization of Ethane creates a unit of the original molecule. The polymerization of Ethane creates relatively large plastic molecules that are far too big to be attacked by the colonies of tiny microbes, which exist everywhere. While there is some contention as to how small a molecule must be before it can be ingested, mineralized and biodegraded by the microbes- the lower end of the range of 5,000 to 40,000 MW is widely in the science world.

36.3 Why don't Ordinary Plastic Disintegrate?

The answer is quite simple. In the production of polyolefin plastic (plastic that float- polyethylene and polystyrene), resin are polymerized and then highly stabilized into very large bio-inert carbon chains. While over a long period of time they will naturally oxidize and degrade, the time frame is simply not convenient for our societal needs. In the development and production of light weight, versatile and tough plastic material made for millions of different uses: industry has very successfully achieved higher performance at lighter and lighter weights (less virgin resources, less energy, less cost) without compromising any of the good characteristics of the plastic materials.

The light weight, ubiquity, low cost, stability and lack of toxicity make plastic a very energy efficient and effective material- all qualities that continue to drive the displacement of many other packaging alternatives. It is the size and stability of the plastic molecule that cause it to be 'bio-inert" a good characteristics during use but a property that works against it after disposal.

The most frequently asked question: *"If ordinary plastic are synthetic material made from petroleum can they really be made to biodegrade?"*

The response is a definitive "YES". This is an answer that many people find quite surprising but the fact is, the origins of plastic are "organic" and as such, it is not surprising that nature can also safely return it to the natural bio-cycle.

36.4 Age of Plastics

The real issue is "ordinary plastics are bio-inert" and will persist as landfill or litter for a very long time.

Though plastics as we know them today are a relatively recent invention but they have become an important part of modern life. Today, our whole world seems to be wrapped in plastic. Plastics have become an important part of modern life and are used in different sectors of applications like packaging, building materials, consumer products and much more. Almost every product we buy most of the food we eat and many of the liquids we drink come encased in plastics. Today, the total volume of plastics produced worldwide has surpassed that of steel and continues to increase and without a doubt, we have entered in the *Age of Plastics* **Growing of Bio-degradable** Demand for materials like plastics is continually "green" plastics in plants growing and will not be abated due to their exotic properties like: low cost, durability, strength, lightweight and flexibility etc. The plastics industry world over includes a large number of production and distribution facilities employing millions workers and contributing economically in a big way. Packaging is the largest market for plastics, accounting for over a third of the consumption of raw plastic materials. Each year about 100 million tons of plastics are produced worldwide. Demand for plastics in India reached about 4.3 million tons in the year 2001-02 and about 8 million tons in the year 2006-07. Currently, however, the per capita consumption of plastics in India is only about 3 kg compared to 30-40 kg in the developed countries. The present market in India is of about Rs. 25,000 crores. The magnitudes of the plastics industry as well as its durability (or non-degradability), however, are proving to be major environmental problems.

36.5 Hazards of Plastics

Plastic is one of the few new chemical materials, which pose environmental problem. Polyethylene, polyvinyl chloride, polystyrene is largely used in the manufacturing of plastics. Synthetic polymers are easily moulded into complex shapes, have high chemical resistance, and are more or less elastic. Some can be formed into fibres or thin transparent films. These properties have made them popular in many durable or disposable goods and for packaging materials. These materials have molecular weight ranging from several thousands to 150000. Excessive molecular size seems to be mainly responsible for the resistance of these chemicals to biodegradation and their persistence in soil environment for a long time. Plastic in the environment is regarded to be more an aesthetic nuisance than a hazard, since the material is biologically quite inert. The plastic industry in the US alone is $ 50 billion per year and is obviously a tempting market for biotechnological enterprises. Biotechnological processes are being developed as an alternative to existing route or to get new biodegradable biopolymers. 20 per cent of solid municipal wastes in US are plastic. Non-degradable plastics accumulate at the rate of 25 million tonnes per year.

Though, it is difficult to imagine everyday life without plastics, but the sustainability of their production has increasingly been called into question. Conventional plastics are persistent in the environment; improperly disposed plastic materials are a significant source of environmental (in all forms soil/land, water and air) pollution, potentially harming life. The plastic sheets or bags do not allow water and air to go into earth which causes reduction in fertility status of soil, preventing degradation of other normal substances, depletion of undergroundwater source and danger to animal life. In the seas too, plastic rubbish – from ropes and nets to the plastic bands from beer packs - choke and entangle the marine mammals. Moreover, fossil fuels provide both the power and the raw materials that transform crude oil into common plastics such as polystyrene, polyethylene and polypropylene. Thus, manufacturing plastic in petrochemical factories consume about 270 million tons of oil and gas every year worldwide. The pressures of increasing non degradable waste and diminishing resources have lead many to try to rediscover natural polymers and put them to use as materials for manufacture and industry.

36.6 Plastic Bag and Bottles: CO_2 Emission

A number of people have asked about the implications of using plastic bags on the personal carbon footprint as well as on the environment in general. There are some comparisons between paper bags and plastic bags available which clearly show that it all depends on how many times these plastic or paper bags are being used.

Littering is probably the severest problem related to plastic bags. Nevertheless let's now have a look at the carbon dioxide (CO_2) emissions for the production and incineration of plastic bags.

The carbon footprint of plastic (LDPE or PET, polyethylene) is **about 6 kg CO_2 per kg of plastic**. If you know the weight of your plastic bags, you can multiply it with the number of plastic bag you are using per year. Then you can easily calculate the carbon dioxide emitted by your own usage of plastic bags. See below for some background information.

☆ The production of 1 kg of polyethylene (PET or LDPE), requires the equivalent of 2 kg of oil for energy and raw material Polyethylene PE is the most commonly used plastic for plastic bags.

☆ Burning 1 kg of oil creates about 3 kg of carbon dioxide In other words: Per kg of plastic, about 6 kg carbondioxide is created during production and incineration.

☆ A plastic bag has a weight in the range of about 8 g to 60 g depending on size and thickness. For the further calculation, it now depends on which weight for a plastic bag you actually use. A common plastic carrying bag in our household had a weight between 25 g and 40 g. So I took the average of 32.5 g.

Take the above relation between kg plastics and kg of carbon dioxide, and you get about 200 g carbon dioxide for 32.5 g of plastic, which is the equivalent of the average plastic carrying bag in our household or in other words: for 5 plastic bags you get 1 kg of CO_2.

36.7 Extraction/Production of Petro-plastic and Bio-plastic

36.7.1 Petro-plastic

Petroplastic is derived from crude oil. During the refining process, hydrocarbons are extracted from the oil to produce various everyday chemicals - octane is extracted and refined to produce auto fuel, nonane and hexadecane to produce diesel, kerosene, which are used in making jet fuel. In fact, there are dozens of chemicals, which are used in most of the products that industrial societies have become dependent on - petroplastic is just one product derived from it.

Millions of years ago, places, which currently have abundant reserves of oil or natural gas, were lush rain forests with rich healthy ecosystems. For various reasons this process was stopped and everything eventually died and was reduced to base organic compounds; ending up in underground reservoirs. Heat and pressure caused these compounds to combine into different types hydrocarbon chains. The variance in temperature, porosity, and pressure determines how it achieved its current state - oil shale exists where rock is very porous; areas of high heat and pressure have abundant natural gas; and somewhere in between is where "sweet crude" exists. The energy density of oil is about 40 to 1 - a simple example is it takes 1 gallon of gas to extract 40 gallons of gas. This fact is why oil is so attractive.

36.7.2 Bio-plastic

Bio-plastic is derived from corn or potato starches. The endosperm of corn, the center white in the kernel, contains the starch, which is extracted by soaking, grinding and washing. This is then combined with plasticizers (sorbitol and glycerin in varying amounts) to produce a thermoplastic. When harvested, the left over material (husks, ears, stalks) are composted and returned to the soil and supplemented with fertilizers. Conversely, corn requires clean water, fertilizer and harvesting equipment all, which require oil today. As oil becomes scarcer, corn production will become difficult but not impossible - unless alternative production techniques can be implemented. Plastics resin is a byproduct of oil - its production has little impact on other uses of oil. Does the production of bio-plastic have an impact on corn or potato for consumption or feed stock? Using the ethanol analogy, it may have minimal impact. According to the renewable fuels association concerns about the impact on food are unfounded. In fact the correlation between food costs had more about scarcity of oil rather than its use in biofuel and bio-plastic. It appears that much of the land devoted to oil alternatives were unused until recently.

36.8 The Solutions for Plastic

Aiming for biodegradable and esuriently products –Physical and chemical methods of pollution control were always in the forefront because they were easy to understand, easy to control and were reproducible. Biodegradation, the real mechanism of nature of balancing the material, was always found to be incompletely understood, unpredictable and uncontrollable if we have to adapt it in the form of biological treatment methods. A better option then is to modify our materials, processes and products in such a way that we can rely upon the biodegradation in nature and recalcitrance, bioaccumulation problems are overcome. We are slowly changing our philosophy and are not merely targeting for clean up or removal of pollutant but are aiming for prevention of pollution or facilitating biodegradation. Manufacturing processes are rapidly changing and biodegradable products are fast replacing man-made, difficult to degradeproducts. Objectives then are to improve upon the method of production, searching for alternative raw materials, recycling, and conversion to suitable forms of certain wastes, so that we do not add any material, waste in nature which nature cannot take care of. "Biodegradation mechanisms" which occur in the soil, aquatic environment though slow are important for us, as they do not involve any cost of treatment when it occurs naturally, and are safer and bring about complete degradation and not mere conversion. Hence incorporating biodegradability is an obvious approach while carrying out production of different items.

36.9 Types of Degradable Plastic

It is important to distinguish between the different types of biodegradable plastic, as their costs and uses are very different. The two main types are oxo-biodegradable and hydro-biodegradable. In both cases degradation begins with a chemical process (oxidation and hydrolysis respectively), followed by a biological process. Both types emit CO_2 as they degrade, but hydro-biodegradable can also emit methane. Both types are compostable, but only oxo-biodegradable can be economically recycled. Hydro-biodegradable is much more expensive than oxo-biodegradable.

36.9.1 Oxo-biodegradable Plastic

This new technology produces plastic which degrades by a process of OXO-degradation. The technology is based on a very small amount of pro-degradant additive being introduced into the manufacturing process, thereby changing the behavior of the plastic. Degradation begins when the programmed service life is over (as controlled by the additive formulation) and the product is no longer required.

There is little or no additional cost involved in products made with this technology, which can be made with the same machinery and workforce as conventional plastic products.

The plastic does not just fragment, but will be consumed by bacteria and fungi after the additive has reduced the molecular structure to a level which permits living micro-organisms access to the carbon and hydrogen. It is therefore "biodegradable." This process continues until the material has biodegraded to nothing more than CO_2, water, and humus, and it does not leave fragments of petro-polymers in the soil. Oxo-biodegradable plastic passes all the usual ecotoxicity tests, including seed germination, plant growth and organism survival (daphnia, earthworms).

Oxo-biodegradable film has been certified as safe for long-term contact with any food type at temperatures up to 40°C, and oxo-biodegradable bags are being bought and distributed by the UK Soil Association, and used for direct contact with organic

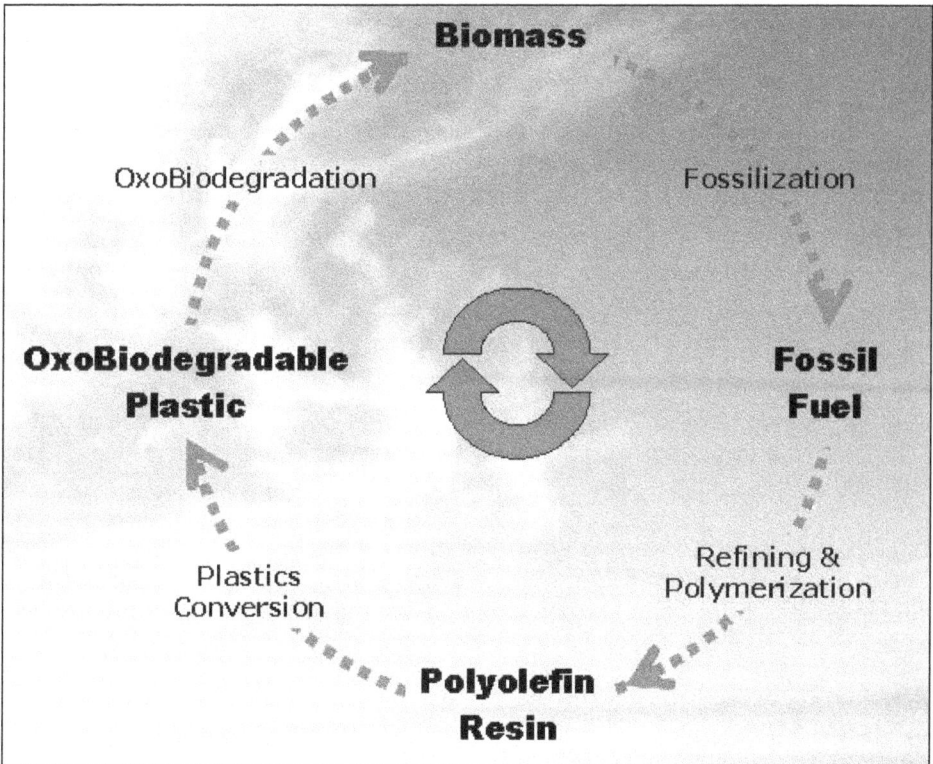

Figure 36.1: Cycle of Oxo-biodegradable Plastic.

food products. Oxo-biodegradable plastic is ideal for frozen food packaging, as it can be kept for extended periods at low temperature, and will then quickly degrade when it becomes a waste product at normal temperatures the length of time it takes for oxo-biodegradable products to degrade can be 'programmed' at the time of manufacture and can be as little as a few months or as much as a few years. They are protected from degradation by special antioxidants until ready for use, and storage-life will be extended if the products are kept in cool, dark conditions.

36.9.2 Fossil Resources

Oxo-biodegradable plastics are currently made from naptha, which is a by-product of oil refining, and oil is of course a finite resource. However, this by-product arises because the world needs fuels and oils for engines, and would arise whether or not the by-product were used to make plastic goods.

Unless the oil is left under the ground, carbon dioxide will inevitably be released, but until other fuels and lubricants have been developed for engines, it makes good environmental sense to use the by-product, instead of wasting it by "flare-off" at the refinery and using scarce agricultural resources to make plastics.

A life cycle assessment was carried out in January 2005 by GUA – (Gesellschaft für umfassende Analysen) of Vienna which shows that:

"Plastic products are made of energy resources. Additionally, their production needs further energy resources. Nevertheless, plastic products frequently enable energy savings from the perspective of the energy balance of the total life cycle compared to the energy balance of an alternative material. Examples for such energy savings by plastic products are:

    ☆ Substitution of material which consume much more energy for production of the same functional units (*e.g.* glass).

    ☆ Performance a certain function with much less material(*e.g.* packaging)

    ☆ Fuel saving because of reduction in mass(transport)

    ☆ Energy saving due to thermal insulation (where insulation with other material would be less effective, technically complicated or too expensive)

    ☆ Saving of resources by avoiding loss or damage of packed products.

Recently, interest has been shown in manufacturing sugar derived polyethylenes. These, like fossil-derived PE, are not biodegradable, but they can be made oxo-biodegradable in the same way as the latter, by the addition of a pro-degradant additive.

36.9.3 Hydro-biodegradable Plastics

Hydro-biodegradation is initiated by hydrolysis. Some plastics in this category have high starch content and it is sometimes said that this justifies the claim that they are made from renewable resources. However, many of them contain up to 50 per cent of synthetic plastic derived from oil, and others (*e.g.* some aliphatic polyester) are entirely based on oil-derived intermediates. Genetically-modified crops may also have been used in the manufacture of hydro-biodegradable plastics.

Hydro-biodegradable plastics are not genuinely "renewable" because the process of making them from crops is itself a significant user of fossil-fuel energy and a producer therefore of greenhouse gases. Fossil fuels are burnt in the autoclave used to ferment and polymerise material synthesized from biochemically produced intermediates (*e.g.* polylactic acid from carbohydrates etc); and by the agricultural machinery and road vehicles employed; also by the manufacture and transport of fertilizers and pesticides. They are sometimes described as made from "non-food" crops, but are in fact usually made from food crops.

Oxo-bio plastics degrade in the upper layers of a landfill, but they are completely inert deeper in the landfill in the absence of oxygen. They do not emit methane at any stage. Paper bags use 300 per cent more energy to produce, they are bulky and heavy and are not strong enough, especially when wet. They will also emit methane in landfill. For the reasons given under "Composting," compostability of plastics is an irrelevance.

36.9.4 Photo-degradable Plastics

These react to ultra-violet light, but unless they are also oxo-biodegradable they will not degrade in a landfill, a sewer, or other dark environment, or if heavily overprinted.

Comparison of Oxo-biodegrdable Plastic and Hydro-biodegradable Plastic

Table 36.1: Comparison of Oxo-biodegradable Plastic and Hydro-biodegradable Plastic

Oxo-Biodegradable Plastic	Hydro Biodegradable Plastic
Usually made from a by-product of oil refining	Made from fossil fuel derived polymers and starch
Can be recycled as part of a normal plastic waste stream	Damages recycle stream unless extracted from feedstock
Can be made from recycled plastic	Cannot made from recyclate
Emits CO_2 lowly while degrading and forms biomass	Emits CO_2 rapidly while degrading
Inert deep in land fill	Can emit methane in landfill
Can use same machinery as for conventional plastic	Needs special machinery
Suitable for use in high speed machinery	Not suitable
Can be compostable	Compostable
Litter or no on-cost	Four or five times more expensive than conventional plastic
Same strength as conventional plastic	Weaker than conventional plastic
Same weight as conventional plastic	Heavier
Leak proof	Prone to leakage
Degradable anywhere on land or sea	Degradable only in high-microbial environment
Time to degrade can be set at manufacture	Cannot be controlled
Safe for food contact	Safe for food contact
No chlorines or heavy metals	No chlorines or heavy metals
Can be incinerated with high energy recovery	Can be incinerated, but lower calorific values
Production uses no fertilizers or pesticides or water	Production uses fertilizers, pesticides and water

36.10 Conclusion

Petroplastic 'photo degrades' in sunlight. The ultraviolet light from the sun causes polymer chains to become brittle and rack under stress. Unfortunately, natural microbes responsible for natural decomposition do not recognize these polymer chains as food. We can thus presume that the plastic does not degrade to base organic compounds - it works by making polymer chains brittle without ultraviolet light.

Bio-plastic, being polymers of glucose (specifically amylose and amylopectin) are considered food by decomposing microorganisms - no special magic or unnatural remedies required.

References

Athalye, A.S. 1992 *Plastic in packaging* (1992)

J.Urbanski, J.,Czerwinski, W., K.Jinicka, H.Zowall, 1977. *Handbook of Analysis of Synthetic Polymer and Plastic.*

Charles A. Harper, *Handbook of Plastic, Elastomer and Composites* (1996)

Crompton, T.R. 1984 *The Analysis of Plastic,* (1984)

Miller E. 1981, *Plastic Product Design-Handbook* (1981)

Anonymous 2010 *Dictionary of Plastic Packaging*

www.bioplastic.com

www.oxobiodegradable.com

www.oxohydro.com

www.plastic.com

Index